AESS Interdisciplinary Environmental Studies and Sciences Series

Series Editor

Wil Burns
Forum for Climate Engineering Assessment
School of International Service
American University
Washington, DC, USA

Environmental professionals and scholars need resources that can help them to resolve interdisciplinary issues intrinsic to environmental management, governance, and research. The AESS branded book series draws upon a range of disciplinary fields pertinent to addressing environmental issues, including the physical and biological sciences, social sciences, engineering, economics, sustainability planning, and public policy. The rising importance of the interdisciplinary approach is evident in the growth of interdisciplinary academic environmental programs, such Environmental Studies and Sciences (ES&S), and related 'sustainability studies.'

The growth of interdisciplinary environmental education and professions, however, has yet to be accompanied by the complementary development of a vigorous and relevant interdisciplinary environmental literature. This series addresses this by publishing books and monographs grounded in interdisciplinary approaches to issues. It supports teaching and experiential learning in ES&S and sustainability studies programs, as well as those engaged in professional environmental occupations in both public and private sectors.

The series is designed to foster development of publications with clear and creative integration of the physical and biological sciences with other disciplines in the quest to address serious environmental problems. We will seek to subject submitted manuscripts to rigorous peer review by academics and professionals who share our interdisciplinary perspectives. The series will also be managed by an Editorial board of national and internationally recognized environmental academics and practitioners from a broad array of environmentally relevant disciplines who also embrace an interdisciplinary orientation.

More information about this series at http://www.springer.com/series/13637

Neil H. Kessler

Ontology and Closeness in Human-Nature Relationships

Beyond Dualisms, Materialism
and Posthumanism

Neil H. Kessler
Department of Natural Resources and the Environment
University of New Hampshire
Durham, NH, USA

ISSN 2509-9787   ISSN 2509-9795 (electronic)
AESS Interdisciplinary Environmental Studies and Sciences Series
ISBN 978-3-030-07583-5   ISBN 978-3-319-99274-7 (eBook)
https://doi.org/10.1007/978-3-319-99274-7

This Springer imprint is published by the registered company Springer Nature Switzerland AG
The registered company address is: Gewerbestrasse 11, 6330 Cham, Switzerland

Preface

I was sitting in a rented cabin on the coast of Maine, its multi-paned glass windows wrapping me in a sea of mist-soaked spruces and oaks when, in the course of my research, I happened upon a paper by Peter Martin entitled "Caring for the Environment" and published in the Australian Journal of Environmental Education in 2007. In it, he explores the applicability of Nel Noddings' care theory to environmental ethics, and through that, to environmental education. As I read it, I found myself wholeheartedly agreeing with Martin's attempt to adapt Noddings' particular, relational ethic to the human-nature relationship. In fact, I was thinking of doing that very thing in my own research. But after reading his essay, I realized that something was missing from it.

In particular, at one point Martin is describing two other authors' critiques of ecofeminist Karen Warren's first-person account of a woman having a relationship with a cliff, which she describes in her essay "The Power and the Promise of Ecological Feminism", published in the journal *Environmental Ethics* in 1990. To their suggestions that this person had either "fantasized" the relationship or was being "anthropomorphic," Martin responds on page 60 by saying, "Suggesting that people's relationships with entities in the environment have structural similarities to interpersonal relationships is clearly open to criticism." He supports this statement by correctly pointing out that empirical support for such a similarity has been scant. But, instead of going on to suggest that it is true nonetheless and requires more, or a different kind of, empirical inquiry, instead he seems to retreat by saying, also on page 60, "My purpose here is [only] to raise the *possibility* of such a likeness becoming a pedagogical *tool* [emphasis added] or means of informing educational practice." In other words, noting the difficulty of making the case for human-nature relationships to have "structural similarities to interpersonal relationships," he settles for the relationship being a metaphor and, in so doing, inadvertently undercuts the possibility of the real relationship that he himself notes, elsewhere in the essay, is crucial to development of care for the environment.

In following the trajectory of Martin's argumentation, I realized he had gone to the end of the same intellectual *cul-de-sac* that so many other modern scholars have. That is, when attempting to muster an argument for similarities between interhuman and human-nature relationships, there is a simple lack of evidence in modern societies from which to draw such parallels. But, I suggest that this is not because these similarities don't exist. Instead, it's because few serious scholars dare to look for them within the modern scientific and educational context. Martin puts it mildly when he describes the attempt to draw such parallels as "open to criticism." In my mind, to attempt such a thing is to risk intellectual heresy. And maybe it's because I have had enough relational experiences in my own life with more-than-human beings and had no place to "put" those experiences except in the realm of spirituality or unsubstantiated belief (instead of in what common sense told me was simple reality and knowledge) that it was exactly in answering the question: "Can we have a real, caring relationship with nature?" that I realized scholarship was sorely needed. In coming to that conclusion, I also understood that answering that question necessitated first answering another. That is, "What are more-than-human beings like, or, what do we take them to be like?" Only after answering *those* questions could we know whether and what kind of relationships we are having with more-than-human beings or that it would be possible to have with them.

In the end, this book is inspired by these foundational questions—as much as it is by the sea-cloaked spruces and the Osprey out on the point of land that weekend in Maine, my heart leaping the following day as I got her, perched on the tallest tree out on the point, to respond when I broke out the Osprey call that I'd been working on for years. It's a whistle whose air comes from deep down in the diaphragm, the lips a simple reed to shape that air, to blow it out into that green, stone-shouldered place with a piercing, descending cry. And though she offered only one response to me before returning to her conversation with another of her kin across the inlet, when I struck that shrill cord for the first time, she turned and looked down, ratcheting out a long call of...annoyance? No matter. She *noticed* me—as I had her from far up the long, private driveway that I trespassingly descended to reach the place where she was—where she surveyed with such elegant fierceness everything tethered so meekly to the Earth beneath her. It is from that place of meeting that this book begins.

Durham, NH, USA   Neil H. Kessler

Acknowledgments

First, I would also like to thank my greatest teacher, Martín Prechtel. His strength and tenacity literally pried open my modern mind to thinking about the human-nature relationship in the ways I have attempted to show in these pages. This book would have been inconceivable without his kindness, generosity, and insistence on trying to crack the hard shell of the modern world that blinds us to the possibilities of true belonging. I think of this book as one, long response to his ornate exhortations.

I'd also like to express deep gratitude to my family. My wife, Mariya Shnaydman, has given so much of her time and care and love just so that this book could come to be. I literally could not have done it without her. And to my sweet son and daughter. They have tolerated my double-mindedness as I worked feverishly to put this book together. It is their love of nature and unfiltered ability to see the magic there that keeps me going. It is toward their future that this book takes gentle yet forthright aim.

I would also like to thank the members of my doctoral committee, without whose intellectual rigor and encouragement the research for this book would've never come to be. In particular, I'd like to thank my committee chair, Dr. John Carroll, who gave me the latitude to write whatever I felt it was important to write, and Dr. Barbara Houston, whose encouragement, so long ago, to "fail boldly" has been a spur pushing me beyond any "normalizing" limits. Her kindness and uncompromising intellect is the stuff of a great teacher and a friend.

In particular, I also want to thank Dori Mercadante, whose unflappability in the presence of great and wild ideas has been a steadying and inspiring influence throughout the writing of this book and long before it. I would not think the way I do about the natural world if it wasn't for her willingness to always ask the next, more difficult question and to not let me off the hook when providing an answer.

I would like to bow to two great and wild horses, my friends Eleanor and Marilyn, whom I a met so long ago standing on their powerful hooves at the edge of a hot springs, secreted there against the sun-tattooed cliffs of the high New Mexico desert. It is the text within the text within the text that I send on wide wings to them. I am also grateful to the wonderful people at the Wild Rockies Field Institute in

Missoula, Montana. If not for the opportunity to teach their one-of-a-kind field-immersive courses, I never would have traveled as far down this wild road as I have. In particular, I'd like to thank Bethany Swanson, whose mix of affability, kindness, and care for this great, natural world have long been an inspiration.

Finally, I would like thank all the little stones, sitting alone in the moonlight. May you not be so lonely, unless that is what you desire to be. I'd like to thank the iron-black stallion in the field in Paonia, Colorado, for shooting me straight through with the power of his voice. And then, I'd like to thank the birds. All of them. The pigeons on the city streets with their maple-bud-red legs and shimmering necks. The poor starlings, who've done nothing more than glitter like a Milky Way-encrusted sky as they follow us humans from place to place. The chickadees who refuse to be banded easily. The red-tailed hawks, inscribing their brawny weightlessness on the dawnlit skies. The red-bellied woodpecker couple up the street from my house, who showed me so many things that first spring in my neighborhood. And all the tit-mouses, their flint-capped tenacity and gregariousness a thread that weaves together the woods around my house on a daily basis. It is because of all of you that I have the desire to speak so that someday it might become common, modern knowledge in the world that with you we *can* speak.

Contents

Part I Understandings of Human-Nature Relationships

1 Ontology and Human-Nature Relationships 3
   1.1 Environmental Problems as Problems of Relationship 3
   1.2 Ontology and Accuracy 6
   1.3 Human/Nature Dualisms 9
   1.4 Modern Human Experiences of Closeness 10
   1.5 An Ecofeminist/Pragmatist Lens 15
   1.6 Definitions/Points of Clarification 16
   References 19

2 Ecofeminist Dualisms 23
   2.1 Influence of Dualisms on Human-Nature Relationship Theories ... 23
   2.2 Types of Dualisms 26
   2.3 Monist Materialism 30
   2.4 Material and More-Than-Material: A Working Distinction 31
   References 33

3 Posthumanism's Material Problem 35
   3.1 Deleuzian *Becoming* 36
   3.2 Origins of Human Exceptionalism 44
   3.3 New Materialism's Old Roots 48
   3.4 Asymmetry 56
   3.5 Justifications for Materialism 62
   3.6 Midgley and Materialism 68
   3.7 Relations, *Relata*, and the Loss of the Self 70
   3.8 Relations of Matter or Relations of Selves? 73
   3.9 Ecofeminist Selves vs. Posthumanist "Selves" 78
   3.10 Agency Without Agents 80
   3.11 Conclusion 86
   References 86

Part II   Dualism and Relational Structure

4   Human-Nature Relationship Model 91
   4.1   Parallels Between Interhuman and Human-Nature
          Relationships 91
   4.2   Existing Human-Nature Relationship Theories................. 93
   4.3   Interhuman Closeness and Interdependence Theory 96
   4.4   Interdependence Theory Applied to a Human-Nature
          Relational Encounter 99
   References.. 106

5   Dualist Effects on Structure and Dynamics 107
   5.1   Conservation Psychology................................. 107
   5.2   Place Theories 111
          5.2.1   Definitions of "Place" 112
          5.2.2   Instrumentalization 112
          5.2.3   More Extreme Forms of More-Than-Human
                  Reduction..................................... 115
          5.2.4   Is Unilateral Attachment Possible? 117
          5.2.5   Problems with Relational Structure and Dynamics....... 118
          5.2.6   Acknowledging the Absence of More-Than-Human
                  Beings from Place Theories........................ 120
          5.2.7   Conclusion.................................... 120
   5.3   Connection to Nature Theories 120
          5.3.1   The "Nature" of Connection 121
          5.3.2   External Connection............................. 122
          5.3.3   Internalized Connections 123
          5.3.4   Inclusion of More-than-Human Other in Human Self..... 124
          5.3.5   Collapse of Relational Structure and Dynamics 127
   5.4   Sustainability Theories................................... 129
          5.4.1   Human "Nature" 131
          5.4.2   Loss of More-than-Human Identity and Individuality..... 137
          5.4.3   What Ought to Be Sustained? 140
          5.4.4   Human Activity as Environmental Activity............. 147
          5.4.5   Materialism of Biocentric/Ecocentric Relational
                  Approaches 151
          5.4.6   Posthumanist Considerations...................... 157
   5.5   Common Worlds Pedagogy 159
   5.6   Critical Animal Studies 167
          5.6.1   Sentience 168
          5.6.2   Veganism 170
          5.6.3   Equating Animal with Nonhuman 173
          5.6.4   Conclusion.................................... 174
   References.. 175

Part III   Human-Nature Relational Ontology

6   Foundations of Human-Nature Relational Ontology 183
    6.1   A Horse in Colorado 183
    6.2   Relations as Ontologically Basic........................... 184
    6.3   Characteristics of Ontologically Basic Relations 189
    6.4   Human-Nature Relations as Ontologically Basic 192
    References.. 193

7   Relational Perception and Knowledge 195
    7.1   Perceiving Closeness in Human-Nature Relationships 195
    7.2   Peircian Feelings 199
    7.3   Poetic Knowledge 202
    7.4   Personal Acquaintance Knowledge 204
    7.5   Propositional Knowledge................................. 207
          7.5.1   Objectivity 208
          7.5.2   Reliability...................................... 214
    References.. 219

8   Material and More-than-Material Considerations 221
    8.1   Materialism's Support for More-than-Human Relational
          Qualities and Capacities.................................. 222
    8.2   Evidence for the More-than-Material?....................... 223
    8.3   Ruminations on More-than-Material Relational Elements 226
    References.. 229

Part IV   Vectors of Interdependence

9   Feelings... 233
    9.1   Feelings As Purely Material............................... 234
    9.2   Feelings As Purely Internal 238
    9.3   Animal Feelings.. 242
          9.3.1   The Denial of Feelings to Animals Is Historically
                  Recent 242
          9.3.2   Examples of Animal Feelings 245
          9.3.3   Insect Feelings 249
    9.4   Plant Feelings?.. 251
          9.4.1   Feeling and Electricity............................ 251
          9.4.2   Electrical Plants 253
          9.4.3   Plant-Animal Electrical Parallels.................... 255
          9.4.4   Possibility of Plant Feelings 257
          9.4.5   The "Kook-Fringe" 261
    9.5   Feelings in "Inanimate Objects" 265
          9.5.1   Bird-David's Animism........................... 265
          9.5.2   Milton's Loved Nature........................... 270
          9.5.3   Feeling With?.................................. 272

9.5.4   Modern Human Perception and Feelings
in the "Inanimate" 275
9.6   A Re-Envisioned Environmental Conservation Example 280
References... 282

10   Thoughts 287
10.1   The Mind-Brain Problem............................. 288
10.1.1   Supervenience 289
10.1.2   Physical Realizationism........................ 290
10.1.3   Emergence 292
10.2   "If I Only Had a Brain," the Plant Thinks... 301
10.2.1   Plant Intelligence 301
10.2.2   Explanations for Plant Intelligence 302
10.2.3   Critique of Elements of Explanatory Theories......... 306
10.2.4   The Possibility of Mental Experiences in Plants 312
10.2.5   Beyond Basic Mental Experience – Can Plants
Be Conscious? 315
10.2.6   Might Plants Be Consciously Aware? 319
10.3   Thoughts in the "Inanimate"? 321
10.4   Exchange of Thoughts 321
References... 325

Part V   Conclusion

11   An Example of Modern Closeness............................ 331
Reference 333

Index... 335

List of Figures

Fig. 3.1 Posthumanist/Materialist understanding of progression
from existing/pre-Enlightenment reality through Cartesian
dualism to resulting distortion of reality   45

Fig. 3.2 Relational understanding of progression from
existing/pre-Enlightenment reality through Cartesian dualism
to resulting distortion of reality   47

Fig. 4.1 Interdependence Theory of Close Interhuman Relationships
showing interaction of feelings, thoughts, and actions.
(Adapted from Kelley et al. (1983)).   97

Fig. 4.2 Initial interaction between Walker and Blue. Shows link
between Blue's feelings and Walker's thoughts about them   101

Fig. 4.3 Interdependence of Walker and Blue. Shows comingling
of thoughts and feelings between Walker and Blue
when Blue is in love with his mate   102

Fig. 4.4 Interaction of Walker and Blue after removal of Blue's mate.
Interaction shows emotional and cognitive interdependence
leading to disgust and hatred   103

Fig. 5.1 Modified interdependence diagram from Fig. 4.1
with all thought and feeling-based, more-than-human
reciprocity removed   109

Fig. 5.2 Modified interdependence diagram from Fig. 4.1 with thoughts
and feelings removed entirely   111

Fig. 5.3 "Three pillars" view of sustainability   147

Fig. 5.4 Environment as ontological context for economic
and social pillars   148

Fig. 5.5 Relational view of the three pillars of sustainability   149

Fig. 6.1 Aristotle's Substance as ontologically independent,
with relations that depend upon it for existence 188
Fig. 6.2 Relations with superseding ontology, and substances
that depend upon them for their existence.................... 189

Fig. 9.1 Materialist ontological representation
of the sharing of feelings 239
Fig. 9.2 Possible alignments of epistemology and ontology for theories
of animism 269

Fig. 10.1 Theorized process of "emergence" of the human mind.......... 294

About the Author

Neil H. Kessler is an adjunct professor in the Department of Natural Resources at the University of New Hampshire. He earned his PhD in natural resources, with a focus on environmental philosophy. He has extensive experience teaching environmental policy, ecology, and ethics, particularly in backcountry settings such as the rain forests of Southeast Alaska. It is in wild places such as this that he honed the ideas for this book. When he's not teaching, Neil serves on his town's Conservation Commission helping craft regulations and promoting nature overall. Otherwise, he likes to blacksmith, garden, and spend time with his wife and two young children.

Part I
Understandings of Human-Nature Relationships

Chapter 1
Ontology and Human-Nature Relationships

1.1 Environmental Problems as Problems of Relationship

Fields and disciplines centered on natural resources and the environment focus on the function of environmental systems and their interactions with human society. One of the chief reasons for this focus is solving environmental problems. It may appear obvious, but in order to offer solutions, it's crucial to understand exactly what the problem is. And though issues like climate change, loss of biodiversity, and other ecological catastrophes are unquestionably problematic, I suggest that to take them as the *root* problems of modern societies' interactions with the environment is to miss what I take them to be more foundationally—that is, *symptoms* of a much deeper and more intimate problem that, when compounded into larger, global sets of interactions, delivers the destructive environmental changes seen in modern societies. What is this deeper, more intimate problem? It is that modern humans have faulty relationships with more-than-human beings.

In some sense, that modern humans are having a flawed or faulty relationship with the more-than-human world is clear. It's the reason that all the fields and disciplines focused on environmental problems seek in their own way to alter how humans interact with the environment. Further, it's hardly controversial to state that environmental problems are anthropogenic (Roberts, 2010; Vitousek, Mooney, Lubchenco, & Melillo, 1997). This means that at a material or physical level, it's unavoidable that a flawed human-nature relationship is occurring. But, while this is largely uncontestable, it's still possible not to *source* these problems in a flaw or set of flaws in the structure and dynamics of each individual interaction—or *relationship*—between human and more-than-human.

For example, some see the solution to large-scale environmental problems as primarily mathematical, and prescribe the reduction of various consumptive human behaviors as the solution. In this logic, if humans "drove less" or "recycled more," environmental degradation would be reduced or eliminated. Such an approach is the

© Springer Nature Switzerland AG 2019
N. H. Kessler, *Ontology and Closeness in Human-Nature Relationships*,
AESS Interdisciplinary Environmental Studies and Sciences Series,
https://doi.org/10.1007/978-3-319-99274-7_1

primary one espoused in the sustainability and sustainable development literature (e.g., Bruntland, 1990; Fiksel, 2006; Wackernagel & Reese, 1998). In this construal of environmental problems, once human impacts drop below some defined threshold for sustainability, they no longer exist. But, such a perspective assumes nothing inherently flawed in the way that a particular human relates with a particular more-than-human being. Such an approach never examines what is happening at the individual level of interaction, a level that eventually accumulates into larger-scale problems.

As an example of problems at an individual level, imagine a college professor taking a group of students into the field for a dendrology lab. As the professor leads the students into a forested area to study deciduous leaf structure and function, she has two choices. She can go over to a tree, pick off a leaf and bring it to the students to show them its various features. Or, she can bring students to the tree, pull the branch gently toward the gathered students and, leaving the leaf attached, deliver the same information. While the difference in approach may seem insignificant, I suggest that the leaf-picking approach is a microcosm of the relationship with more-than-human beings that directly produces large-scale environmental destruction. One either ignores the tree as a relational Self and imposes one's will—perhaps calculating that, at a material level, one leaf picked is not hurtful or destructive—or one does not. I pause here to say that when I use the term Self, I refer to an individual-in-relation that is worth consideration as a relational partner. This Self is a subject with his or her own purpose (teleology), agency and individuality in a relational exchange. I'll define this term more fully at the end of this chapter.

To return to the tree in the dendrology lab, in order to pick the leaf one must be willing first—at this personal and intimate level—to discount a more-than-human being as a Self, or to at least ignore or hide from the legitimacy of her existence as a Self. Once such an ability and practice is in place, each larger impact can also be justified. Because individual more-than-human beings as Selves either don't exist or don't matter in an individual way, arguments can always be presented for enlarging impacts. Metrics like "carrying capacity" can be presented to justify impacts up to a certain scale as can a certain level of destruction as long as "ecosystem goods and services" aren't substantively (read as: mathematically) diminished. Again, the individual more-than-human being is nowhere to be found in such constructs. Once arguments like this are activated, the stage is set for cumulative impact scenarios such as the "tragedy of the commons" (Hardin, 2009). As one can see, both the individual and large-scale level of impact require a centering of the action on human desire and a substantial, if not total, disregard for the more-than-human beings affected. In contrast, if the individual tree were treated as a relational Self, then large-scale deforestation would be almost impossible to achieve—and tantamount to genocide. Thus, the disregard for the more-than-human world *has to* begin at the individual level. If this is true, then ultimately, environmental problems are relational, personal, and absolutely intimate in their origins.

Some might still counter my suggestion by stating the quantitative "fact" that if humans did consume less, environmental problems would disappear. To counter this, I'd first suggest that the problem isn't gone, it's simply dormant. Second, I point out that if the problem really were mathematical, then all the public service and environmental education campaigns to inform the public of what changes need to be made in our habits of consumption would meet with unmitigated success. Yet, this has not been the case (Kollmuss & Agyeman, 2002). The fact that we know what we need to do and choose not to do it means that the problem is not wholly, or even primarily, mathematical. Of course some might respond by saying that the problem is still wholly human—*we* don't reduce our levels of consumption of fossil fuels, for instance, because *we* want to drive wherever and in whatever vehicle we desire. But, such a desire itself is already "infected" with a disregard for the effects of such human actions on more-than-human beings. It still originates in a relationship humans have with more-than-human beings that makes it so that humans don't want to, don't have to, or don't think it right or important to, relate in any fundamentally different way with more-than-human beings. To put it plainly, if modern humans think the more-than-human beings with whom they interact are not relational Selves, then the way is paved (pardon the pun) for the relatively unproblematic (to the human in the very short term, at least) picking of the leaf from the tree.

To reinforce my suggestion that environmental problems are problems of relationship, I note that this suggestion is echoed by others. Foster (2000), for example, says of the "most contemporary social-scientific analyses of environmental problems" that they are "centered on what is now widely believed to be a global crisis in the human relation to the earth" (p. 16). Bell and Russell (2000) also encapsulate the issue nicely when they say, "Of primary concern and interest to us are relationships among humans and the 'more-than-human world' (Abram, 1997), the ways in which those relationships are constituted and prescribed in modern industrial society, and the implications and consequences of those constructs" (p. 190). Lastly, one of environmental education's two founding documents, the Belgrade Charter (UNESCO, 1975), has as its ultimate goal for the kind of environmental action fostered by environmental education, "To improve…the relationship of humanity with nature" (p. 14). What that relationship is and ought to be, and how ontological perspectives on humans and more-than-human beings influences that determination, forms the substance of this volume.

I'm also aware that by using the term "faulty," implicit within this is the suggestion that there are correct and incorrect ways to understand more-than-human beings and our interactions with them. This may not comport well with contentions in the environmental ethics field, amongst others, that put the issue of how humans interact with more-than-human beings on an ethical footing—by characterizing it as a difference in human beliefs or values. But, if the issue turns on a rooting in well-argued ethical stances, each view of how more-than-human beings are and ought to be treated is of equal validity to every other well-argued and supported view, regardless of how different they are from one another. This is generally the landscape of the academic discourse on human-nature relationships today. But, what is missed in such a framework of consideration is that the determination of what human-nature relationships are is taken to originate in human conceptualizations of those relationships alone, not where I believe they must originate: within a human-nature relational context where the nature and functioning of the human-nature relationship is determined by both the humans and more-than-human beings participating in it. If, at this juncture, the reader is perplexed by how developing an understanding of the nature of human-nature relationships within this context is to proceed, it's precisely because we don't enter into dialogue with more-than-human beings to find out what would happen to influence our ideas about, and behaviors within, those more mutu-alistic arrangements. Also, because modern humans tend to assume that more-than-humans cannot enter substantively into such a dialogue, neither do we expend the effort to learn what languages they might speak if we do take the time to *ask* for their input and *listen* to the answer. I leave what such a dialogue might look like intentionally vague here because one's relationship with the more-than-human world is particular and intimate, as I suggested at the outset, and will take on a form different for each human being. Neither, however, do I mean that such dialogues are purely material, as that comports with the general notion of ecology, where every-*thing* is connected. The point here is that the "things" connected in ecology are Selves, and thus a purely material interchange such as an ecological view is not a truly *relational* one, and it is the relational one that I am arguing is missing these Self-based elements, and thus is faulty. Again I will define what I mean by Selves and also by *relations* of this sort more fully at the end of this chapter.

If determination of what human-nature relationships are like is to originate in the context of the relationship itself, then one of our first tasks must be to define what we take humans and more-than-human beings to be like, and thus what we are capable of in our interactions with each other. Only when we know what *can* happen can we determine what *ought to* happen. In other words, the ethical determination must be preceded by an ontological one. Ontology is the study of "…the nature or essence of being or existence" ("Ontology", 2018). Thus, what human-nature rela-tionships *ought to* be is not a matter first of values, but of perception of humans and more-than-human beings as we interrelate in the reality *out there*—outside of any either pre-determined or *post hoc* human conceptualization of what is actually hap-pening between a human being and more-than-human being in any particular inter-action. Ultimately, this is why I suggest that modern conceptions of human-nature relationships are generally flawed: because modern humans are making *mistakes in perception of the human and more-than-human world*. This means that we are not being accurate in assessing the possibilities for interaction between us.

1.2 Ontology and Accuracy

To expand upon the notion that there may be mistakes in perception leading to inac-curate assessment of the possibilities for human-nature relationships, I begin by asserting that in the research conducted for this book, nearly all of the evidence found that supports the claim that substantive human-nature relational experiences such as closeness are highly improbable carried with it a pre-existing assumption that this was the case. This implies that modern views of the relational qualities and capacities of both human and more-than-human beings may not be justified by or grounded in elucidation of facts about human relationships with the more-than-human world, but rather by pre-existing ontological commitments that filter relational experiences and their interpretation through a lens that obscures their actual features. Further, the assumption that substantive relational experiences like closeness are nearly impossible is itself rooted in another assumption, that most more-than-human beings don't have the capacities and qualities to contribute to the kind of closeness humans experience with each other. If such assumptions exist prior to any attempts to interpret human-nature relational experiences, then any conclusions drawn about limited human-nature relational possibilities run the risk of following a self-reinforcing, circular pattern of thought which cannot be accepted uncritically.

This volume will establish the basis for rejecting assumptions about the improbability of substantive human-nature interchange, and in the process will re-examine ontological questions surrounding just what qualities and capacities beings have and how they interact to influence possibilities for relationship. If what I suggest in the section above is true, that all environmental problems are problems of relationship, then one implication of any ontological remediation work is that possibilities for vastly improved human-nature relationships should become available. Some might respond by suggesting that a revision of human-nature relational ontology is no guarantee of the end of abusive environmental relationships, given that humans know that they can be close with one another, yet still treat each other very poorly. Though it's true that such poor treatment of other humans occurs, it's rare that such treatment is justified on ontological or ethical grounds. While calling out certain behaviors as unethical may not inoculate humans from poor treatment by their fellow humans, it still remains ethically unjustifiable, creating a certain leverage for change. I believe that a commensurate shift in the ontology, and then later the ethics of human-nature relationships, will create similar leverage and spur a more energetic and urgent search for solutions rooted in something more than an anthropocentric desire for survival.

As mentioned above, another implication of the possibility that pre-existing assumptions may be skewing our interpretations of experience (and therefore our concepts of ontology) is that it should no longer be viable to attribute radical differences in these interpretations only to variations in human beliefs, values, ideals, social or psychological constructions, epistemic spaces, or worldviews. Instead, it may be that some are working with a set of flawed assumptions while others are not. To underscore this difference, I note that it is a hallmark of many indigenous cultures for humans to have close relationships with more-than-human beings (Cocks, 2006; Pierotti & Wildcat, 2000; Salmon, 2000). For instance, Australian Aborigine Bill Neidjie (1989) says, "all that animal [animal, bird or snake] same like us. Our *friend* that" (p. 139). I pause to note that, though I use a few examples of indigenous thinking in this chapter, in the rest of the book I will refer very little to these kinds of views on human-nature relationships in an explicit sense. I take this tack intentionally, though not to obfuscate the heavy contribution of indigenous thinking to my own (I will address concerns about cultural appropriation at the end of this chapter). I have had no greater teacher than Martín Prechtel, who has taught me so much and so deeply that this book would be inconceivable without the generosity and kindness he has shown me and others in sharing with us some small part of what he knows.

What I've come to realize with his help, though, is that indigenous knowledge of what's possible between humans and the more-than-human world is so consistently characterized by modern thinkers as originating from a different worldview (Aikenhead & Jegede, 1999; Toledo, 2001) that a sense of irreconcilability with a modern worldview seems unavoidable. This, in turn and at some root level, serves to forestall further examination. For example, indigenous humans believe that "everything is alive" (Castellano, 2000) while most modern humans do not, at least not in such a literal manner. In another example, Burton (2002) contrasts the indigenous Lakota view of Bear's Lodge/Devil's Tower National Monument as a sacred place for renewal of the tribe's relationship to the land with that of modern public lands manager's view that it is to be "managed" for outdoor recreation, conservation and "multiple uses" (p. 5). These are two very different views that appear ontologically incompatible. The modern response to that incompatibility is to source each in different sets of beliefs, values, etc. and then to attempt to construct an ontology that can accommodate these radically disparate views. But, what if I were to say that indigenous humans *know* that everything is alive and modern humans do not? *Then* the issue becomes one of the *accuracy* of interpretation of the human-nature relational experience itself and, by extension, the worldviews—not as ends, but as lenses that may or may not distort interpretation of the relational experience itself. In other words, when positioning the understanding of shared relational experiences as a matter of human beliefs alone and not as knowledge, the modern perspective presumes that the salient features of the human-nature relational experience are ultimately decided in isolation inside the human belief-system, only to be subsequently projected outward onto some ill-defined, *tabula rasa* of the more-than-human world that will accept any human-internally coherent definition, no matter how dissimilar one definition is from another.

But, whether or not a stone sitting in the moonlight is *alive* is not a thing a human can bestow with her beliefs. The only way that the stone's existential aliveness depends on a human believing it is within an ontological idealist position, and this will not do. Down such a path lies ever more extreme forms of anthropocentrism which should be flatly rejected. Even posthumanism's intra-action (Barad, 2007), which rightly conscripts supposed atomistic individuals into an inextricable interdependence where events and yes, even characteristics of the individual, are mutually and dynamically co-constituted, cannot bestow a beating heart, soul, or actual aliveness to the stone. I'll explore posthumanism's limitations in this and other areas in much greater depth in Chap. 3, so will only state here that to be part of an intra-action, even inextricably, does not mean that the existence of certain qualities and capacities of a discernible individual within that intra-action (qualities and capacities that inhere in the individual-in-relation, and that are portable to other intra-actions by that individual) are only and wholly constituted by a belief in them by some human co-relator. The anthropocentrism and idealism of such a notion is clear. I don't think posthumanists intend this idealism (theirs appears only to be an attempt to account for inextricably mutualistic influences in a mostly material sense) but this is the logical extension of any theory that sources the portrayals of the nature of human-nature relationships in beliefs and worldviews that can produce radically different versions of the supposed same reality. Ultimately, there is a stone out there quite apart from this human perceiver and her beliefs and worldview. In the end, what the human believes or holds as a worldview is irrelevant. In the stone, the same quality of being alive, amongst other qualities and capacities, either exists or it does not.

One conclusion that can be drawn from this is that if the stone *is* alive, then the modern human *post hoc* conceptualization that it is not alive is a *mistake in thinking*. It is a failure of knowledge. It is, therefore, a *less accurate* ontological view of the stone. Seeing the differences in human thinking this way, the irreconcilability heretofore thought insurmountable is diffused. Resolving these disparities becomes an exercise in discovering possibilities and correcting for mistakes in perception, not in changing one or the other group's foundational values or beliefs. If I am correct in my analysis of the roots of this issue, and that closeness in human-nature relationships is possible, the work of more clearly understanding the reality of these relationships lies not in offering indigenous worldviews as alternatives, but instead in penetrating to the depths of the mistakes in modern conceptualizations and correcting them. From the indigenous view is only injected the possibility that those in the more-than-human world *can* have the qualities and capacities necessary to support substantive relationships, including close ones, as we understand them when they occur between human beings. If true, an analysis of modern ontologies should reveal where close relationships have been missed, rejected, or hidden from view.

1.3 Human/Nature Dualisms

In the section above, I referred to pre-existing assumptions built into modernist ontologies that characterize closeness in human-nature relationships as highly unlikely. I trace the roots of such assumptions to what ecofeminists have termed "human/nature dualisms" (Plumwood, 1993). A common term in the discourse on human-nature relationships, a *dualism* is the ontological notion of reality being made up of two foundationally distinct elements. Descartes is one of the best known dualists, believing that mind and body were ontologically or metaphysically distinct. His brand of dualism is most often referred to as a "substance dualism" or "Cartesian dualism." Ecofeminist and other theorists use the term "dualism," however, to describe how two things which may or may not be ontologically distinct are *post hoc* conceptualized in such a way that one half of the dualized pair is positioned as inferior to the other (p. 47). A human/nature dualism is one where the more-than-human world is conceptualized as somehow inferior and subordinate to the human. Though posthumanist theories are right, for example, to note that dualisms are at work in anthropocentric conceptions of human-nature relationships, the dualisms which those theories attempt to subvert are Cartesian dualisms, where the split between the material and more-than-material is seen as the foundation for the privileging of humans. In ecofeminist dualisms, however, it is the split and imposed imbalance between human and more-than-human that is seen as the foundation of anthropocentrism. I'll discuss the implications of this difference in Chap. 3, so only point out here that my view aligns with that of ecofeminism, where human/nature dualisms are the primary causes of anthropocentrism. I even identify the monist materialism that underpins the strongest forms of posthumanism as dualist in this ecofeminist sense. I'll discuss this more later in this chapter and then extensively in Chap. 3. Since I refer quite frequently to the human/nature, ecofeminist sense of dualism throughout the book, going forward I add no qualifier to it and simply refer to it is "dualism." Otherwise, when referring to Cartesian dualism, I'll identify it as such with the "Cartesian" qualifier.

Throughout this volume I'll work to demonstrate the ways in which monist materialism and other dualisms operate to distort, skew or diminish experiences of human-nature relationships. I'll examine how, in such action, dualisms marginalize, reduce or negate possibilities for closeness and other relational elements that make any relationship personal, valuable, and determinative of how one decides to live one's life. By both affecting the ways in which modern humans interpret their human-nature relational experiences and the ways in which those interpretations affect subsequent relational engagement, dualisms and the ontological conceptualizations they undergird form a self-fulfilling, self-reinforcing, and ultimately hegemonic lens through which human-nature relational experiences are filtered in the modern world. Under dualism's withering glare, possibilities for close human-nature relationships stand little chance of survival. By identifying these dualisms and sweeping them aside, my hope is to open up modern ontologies to include the possibilities that human-nature closeness both exists and can be fostered.

1.4   Modern Human Experiences of Closeness

Why does this book focus on closeness? It does so because, despite modern ontological commitments that limit or distort interpretations of close relational experiences as such, I note that some number of modern humans still have powerful experiences of closeness with the more-than-human world. If true, because those experiences are afforded so little serious attention in scientific and academic circles, or when they are considered, are so thoroughly transformed by the hegemonic dualist lenses that most theorists employ to examine them, that a conceptualization of ontology that accommodates them is sorely needed. In addition, close human-nature relationships will certainly have an ameliorative effect on the large-scale environmental problems humans face today. If my contention that these problems originate in flaws in the particular relationships humans have with more-than-humans, then the formation of close human-nature relationships will counteract such flaws at their root.

Lastly, and perhaps most importantly, these kinds of experiences are *missing* from modern human interactions with the more-than-human world. I don't suggest this as a way of prescribing some addition to more basic human-nature relationships, but instead to install the notion that they are an essential and *inherent* part of any human-nature relationship, and have gone missing from modern ones for the reasons that I'll explore in this book. This idea is not mine, it is that of Martín Prechtel (1998, 2012). And though he would likely not put it in such unadorned terms, the idea that human-nature relationships have deteriorated or devolved to lack this intimacy under modernism and its precursors in the Enlightenment, Ancient Greek civilization, and other historical influences is entirely his. In this book I attempt to respond to his idea by connecting it to both the particular history of dualism in human-nature relationships, and how that dualism acts to initiate, accelerate, and hegemonically reinforce this deterioration.

As I'll discuss in Chap. 3, the effort to separate humans from the more-than-human world is a historical fact openly discussed at the time by the likes of Francis Bacon and René Descartes, amongst other philosophers and scientists of their day. The effectiveness of their efforts to turn more-than-human beings into relationally inaccessible biological or inanimate objects so that they might be used for human purposes is still reverberating through modern ontologies today. Even posthumanists, who claim to have found a way to counteract humanism, ultimately fail to do so because they must resort to a de-Self-ed matter to overcome what they still unwittingly accept as this "little relational problem" humans have with more-than-human beings. That problem being that more-than-human beings supposedly cannot, as *individuals*, reciprocate human relational overtures in any substantive way. While I'll still build a more fulsome definition of a Self at the end of this chapter, here I'll point out that a Self is a relationally relevant and active being, and one that is ontologically irreducible to any more essential, primary or originative material components. It is human and more-than-human Selves, and the ontologically irreducible relations in which they are inextricably embedded, that need to be once more recognized in modern human-nature relationships. This is entirely achievable when dualisms are expunged from the conceptual lenses we employ to examine and theorize about the ontology of such things.

I take the fact that modern human beings still have experiences of closeness with the more-than-human world, despite the fact that there are great societal disincentives to interpret them as such, and even more to express such interpretations to others, as support for both their reality and their essentiality. Examples of individual, modern human experiences of closeness with the more-than-human world can be fairly easily found. As a first example, I note what Nobel Prize winning geneticist Barbara McClintock says of working with corn that "it surprised me because I actually felt as if I were right down there [with the corn] and these were my friends" (Keller, 1983, p. 117). Of those plants she says, "I know them intimately and I find it a great pleasure to know them" (p. 198). Nature writer and philosopher Barry Lopez (1992) gives another example when he says, "It has been my privilege to travel, to see a lot of country, and in those travels I have learned of several ways to become intimate with the land" (p. 14). As to what kind of intimacy Lopez means, he tells us it is the kind of intimacy "one would [cultivate] with a human being" (p. 13).

As one example of an opportunity for intimacy that was missed, but whose ele-ments presented themselves, Pacini-Ketchabaw and Nxumalo (2016) describe an encounter between humans at a child care center and a mother racoon:

> Captivated by the raccoons' humanlike behaviours, a group of children and two educators watch this charming raccoon family through the classroom window. All of a sudden, the mother raccoon turns her head toward the window as if she is letting these curious humans know she is aware of them. Responding to this motion, one of the children places his hand on the window to gesture hello. The mother raccoon leaves her youngsters and approaches the window. Without hesitation, she raises her paw to meet the child's hand through the glass. Silently, the child and the raccoon gaze into each other's eyes. The other children and the educators look at each other with surprised expressions. No one moves until the raccoon walks away from the window and joins her youngsters. (p. 156)

In this example, the *simpatico* between raccoon and human strikes me as a pow-erful instance of the particular, more-than-material, relational, ontologically basic encounter between human and raccoon that can lead to close relationship. But, Pacini-Ketchabaw and Nxumalo don't see it this way. They do not offer it as an example that flies in the face of humanist ontology (which they do discuss in their article) or of it being a clear gesture of friendship. Instead, they take it as an exam-ple of how the humanist nature/culture divide is being broken down *spatially* by the raccoons. Though their description of it clearly communicates the power of the gesture, they only end up arguing that it is an instance of raccoons failing to com-port with humanist ideas of the supposed *physical* separation between human and raccoon, and otherwise say nothing more about the experience. Here was their opportunity as posthumanists to interpret the moment with the boy as a moment of intimacy, or at least a moment within which the foundation of the conditions neces-sary for intimacy to emerge were laid. Instead they run, intellectually, as far from those two hands connected through the glass as they can get. They literally fail to mention its import even though, in the language they use to describe it, they see the power of it plainly enough. And while they are right at a material level about space and separation, I suggest that by taking this material-only view of the encounter, the authors miss something essential about human-nature relationships and the posthu-manist potential for a radical re-envisioning of them.

I'll explore this misplaced focus on the material in posthumanism extensively in Chap. 3, but will conclude here by noting that in such a clear and arresting instance of the potential move to intimacy, for all the modern, posthuman talk of bridging that gap, this was an opportunity that required no materialist gymnastics to pull off. It was the reaching out, in a gesture of mutual understandability, by the raccoon— and that was clearly received that way by the child whose hand the racoon reached out to touch. If dualist modern thinking in posthuman educators subverted that ges-ture to the preference for an abstracted circumlocution to get to some point about physical human space, then it shows the lengths to which modern humans (uncon-ciously or otherwise), so accepting of and loyal to their own dualist reductions of the capacities for intimacy in raccoons and humans—or so insistent on putting it on an impersonal, material footing—are willing to go to ignore a straightforwardly understandable gesture. A gesture that would inspire, almost reflexively, the desire to reach back out in in response. If a posthumanist would counter my suggestion by conceding that this could have led to something more, but that all relationships, close or otherwise, still begin in the material, I would say that this is not a settled fact, but an assumption. It is true that all relationships *require* the material, otherwise where would such relationships express themselves? But, to insist that they begin there, and get their relational power there, is based purely in an insistence on monist materialism as a starting point. It is not proof that such a view of ontology should be the end-point.

Elsewhere, the literature examining the orientation of modern children to more-than-human beings is particularly relevant here, as children often engage with more-than-human beings as close relational partners. For example, Loughland, Reid, and Petocz (2002) found that some students, and primary school-aged students more than older, secondary-aged ones, had experiences with more-than-human beings sufficient to define the belief that "people and the environment are…in a mutually sustaining relationship" (p. 194). The authors offer example statements from children's self reports such as "we should care for [nature] and it will look after us" and the environment is a "a place where we coexist in harmony" (p. 195). Chawla (1990) has also worked extensively with children's interactions with more-than-human nature. In one study she notes various intense feelings of communion between the child and the more-than-human world. She describes some of these feelings as a "mutual sense of belonging" (p. 21) and quotes one autobiography where the author describes his childhood experience by saying,

> The woods befriended me…I was usually with a group of boys as we explored the woods, but I tended to wander away to be alone for a time for in that way I could sense the strength of the quiet and the aliveness of the woods. (p. 20)

While one can discern some ambiguity in the author's description of the woods befriending him, it seems that his intention was not to describe any projection of his own feelings, but instead what he felt from the woods in the sense that it was the *woods'* strong quietness and aliveness. Another example comes from famed art historian Bernard Berenson (1949), who remembered that, as a child, one morning he "climbed up a tree stump and felt suddenly immersed in Itness. I did not call it by that name, I had no need for words. It and I were one" (p. 18). Louv (2005), in an interview with a young poet, describes her as she recounts to him the destruction of her "special part of the woods" when she was a child: "[the] young poet's face flushed. Her voice thickened. [And she said,] 'And then they just cut the woods down. It was like they cut down part of me.'" (p. 14). Lepidopterist Robert Michael Pyle (1993) says of his childhood,

> My own point of intimate contact with the land was a ditch…From the time I was six, this weedy watercourse had been my sanctuary, playground, and sulking walk…Even if they don't know 'my ditch,' most people I speak with seem to have a ditch somewhere…These are places of initiation, where the borders between ourselves and other creatures break down…It is through *close* and intimate contact with a particular patch of ground that we learn to respond to the earth, to see what really matters. (p. 4)

Hoffman (1992) refers to an account from the childhood of a 33-year-old German woman, who said,

I always felt that nature had a definite soul. In our backyard an old maple tree stood...I
would hug this old tree, and I always felt that it spoke to me...Not only the trees could
speak to me, but also the plants, streams, and even the stones...When I would find an especially beautiful rock on the road, I would take it, feel it, observe it, smell it, taste it and then
listen to its voice... (p. 24)

What I find particularly interesting about these childhood experiences is that
younger children tend to feel more of a relational orientation than older children
(Loughland, Reid, Walker, & Petocz, 2003). While some attribute this to a younger
child's greater imagination (Gleason, Jarudi, & Cheek, 2003) or as an error in thinking (Bullock, 1985), I suggest the opposite, that young children's perceptual capacities may be far more *open* and *accepting* of external reality. It is only as they grow
and learn what their societies or cultures conceptualize to be true, possible, and
acceptable in regard to the human-nature relationship that they learn to compart-
mentalize and categorize their perceptions as true or not true—as reflective of experience or as *post hoc* conceptualization of the original experience as projection.
Thus, the older one gets, the *harder* one's imagination works to hide from consciousness the relational experiences humans are having, or could have, with more-
than-human beings. Loughland et al. (2003) find an interesting statistic that could
be seen to support my assertion. They say that "increasing [environmental] knowledge reduces the odds of the 'relation' [as opposed to 'object'] conception by a
small but statistically significant amount!" (p. 13). While they attribute this effect to
"environmental knowledge...being learned in such a way that relation concepts are
not being developed" (p. 13), I suggest instead that relation concepts are innate, and
felt intuitively by children directly from the more-than-human beings with whom
they are interrelating. It is only by later acquiring concepts about the improbability
of relationality are such innate, relational constructs obscured.

I will explore in more detail the means by which this might occur later in the
book, so conclude here by suggesting again that a close relational experience may
be what is happening at a perceptual/experiential level, with objectification of more-
than-human beings a *post hoc* alteration that distorts or reduces the power and complexity of that reality. If these close relationships do exist or have the potential to
exist, then imagining that they don't or can't is a *mistake in thinking*. The remainder
of the anecdote from the German woman in Hoffman (1992) brings this possibility
home most forcefully. She recounts the methodical destruction of her close interactions with more-than-human beings by saying,

Then school began, and everything changed. Because of my intense involvement with
nature, I couldn't relate well to other children...[O]ne day when my classmates saw me
talking to a big chestnut tree in front of the schoolyard...they told the teacher, who requested
a meeting with my parents the next day...My parents recounted the conversation to me and
clearly showed how ashamed they were "to have such a crazy child." From that day onward,
my magic was systematically ruined or destroyed...So it happened, that I started believing
that nature was mute and couldn't speak to me. (p. 25)

What's clear from these examples is that many modern human beings are having,
or have had, close relational experiences with more-than-human beings. The critical
step will be to see how these experiences are, and ought to be, interpreted.

1.5 An Ecofeminist/Pragmatist Lens

If my core assertion is true—that conceptual limitations placed on possibilities for closeness in human-nature relationships is rooted in assumption and not evidence—then the nature of the close human-nature relational experiences such as those described above has yet to be settled. One means by which this book makes conceptual "room" for entertaining the possibility that they are just exactly what their human experiencers suggest they are—that the German woman in Hoffman's text *was* intensely involved with nature and the woods *did* befriend the author in Chawla's (1990) account—is by leveraging the philosophical orientations of both ecofeminism and American pragmatism (hereafter "pragmatism"). Both schools of thought source truth in what ecofeminists call "particular…experiences" (Warren, 1990, p. 113) and what pragmatists call "primary experience" (Dewey, 1929, p. 3).

The key feature of both is that truth does not begin in theory but in experiences that occur prior to any conceptualization about them. In these schools of thought, theories are only "good" if they comport with the phenomena *as experienced*, and must be continually checked against that experience to ensure their validity. Since in their philosophical contexts, experience is the vessel of truth, then if some humans are having close relational experiences with more-than-human beings, we can allow for the possibility that this is exactly what *is* happening, and that the experience is the best and truest source for this information. Is it possible that a human can be mistaken in this—perhaps by experiencing it as a hallucination or some other internally-sourced distortion? Yes, but this is only one possibility. In an ontology that does not *a priori* deny the possibility of such experiences, that they may not be a hallucination or other distortion is of at least an equally legitimate possibility. It then becomes a matter of vetting experiences and using the capacities that humans and their relational counterparts in the more-than-human world have to determine the origin of such interpretations of experience. Without getting caught in esoteric explorations of epistemology, we can acknowledge that using tools to determine the veracity of a description of experience is how one arrives at truths. When dualisms are swept aside, the means one uses to determine such truths can be markedly different from the monist materialist methods often employed in both humanist academic and scientific explorations and in posthumanism. Feeling the Other in a relational, Peircian (1960) sense is one possibility I explore in Chap. 7. If one's ontology is well constructed, nothing precludes one's experience of "closeness with" members of the more-than-human world from being the best and most accurate description of what is occurring. Of course, what exactly constitutes these close experiences is the crux of how these relationships should come to be understood, but that they can be close in an interhuman sense is, at this juncture, as yet unproblematic and unsettled. In the next chapter. I'll explore in more depth the influence of dualisms on understandings of closeness experiences, and how they work to distort those understandings.

1.6 Definitions/Points of Clarification

Before moving on to the next chapter there are some points that require clarification and terms to define. The first is whether I consider this book to be a work of posthumanism, understood loosely as a school of thought focused on refuting human exceptionalism in order to install a more egalitarian notion of human-nature relationships. While my work in this volume certainly fits within that general description, at the core of the most influential posthumanist ontologies lies a commitment to new materialism that I reject as dualist. Chapter 3 contains an exploration of this and other features of posthumanism that I contend work against our shared goals. Because of this, I cannot count this as a work of posthumanism.

As another point of clarification, the term "human-nature relationship" used throughout this work appears in the literature of various environmentally related disciplines and fields. Moving forward, any reference to the "human-nature relationship literature" is not discipline-specific, but instead comprised of the literature formed from a nexus of fields and disciplines that, through their own analytic lenses, address human-nature interactions. For example, there are the fields of environmental studies, environmental management, and environmental education, the disciplines and subdisciplines of natural resources, recreation management, conservation psychology, environmental ethics, ecology and so on. And finally, there are numerous theories emerging from these and other fields and disciplines, of which I'll consider several in this book. There is *place attachment* theory from the sociology discipline, *sustainability* and *sustainable development* theories emerging from an interdisciplinary framework. There are posthumanist fields such as *common worlds pedagogy and critical animal studies.* Several human-nature relationship theories focus on affective dimensions, such as emotional affinity theory (Kals, Shumacher, & Montada, 1999; Müller, Kals, & Pansa, 2009) and significant life experience theory (Cachelin, Paisley, & Blanchard, 2009; Chawla, 1998; Tanner, 1980). And finally, various "connectivity to nature" theories (Dutcher, Finley, Luloff, & Johnson, 2007; Hinds & Sparks, 2008; Mayer & Frantz, 2004; Nisbet, Zelenski, & Murphy, 2008; Schultz, 2002) address both cognitive and affective dimensions of the human-in-relationship with more-than-human beings. Ultimately, then, when I discuss the "human-nature relationship" and its literature, it is to this diverse body of scholarship that I refer.

Another term I use liberally throughout this work is that of the Self with 'S' capitalized. First, I trace the inspiration for this Self to ecofeminist theorist Plumwood's (1993) "relational self" (p. 155). Here, I define that Self is an individual-in-relation that is worth consideration as a full relational partner in the sense that she is in relationships, exerts an influence over those relationships, is aware of them, and can contribute thoughts, feelings and actions that contribute or detract from the functioning of those relationships. She is a subject with her own *telos*, agency and individuality in a relational exchange. She has the capacity to be both and "I" and a "Thou" in the sense that is defined by Martin Buber (1923/1970)—that is, she "fills the firmament" (p. 59). I'll explore this sense of Buberian I/Thou later in the book as well. While this Self can never be separated from her relations, neither is she wholly defined by any single one of them, and persists as an identity through space and time. She carries her Self-ness with her into a variety of relationships, the latter of whose structure and dynamics help shape this Self in the matrix of their ever-shifting *now* of those relationships, yet this Self persists as a unique and identifiable thing. In this sense, I take a Self to be ontologically irreducible as an entity. And, while a Self co-occurs with a material body, the entirety of her being is not determined by that body. If one were to note that humans qualify for this and thus it is ultimately anthropocentric, I'd only counter here by saying it's only anthropocentric if we begin in the assumption that humans have exclusive possession over such qualities instead of numbering ourselves amongst many beings that do. I'll augment this definition of Self and call out particular features of it as the situation arises throughout the remainder of the book.

A working definition of what I take relations or relationships to be is also in order. In posthumanism, the tendency is to look at relations in their most foundational material forms as the interaction of matter such that real effects are created (Barad, 2007). In this sense, it is a microstructural material relationality that feeds "upward" and helps form both the relational matrix of beings, and the intra-actions that are the beings themselves. In my use of the term *relations* or *relationship*, instead I see it as the substantive exchange between two or more beings that are Selves. Interhuman relationships fit this description, but throughout this work, I will argue that all manner of beings can have these kinds of relationships. These relations are…for lack of a better way of putting it, *personal* and therefore meaningful in a relationally "substantive" way. The particular substance of the relationship fits well with the ecofeminist sense of relations when they discuss elements of human-nature relationships. Their relations tend to be underpinned by elements of care, reciprocity, agency, teleology (Plumwood, 1993; Warren, 1990) and other qualities and capacities of Selves and the relations in which they find themselves. My sense of relations incorporates these ecofeminist features and further, I take these relationships to be ontologically basic since I don't exclude any beings from being Selves, and I take all manner of Selves to have the potential to enter into these kinds of relations. My use of various forms of the term relations—relationality, relationship, relationalism, etc.—all fit within the definition here.

In addition, some clarification of my use of the terms *human*, *more-than-human* and *nature*, is warranted. In this book I have chosen, for human beings, to use the term *humans* to refer to multiple human individuals and *human*, singular, as either an adjective to describe some quality that is unique to humans, as a noun to refer to a single human, or as an overarching noun to describe something about human beings as a group, such as, "It is distinct to the human." In contrast, when referring to all beings that are not human, I use the term *more-than-human* or *more-than-human being* to designate an individual, and *more-than-human beings* to designate multiple individuals. I use the term *more-than-human* as the adjectival form. At times, I also use *nature* as a designation for more-than-human beings equivalent to my use of *human* as an overarching noun, but I avoid using it to mean multiple more-than-human beings. I do this because using *nature* as a plural means the potential of failing to recognize more-than-human beings as individuals in a way equal with humans as individuals. Coupling *humans* and *nature* collectivizes more-than-human beings as a monolithic "other" category, not as a group or groups of individual more-than-human beings or Selves with whom humans are in actual relationship. In combination, then, I use *human* and *nature* as overarching nouns or as adjectives, *humans* and *more-than-human beings* to refer to multiple individuals of each, and *human* and *more-than-human or more-than-human being* to refer to single individuals of each. In articulating this, I also recognize that the cultural convention of using the term *nature* as a plural form of more-than-human beings is so strong in modern discourse that I feel confident that I will not catch every instance of my own use of the collective term in the book. My hope is that my recognition of the difference here, and my rigorous yet not infallible attempt to uphold it, will be sufficient to underscore a recognition of these nuances.

One other point of clarification is, I recognize that the use of *human* and *nature* has been criticized as creating a false divide between humans and more-than-human beings where humans are not seen as part of an overarching nature (deftly explored by Cronon, 1996, see also Gerber, 1997). While I participate in this false dichotomy by using nature to mean "more-than-human beings," and while it's true that the border between human and more-than-human is far murkier than some suggest, I hold that this distinction is a legitimate reference point for considering the ways in which these very conceptual divisions have real impacts on the human-nature relationships that form the subject of this book. If human exceptionalism is predicated on keeping humans both superior and distinct from the rest of the natural world, and if this book is dedicated to exploring the ontological roots of what modern humans take those that we consider not to be "us" to be like, it only makes sense to explore the qualities and capacities of those others to see if the divide is real and the "nature" of it accurately portrayed.

Earlier in the chapter I spoke to this book's relationship with indigenous ideas and knowledge. For some, this raises a concern about cultural appropriation. In response, I'll say first that the danger of engaging in such a practice given the subject matter is real. However, I don't believe I am engaging in cultural appropriation for two reasons. First, my approach purposely *avoids* the practice by insisting on analyses of modernist ontologies through *their own sets of cultural and philosophical lenses*. I take this approach in order to see whether the modern rejection of the possibility of close human-nature relational experiences can be identified purely through analysis of mistakes in perception and conception that modern societies make. This stands in contrast to looking at reality through another set of human-created (i.e., cultural) lenses.

Second, by proposing that failures to see closeness in human-nature relationships do not originate in human-originated culture, or in isolated human values and beliefs created therein, in this instance what could be appropriated does not belong to a culture and does not have cultural origins or sets of beliefs. It is only when the underpinnings of human-nature relationships are thought to be *preceded* by cultural views—a belief I explicitly reject early in this chapter—that cultural appropriation is a threat. Instead of my saying, "Let's think like 'them'" I propose to ask the question, "Why do we think the way *we* do?" Given the answers that flow in response to the latter, the work becomes that of identifying any flaws in that thinking which preclude us from seeing certain human-nature relational experiences differently than we modern humans currently and predominantly do.

Further, I argue that experiences humans have with the more-than-human world cannot be "owned" by a culture defined as a group of humans. They are only *participated in* by that culture, and of course a human culture's response to those relationships are unique to that culture and are able to be appropriated. But, again I'm suggesting the opposite in my work when I recommend that modern humans look within our own views and *our own cultural histories* for the basis of misperceptions about the more-than-human world and human relationships with it. Ultimately, cultural appropriation is the exact practice of swapping irreconcilable worldviews that I reject at the outset of my argument that the ontology of human-nature relationships is about accuracy, not worldview. With that out of the way, I will move on to discuss dualisms in more depth in the next chapter.

References

Abram, D. (1997). *The spell of the sensuous: Perception and language in a more-than-human world*. New York: Random House.

Aikenhead, G. S., & Jegede, O. J. (1999). Cross-cultural science education: A cognitive explanation of a cultural phenomenon. *Journal of Research in Science Teaching, 36*(3), 269–287.

Barad, K. (2007). *Meeting the universe halfway: Quantum physics and the entanglement of matter and meaning*. Durham, NC: Duke University Press.

Bell, A. C., & Russell, C. L. (2000). Beyond human, beyond words: Anthropocentrism, critical pedagogy, and the poststructuralist turn. *Canadian Journal of Education/Revue Canadienne De l'Éducation, 25*(3), 188–203.

Berenson, B. (1949). *Sketch for a self-portrait*. Bloomington, IN: Indiana University Press.

Bruntland, G. H., & World Commission on Environment and Development, Australia, & Commission for the Future. (1990). *Our common future*. Melbourne, Australia: Oxford University Press.

Buber, M. (1970). *I and thou*. (W. Kaufmann, Trans.). New York: Charles Scribner's Sons. (Original work published 1923).

Bullock, M. (1985). Animism in childhood thinking: A new look at an old question. *Developmental Psychology, 21*(2), 217–225.

Burton, L. (2002). *Worship and wilderness: Culture, religion, and law in public lands management*. Madison, WI: University of Wisconsin Press.

Cachelin, A., Paisley, K., & Blanchard, A. (2009). Using the significant life experience framework to inform program evaluation: The nature conservancy's wings & water wetlands education program. *The Journal of Environmental Education, 40*(2), 2–14.

Castellano, M. B. (2000). Updating aboriginal traditions of knowledge. In G. J. S. Dei, B. L. Hall, & D. G. Rosenberg (Eds.), *Indigenous knowledges in global contexts: Multiple readings of our world* (pp. 21–36). Toronto, Canada: University of Toronto Press.

Chawla, L. (1990). Ecstatic places. *Children's Environments Quarterly, 7*(4), 18–23.

Chawla, L. (1998). Significant life experiences revisited: A review of research. *Journal of Environmental Education, 29*(3), 11–21.

Cocks, M. (2006). Biocultural diversity: Moving beyond the realm of 'indigenous' and 'local' people. *Human Ecology, 34*(2), 185–200.

Cronon, W. (1996). The trouble with wilderness: Or, getting back to the wrong nature. *Environmental History, 1*(1), 7–28.

Dewey, J. (1929). *Experience and nature*. London: George Allen & Unwin, Ltd..

Dutcher, D. D., Finley, J. C., Luloff, A. E., & Johnson, J. B. (2007). Connectivity with nature as a measure of environmental values. *Environment and Behavior, 39*(4), 474–493.

Fiksel, J. (2006). Sustainability and resilience: Toward a systems approach. *Sustainability: Science Practice and Policy, 2*(2), 14–21.

Foster, J. B. (2000). *Marx's ecology: Materialism and nature*. New York: Monthly Review Press.

Gerber, J. (1997). Beyond dualism–the social construction of nature and the natural and social construction of human beings. *Progress in Human Geography, 21*(1), 1–17.

Gleason, T. R., Jarudi, R. N., & Cheek, J. M. (2003). Imagination, personality, and imaginary companions. *Social Behavior and Personality: An International Journal, 31*(7), 721–737.

Hardin, G. (2009). The tragedy of the commons. *Journal of Natural Resources Policy Research, 1*(3), 243–253.

Hinds, J., & Sparks, P. (2008). Engaging with the natural environment: The role of affective connection and identity. *Journal of Environmental Psychology, 28*(2), 109–120.

Hoffman, E. (1992). *Visions of innocence: Spiritual and inspirational experiences of childhood.* Boston: Shambhala.

Kals, E., Schumacher, D., & Montada, L. (1999). Emotional affinity toward nature as a motivational basis to protect nature. *Environment and Behavior, 31*(2), 178–202.

Keller, E. F. (1983). *A feeling for the organism, 10th anniversary edition: The life and work of Barbara McClintock.* San Francisco: W. H. Freeman and Company.

Kollmuss, A., & Agyeman, J. (2002). Mind the gap: Why do people act environmentally and what are the barriers to pro-environmental behavior? *Environmental Education Research, 8*(3), 239–260.

Lopez, B. (1992). The rediscovery of North America. *Orion, Summer*, 10–16.

Loughland, T., Reid, A., & Petocz, P. (2002). Young people's conceptions of environment: A phenomenographic analysis. *Environmental Education Research, 8*(2), 187–197.

Loughland, T., Reid, A., Walker, K., & Petocz, P. (2003). Factors influencing young People's conceptions of environment. *Environmental Education Research, 9*(1), 3–19.

Louv, R. (2005). *Last child in the woods: Saving our children from nature-deficit disorder*. Chapel Hill, NC: Algonquin Books of Chapel Hill.

Mayer, F. S., & Frantz, C. M. (2004). The connectedness to nature scale: A measure of individuals' feeling in community with nature. *Journal of Environmental Psychology, 24*(4), 503–515.

Müller, M. M., Kals, E., & Pansa, R. (2009). Adolescents' emotional affinity toward nature: A cross-societal study. *Journal of Developmental Processes, 4*(1), 59–69.

Neidjie, B. (1989). *Story about feeling*. Broome, WA: Magabala Books.

Nisbet, E. K., Zelenski, J. M., & Murphy, S. A. (2008). The nature relatedness scale: Linking individuals' connection with nature to environmental concern and behavior. *Environment & Behavior, 41*(5), 715–740.

Ontology. (2018, June). *Oxford English dictionary online*. Retrieved from http://www.oed.com

Pacini-Ketchabaw, V., & Nxumalo, F. (2016). Unruly raccoons and troubled educators: Nature/culture divides in a childcare centre. *Environmental Humanities, 7*(1), 151–168.

Peirce, C. S. (1960). *Collected papers of Charles Sanders Peirce* (C. Hartshorne, P. Weiss, & A. W. Burks, Eds.). Cambridge, MA: Harvard University Press.

Pierotti, R., & Wildcat, D. (2000). Traditional ecological knowledge: The third alternative (commentary). *Ecological Applications, 10*(5), 1333–1340.

Plumwood, V. (1993). *Feminism and the mastery of nature*. London: Routledge.

Prechtel, M. (1998). *Secrets of the talking jaguar*. New York: Tarcher/Putnam.

Prechtel, M. (2012). *The unlikely peace at Cuchumaquic: The parallel lives of people as plants: Keeping the seeds alive.* Berkeley, CA: North Atlantic Books.

Pyle, R. M. (1993). *The thunder tree: Lessons from an urban wildland*. Boston: Houghton Mifflin.

Roberts, J. (2010). *Environmental policy*. London: Routledge.

Salmon, E. (2000). Kincentric ecology: Indigenous perceptions of the human-nature relationship. *Ecological Applications, 10*(5), 1327–1332.

Schultz, P. W. (2002). Inclusion with nature: The psychology of human-nature relations. In P. Schmuck & P. W. Schultz (Eds.), *Psychology of sustainable development* (pp. 61–78). Norwell, MA: Kluwer Academic Publishers.

Tanner, T. (1980). Significant life experiences: A new research area in environmental education. *The Journal of Environmental Education, 11*(4), 20–24.

Toledo, V. M. (2001). Indigenous peoples and biodiversity. *Encyclopedia of Biodiversity, 3*, 451–463.

UNESCO. (1975). *The international workshop on environmental education, Belgrade, Yugoslavia, 13-22 October 1975. Final report ED76/ws/95*. Paris: UNESCO.

Vitousek, P. M., Mooney, H. A., Lubchenco, J., & Melillo, J. M. (1997). Human domination of earth's ecosystems. *Science, 277*(5325), 494–499.

Wackernagel, M., & Rees, W. (1998). *Our ecological footprint: Reducing human impact on the earth*. Vancouver, Canada: New Society Publishers.

Warren, K. J. (1990). The power and the promise of ecological feminism. *Environmental Ethics, 12*(2), 125–146.

Chapter 2
Ecofeminist Dualisms

2.1 Influence of Dualisms on Human-Nature Relationship Theories

As suggested in the previous chapter, the origin of the view that more-than-human beings have little to contribute to closeness experiences in human-nature relationships is to be found in what ecofeminists have identified as *human/nature dualisms*. To reiterate from the previous chapter, in a general sense a *dualism* is the ontological or metaphysical notion of reality being made up of two foundationally distinct elements. This is most often referred to as a "substance dualism" or "Cartesian dualism." Ecofeminists use the term "dualism," however, to describe how two things which may or may not be ontologically distinct are *post hoc* conceptualized to be so in such a way that one half of the dualized pair is positioned as inferior to the other (Plumwood, 1993, p. 47). Again, in this volume I use "dualism" to refer to ecofeminist, human/nature dualisms and "Cartesian dualism" to refer to that type.

As these dualisms relate to the human-nature relationship, Plumwood says that "the western model of human/nature relations has the properties of dualism... [which] results from a certain kind of denied dependency on a subordinated [more-than-human] other...in which the denial of the relation of domination/subordination shape the identity of both" (p. 41). Here, she is describing how the human depends on a subordinated nonhuman, and that this subordination is accomplished and/or justified through *post hoc* dualist conceptualizations of the more-than-human as inferior. In human-nature relationship theories, then, the peripheralization or elimination of the more-than-human isn't the result of the more-than-human actually being inferior or peripheral. It is achieved through the *post hoc* operation of undetected dualistic views of what more-than-human beings are understood to be like, and thus what is taken to be relationally possible between them and human beings. In this book, I take these dualisms to be the most foundationally undermining modes of thought in the human-nature relationship literature. Their pervasiveness and

© Springer Nature Switzerland AG 2019
N. H. Kessler, *Ontology and Closeness in Human-Nature Relationships*,
AESS Interdisciplinary Environmental Studies and Sciences Series,
https://doi.org/10.1007/978-3-319-99274-7_2

persistence there can best be described by Plumwood, who notes that dualisms "may lurk in the background in unexamined and concealed form…[and] form a web or network. One [dualism] passes easily into the other, linked to it by well-traveled pathways of conventional or philosophical assumption" (p. 45). So subtle and pervasive are they in established scientific and academic discourse that I believe them to have formed a hegemonic constellation of thought that reflexively reinforces its own position not through evidence, but by installing, without substantiation yet still positioned as ontologically correct, its own perspective. It then uses that positioning as leverage to marginalize and negate any perspective or line of thought not in accord with its own. Dualisms have become so powerful and deeply rooted in modern thinking that most of the time, those offering theories that carry them are rarely even aware of their existence.

In human-nature relationship theories, when dualisms are present they can reduce more-than-human beings to passive (Merchant, 1980; Plumwood, 1993), material (Barad, 2007; Coole & Frost, 2010; Evernden, 1992; Plumwood, 1993), objects of instrumental-only value (Gaard, 2010; Plumwood, 1993). Given such reductions, it's easy to see how the more-than-human world becomes marginalized. What it's equally easy to see, however, is just how out of keeping such a reduced relational partner is with the Self that occupies the experiences of many modern humans, examples of which I discussed in the previous chapter. I acknowledge here that posthumanists theorists also seek a way to counteract the reduction of more-than-human beings in theories of human-nature relationships, but find their reliance on a sense of relationalism that is purely material and microstructural, and their unwitting perpetuation of dualism in the process of installing it, to work against possibilities that more-than-human beings can be Selves. I'll discuss this extensively in Chap. 3 and also throughout this volume.

What one finds emerging from the operation of dualisms in theories developed to explain close human-nature relationships is a deep tension between the experiences of human-nature relationships *as close* and most modern theories developed to explain them. This is true even in posthumanism, where such inherently relational qualities like closeness are so far removed from the ontology of relations as to be invisible, and a clear path from the posthumanist notion of agential realist relational interchange at a material level to experiences such as closeness largely unarticulated. In its essence, the tension is drawn between actual experiences of closeness between relational Selves and *post hoc* theoretical conceptualizations about them that position humans as acting largely alone, fabricating a facsimile of closeness out of internal cognitive and affective elements and projecting them outward onto the relationally inert more-than-human world. I will explore specific examples of these activities extensively in Chap. 5.

The tension I've just outlined is succinctly articulated by Milton (2002), who describes the former as modern humans "*discover[ing]* the personhood of nature and natural things by perceiving…" and the latter as modern humans believing that they "make nature and natural things *into* persons, that [they] *construct* them as persons" (p. 44). Given this tension, one can also see that human-nature relationship theorists who cleave to the "construction" tack have allowed themselves to be painted into a conceptual corner. By relying heavily on interhuman relational constructs of closeness while adhering to dualistic ontological positions that undermine closeness' most foundational feature: *reciprocal* exchange between relational Selves, they've constructed an unresolvable impasse between experience and theory.

The tension between closeness as experienced and closeness as fabrication can be traced to two possible sources. The first is the "reality" that humans and more-than-human beings simply don't have the qualities and capacities that, combined, can support a close human-nature relationship with similarities to close interhuman relationships. If that's the case, all experiences of closeness by humans are the internal-to-human concept they're predominantly portrayed to be—either overtly, or by unspoken assumption—in the human-nature relationship literature today. If this is the most accurate reading of reality, then one ought to primarily appeal to the psychology disciplines for understanding the conditions under which humans conceptualize their interactions *as relational* with a relationally inert partner. In such a view, when some seek to solve environmental problems, for example, by promoting a feeling of closeness between human and the rest of the more-than-human world, they are actually engaged in an elaborately cloaked form of hypnotism. The human is encouraged to feel as if he is close to nature when true closeness is not possible. The second possibility is that humans and more-than-human beings *do* have the necessary qualities and capacities to engage in close relationships, but the predominant, modern human conceptions of humans, more-than-human beings, and the possibilities for their interaction are somehow flawed. If, as I have already suggested, it is dualisms that inject these flaws, then they are operating in an undetected way to conceptually obfuscate the qualities and capacities of more-than-human beings that would enable their substantive participation in close relationships.

If the first possibility is true, then any further attempts to use close interhuman relationship models in human-nature relationship theory are at best metaphoric. Ecofeminist King (1991) directly addresses the trouble of employing a metaphor in his response to Warren's (1990) first-person narrative of a human having a close relationship with a cliff she is climbing when he asks, "The rock is personified as a partner and friend…[but w]hat…is gained by the metaphors of conversation, partnership and friendship when these are taken out of their human context…?" (King, 1991, p. 85). My answer to King's question is: If the closeness is only a metaphor, then not very much is to be gained. But, it is not yet a settled matter that human/ more-than-human friendship can only be metaphoric. To conclude such a thing requires the dualist assumption that the cliff cannot *be* a friend. Further, it seems at the very least naive to expect all the benefits of the behavior of someone in a real close relationship—the strength, vigor and tenacity of love and care—when no relationship actually exists. That is because if such qualities are present, they are not derived internally, but in the matrix of the exchange between those in the relationship. I suggest that employing such a false relational construct, if it is false, has neither the power to truly improve human-nature relationships, nor the conceptual "gravity" to explain the bonds, love, care and intimacy that *are* being experienced by many humans in relationship with more-than-human beings (see Chap. 4 for more of these qualities and how they are used by human-nature relationship theorists to advance their views). Ultimately, if the closeness is metaphoric, its use by human-nature relationship theorists is of marginal value at best and ought to be discarded.

If, on the other hand and as I suggest as the basis of this book, the second possibility is correct and there *are* flaws in the conceptions of the qualities and capacities that humans and more-than-human beings have, and how they interoperate in the human-nature relationship, then there is valuable work to be done in articulating what those flaws are and correcting for them. It is the substance of this book to undertake such an effort so that prescriptions offered for the improvement of human-nature relationships effectively address the root causes. If at this juncture the reader is wondering exactly which more-than-human beings I take to be capable of being a relational Self, it is every known and knowable thing in the world—from stars to slime molds to Killer Whales to rivers full of stones. For all things in existence, Part IV of the book contains an extensive exploration of both the dualistic denial of, and the alternative possibility of the capacity for, the thoughts and feelings that I will lay out in Chap. 4 as the baseline elements necessary to engage in close relationships.

2.2 Types of Dualisms

Above I suggested that many human-nature relationship theories are underpinned by ontologies that carry ecofeminism's human/nature dualisms. The first important dualism of this type to consider is that of defining a being in the more-than-human world as a *passive object* instead of an "active subject" (Merchant, 1980, p. 400). Plumwood (1993) notes that defining the more-than-human world thusly positions it as "empty, passive and without value or direction of its own" (p. 111). The effect that this passive-object definition has on human-nature relationship theories is that it limits or eliminates more-than-human beings as Selves. For example, Batisse (1982) explores the merits of creating biosphere reserves as an environmental conservation measure and notes three objectives to doing so. First, it is meant to "conserve for present and future use the diversity...of biotic communities...", second to "...provide areas for research...", and third to "...provide facilities for education and training" (p. 102). In all three, the more-than-human world is only defined as material, and only those material elements of instrumental value to human desires and goals. Whatever goals or purposes a being in the more-than-human world has for his own sake are either denied or deemed irrelevant by the dualistic lens through which he is considered. The survival of beings besides humans within such a biosphere reserve—and that as an end unto itself—is ignored. Posthumanism would also stand against defining the more-than-human world in this way. But, even though they would position that world instead as active and containing agency (not in an individual sense but as a general property of matter) in the end more-than-human beings are still positioned as having few relational qualities or capacities in the individual sense of how I defined a Self in the first chapter. Thus, more-than-human beings still become objects instead of subjects, with the result that while they become active objects, the activity cannot be attributed to them as Selves, and so still leaves them vulnerable to instrumentalization for human ends. As Chandler (2013) notes, in posthumanism *all* beings become objects.

The second dualism I consider is *materialism*. Plumwood (1993) suggests that what is dualist is a "materialist reductionism" which "conceiv[es] not only nature and animals but also the human itself in mechanistic and reductionistic terms" (p. 121). One effect such a reductive materialism has is to portray more-than-human beings as "stripped of psychological or mindlike attributes, [where a]ll talk of teleology, of agency, of goals, of striving, of choice and freedom is exorcised... [and] leaves nature as a sphere voided of meaning, a mere endless movement of matter" (pp. 121–122). Evernden (1992) says that "if subjectivity, willing, valuation, and meaning are securely lodged in the domain of humanity, the possibility of encountering anything more than material objects in nature is nil" (p. 108). Here again there is a denial of a Self-ness to more-than-human beings. If one defends such a position by, for example, claiming that things such as teleology, agency and goals are *post hoc* human constructs superimposed on an external reality that does not contain them inherently, I'd suggest that such a stance itself relies on a dualist ontological assumption that they do not exist. This is a claim that it is one of the central purposes of this work to refute. Thus I state here, and will offer support for in Part III (Chap. 8 specifically), that teleology and other elements can exist in more-than-human beings and thus can be perceived by, and relationally affect, human beings.

In addition, though Plumwood above only goes so far as to suggest that the mechanistic nature of reductive materialism is dualist, I take all monist material ontological positions to be dualist. I base this on the fact that though there are myriad forms of materialism, all appear to eliminate more-than-material ontological possibilities for thoughts, feelings, intuition, spiritual entities, etc., through *post hoc* conceptual reduction instead of empirical evidence. For example, in Foster's (2000) excellent account of Epicurean materialism, he eradicates the possibility of spiritual ontological elements through *post hoc* qualification of experiences of "teleology, vitalism, idealism...or whatever one likes to call it" as a postmodern conceptual response to "the one-sided determinism of [Democritus' mechanistic materialism]" (p. 11). Since in a materialist ontology in general, more-than-material elements cannot exist, theorists like Foster's are free to attribute experiences of what might otherwise be taken as more-than-material to some form of internal-to-human, *post hoc* conceptual response to a material, external reality. Within such ontological constraints, then, spiritual elements are at best a benign psychological projection and at worst, outright pathological thinking. Midgley notices this problem for materialism as well when saying,

> If indeed it is true that there are no spiritual forces acting on human beings from the outside - if our spiritual experiences are entirely generated from within – then what goes on in our spiritual life is unavoidably psychological business. (p. 61)

In making his claim, Foster (2000) relies on the materialist ontological constraint that spiritual elements *can* only be internal-to-human concepts as support for his contention that they are. This is circular logic. It is not based in any experience, but in a dualistic line of reasoning that elevates the material by assumption, and from that vantage point, enables materialist concepts to reduce or negate the potential for any more-than-material ontological elements.

Further, in characterizing the two equally implausible ontological extremes as mechanistic materialism and reactionary "idealism...[and] spiritualism" (p. 19), respectively, Foster is not alone. Pragmatist philosopher John Dewey (1929) also sees these as the two ontological extremes for which either a middle ground, or an escape from the spectrum entirely, is the solution. The middle ground Dewey seeks is the chaotic flux of "primary experience" (p. 3) that precedes both of these human-generated, conceptual overlays. For Foster (2000), that primary experience is material. What neither author sees, however, is that a view of reality as primarily chaotic and/or material itself depends upon a pre-existing assumption that this is the "most basic" reality. This is not fact, but assumption, as I'll argue in greater detail in Part III. That it may *seem* like fact to the reader is, in my estimation, testament to the hegemonic force with which materialism and its underwriter, chaos, have pushed their way into the deepest recesses of established, contemporary academic and scientific thought (see Kessler, 2012, for more on this effect). Given materialism's power to supplant not only more-than-material elements, but to act as a foundation for arguing that only humans have most of the qualities we'd attribute to Selves, I suggest that materialism is the most powerful human/nature dualism.

This is true even of the so-called "new materialisms" that are the ontological underpinnings of so much of posthumanist theory. I'll explore new materialism in depth in the next chapter, so will only state here that, notwithstanding its vigorous attempts to subsume all of relational reality under its broadened definition of what is material, new materialism leaves intact the dualist material hierarchy where humans reside at the apex when it comes to the kinds of relational qualities and capacities I explore in this book. New materialist theorists attempt to explain away this dualist remnant of what Warren (1990) terms the "logic of domination" (p. 128) as a non-subordinating type of asymmetry or difference—a supposed plain material fact and value-neutral outgrowth of advances made in the natural sciences. But, those theorists failure to see that endeavors in the natural sciences in general, and the knowledge that the objectifying and detached relational lenses they produce, are direct outgrowths of the logic of domination of nature embodied in the Enlightenment and the Scientific Revolution. To not have accounted for this history is to leave its foundational dualisms intact. Thus, new materialism unwittingly fails to escape the dualisms deeply engrained in its own materialist position. I will discuss monist materialism more in the next section and then more extensively in the next chapter.

Moving beyond materialism, more-than-human beings possessing only *instrumental* (to human ends) value is another dualism. I trace this dualistic form of valuing to a predictable extension of the ontological portrayals of more-than-human beings as objects of a material nature. I mean this in the sense that, given those definitions, it's virtually impossible to argue for ethical consideration of more-than-human beings as ends in themselves. Plumwood (1993) says that within this instrumentalization "the other appears only as a hindrance to or as a resource for the self's own needs, and is defined entirely in relation to [the human's] ends" (p. 145). One of the many examples of the instrumental value of nature in human-nature relationship theories is in the notion of "place identity" (Williams & Vaske, 2003, p. 831), which is described as "an emotional attachment" that is a "repository for [human] emotions and [human] relationships that give meaning and purpose to [human] life" (p. 831). I'll explore this in detail in the "Place Theories" section of Chap. 5. Even before exploring any theories, terms like "natural resources" immediately threaten to eliminate more-than-human beings as Selves. *For whom* are these more-than-human beings resources? The human of course. This kind of dualist valuation largely obscures any other value that more-than-human beings might have and that contributes to a very different kind of human-nature relationship. While posthumanism doesn't subscribe to dualist instrumentalizing directly, both the dualist carryovers lurking in their theories, and their de-individuation of all individuals can still serve to promote instrumentalizing points of view.

Yet another relevant dualism is *individualism*, defined as "the view that selves are essentially solitary, separate, isolated, atomistic individuals defined apart from the social contexts and relationships in which they find themselves" (Warren, 2000, p. 90). In the case of a human/nature dualism, this context includes interactions with more-than-human beings. Its main operation is to sponsor a version of relationships as emerging from the primary and precedent individual. I'll explore this in more detail when it comes in the form of *substantivism* in Part III, so suffice it to say here that in individualism, when a relational experience occurs between human and more-than-human, individualism empowers the human to see herself as the source of that experience—or at least its most salient relational features. It aids the overwhelming human focus in human-nature relationships that I discussed in the first chapter.

The result of defining the more-than-human as a passive, material object with instrumental-only value upon whom individual humans project various feelings of closeness is a decidedly *anthropocentric* view—another dualism—of humans and the resultant human-nature relationship. While some argue that the other dualisms I've discussed above are the result of anthropocentric thinking, such as posthumanist theories in general (Ferrando, 2013), I suggest the opposite: that anthropocentrism is more likely *the result of* these other dualisms. If in a dualist fashion, more-than-human beings are defined as passive, material, relationally limited, and/or instrumental objects devoid of their own significance and/or meaning, agency, and intelligence, then there really is only one relational partner that *can* be an active, purposeful subject with significance and meaning, agency, and intelligence. This is the human. This certainly can produce and reinforce anthropocentric thinking.

Ultimately, all of these dualisms have a self-reinforcing effect in the development of theories to explain human-nature relationships. The more they form the conceptual underpinnings of the definitions of human, more-than-human and their resultant relationship, the more they preclude interpretation of elements of that relationship in any way that falls outside this codified set of dualisms. Thus, even were there to appear contradictory evidence coming through directly to the human in his experience with more-than-human beings, it will go undetected. That, or the *post hoc* conceptual lens through which it's interpreted will either reduce it out of existence or "buttonhook" it instrumentally back to being a function and/or product of the human mind. Such ontological sleight of hand cannot uncritically stand as a valid way to interpret experiences of human-nature relationships. More is required if we are to learn about and try to improve these relationships.

2.3 Monist Materialism

By definition, monist materialism posits that all experienced phenomena conform to the materially explainable. Therefore, anything that doesn't fit comfortably within such a context —experience of the spiritual external to one's Self, for example—is rejected, ignored, or must be transformed to fit the assumption of materialism. If one were to ignore that it is monist materialism under consideration and described *post hoc* conceptual lenses that allow one to reject, ignore and foundationally transform described experiences, those in feminist, ecofeminist and other fields would have no trouble recognizing the hallmarks of human/nature dualism. Yet, for some reason, monist materialism escapes such detection.

One reason for this is the fact that, in some academic circles materialism has been useful in providing an ontological common ground upon which to build theories of human-nature relationships. When explanations for such things as human-animal communication differ so radically between, for example, the indigenous and the modern, one naturally begins to search for *something* that can be taken to be the same between the two versions of reality. Unless we are to allow ourselves to slip into subjective idealism or radical constructivism, that thing is materialism. This understanding of the genesis of posthumanist adherence to a monist materialism is reflected in what Ferrando notes about posthumanism's "new materialism." Of it she says, "New materialisms philosophically arose as a reaction to the representationalist and constructivist radicalizations of late postmodernity…[that were in danger of making] external reality…unapproachable" (pp. 30–31). But, authorizing the material as the only ontological common ground attainable and shunting the deep differences between, for example, indigenous and modernist views over to epistemology, ethics, or matters of belief and values, does not resolve the problem. It only obscures the tension between those that say that there is more *out there* than the material and those that reject such notions. But, that tension exists, and it's because the other method of reconciliation is far more threatening—that of determining that modern ontologies are inaccurate portrayals of reality. I discussed this at length in the first chapter, and so here only tether it to dualisms as the assumption that undergirds monist material approaches.

This is not to say that acknowledging universal material interconnections is necessarily negative. It is certainly a step to improving human-nature relationships.

But, in this book I will argue that monist materialism does not provide an accurate rendering of the ontological basis for, or the sum total of, the nature of those interactions. Therefore, it cannot be the only or even the primary step. I also note that when dualisms such as materialism are moved aside, other commonalities upon which to found an ontological theory can be found. Therefore, to suggest (a) that materialism is the only means to rehabilitate human-nature relationships and (b) because only the material exists we should press forward with monist material ontological theories, is erroneous.

If one still doubts my assertion that monist materialist ontology is dualist, a straightforward test can be applied. To determine whether one is holding to a dualist point of view, one only need to determine whether the potentially subordinated/dualized half of a dyad (a) is granted equal standing to the elevated half, or (b) is so thoroughly discredited that it can be taken as "fact" not to exist. If (a) or (b) can be shown to be true, then the position represented by the first half of the dyad is not dualist. If neither can, then the position is dualist. Those seeking to argue that monist materialism is not dualist would be forced to acknowledge that they don't pass the test in (a), since any more-than-material elements are certainly marginalized if not negated in their theories. What they'd likely do is suggest that (b) is true and that more-than-material elements such as spiritual forces or more-than-material notions of consciousness have been so thoroughly defrocked in academic and scientific circles that it is a "fact" that they don't exist. A quick example of this suggestion can be seen in Bennett's (2010) articulation of the baseline assumptions of her posthumanist, vital materialism when saying that "[t]his materialism…does not believe in G[-]d or spiritual forces" (p. 63). She offers no justification for such a rejection, she simply presents it as fact. I suggest that it is permissible in modern scholarship for her to do so not because it *is* a fact, but because dualist monist materialism has been so effective at marginalizing and negating the possibility of more-than-material elements of reality that no other possibilities are viewed as legitimate. This is a self-reinforcing, circular pattern of thinking that should be discarded pending a full examination of its assumptions and effects. The substance of this volume is dedicated to that examination. In the context of the test, if I'm right in what I say then monist materialists cannot turn to (b) to support any claim to not being dualist. Pending the more extensive examination of monist materialism's roots, which I'll undertake in Chap. 3, monist materialism fails the test and must be provisionally considered dualist.

2.4   Material and More-Than-Material: A Working Distinction

Thus far in this volume I've made multiple references to more-than-material ontological elements without explaining exactly what I take them to be. I'll explore examples of these and material elements, and why I take them to be in either category, later in the book. But, if I have the reader's forbearance to this point that it is not yet settled that all reality is material, then to find some kind of starting place to differentiate between the two, I state simply that I take elements that are not obviously material—that defy easy or intuitive (and near-complete) explanation at a material level—to be potentially more-than-material. For example, feelings, though certainly linked to all manner of physico-chemical operations, at times defy those explanations. Feelings are not just things humans possess, but at least animals as well. For animals, these have been shown to include fear, joy, happiness, shame, embarrassment, resentment, jealousy, rage, anger, love, pleasure, compassion, respect, relief, disgust, sadness, despair, and grief (Bekoff, 2000, p. 861). Also, my use of the term "feelings" does not only encompass an individual being's emotions. I take them to be phenomena by which relational communication takes place. The nature of feelings as such, and why I don't take them to be material in origin despite the growing body of data showing material influences on emotion (LeDoux, 2000; Panksepp, 1998; etc.), will be explored in depth in Parts III and IV.

I also take consciousness, and cognitive elements such as attention, memory, categorization, language, problem solving, intentionality, and agency to have irreducibly more-than-material components. I discuss this in detail in Part IV, so here state that I don't source the mind in the brain either, despite growing material evidence for a tight correlation between the two. I will argue that the finding of material correlates does not imply causation, nor confirm precedent material status to the brain. It only does so when the *a priori* assumption of a monist materialist ontology is deployed.

"Things" that are not necessarily *in* a being, such as relations, love, *telos*, meaning, and spiritual elements such as a soul (in humans or more-than-human beings) and spiritual beings such as deities (or G-d, if one is inclined in a monotheistic direction) are also included in my conception of what is more-than-material. The remainder of the book will consist of an expansion and justification of these categories, and their essentiality to close human-nature relationships.

On the other hand, I take material elements to be those which, through experience, appear to be physical or manifest materially in the world external to oneself, and internal to oneself in the Cartesian sense of "body." Thus, a horse's body is material as is my body. A field across which he gallops is also material. I'm aware that it is the convention of those drawing the boundaries around that which is "material," that energy is also taken to be material, and yet for me, at times it still resists such easy qualification. Energy as electricity, perhaps, ought to be considered material, but what about when one talks of the human body's energy, such as the *qi* in Chinese medicine that is manipulated by such things as acupuncture (Kaptchuk, 2002)? At one level, it is accessed by material means and considered by some to be a tangible, electromagnetic energy. But if a body dies, does the *qi* die with it? Or, it being energy, is it disbursed into the material world in the sense of Joule's law of the conservation of energy, or into another kind of existence such as an afterlife? In the end and at some level, I'm not sure it has nothing to do with the soul.

References

Barad, K. (2007). *Meeting the universe halfway: Quantum physics and the entanglement of matter and meaning*. Durham, NC: Duke University Press.

Batisse, M. (1982). The biosphere reserve: A tool for environmental conservation and management. *Environmental Conservation, 9*(2), 101–110.

Bekoff, M. (2000). Animal emotions: Exploring passionate natures. *Bioscience, 50*(10), 861–870.

Bennett, J. (2010). A vitalist stopover on the way to a new materialism. In D. Coole & S. Frost (Eds.), *New materialisms: Ontology, agency, and politics* (pp. 47–69). Durham, NC: Duke University Press.

Chandler, D. (2013). The world of attachment? The post-humanist challenge to freedom and necessity. *Millennium, 41*(3), 516–534.

Coole, D., & Frost, S. (2010). Introducing the new materialisms. In D. Coole & S. Frost (Eds.), *New materialisms: Ontology, agency, and politics* (pp. 1–43). Durham, NC: Duke University Press.

Dewey, J. (1929). *Experience and nature*. London: George Allen & Unwin, Ltd.

Evernden, L. L. N. (1992). *The social creation of nature*. Baltimore: JHU Press.

Ferrando, F. (2013). Posthumanism, transhumanism, antihumanism, metahumanism, and new materialisms. *Existenz, 8*(2), 26–32.

Foster, J. B. (2000). *Marx's ecology: Materialism and nature*. New York: Monthly Review Press.

Gaard, G. (2010). *Ecofeminism*. Philadelphia: Temple University Press.

Kaptchuk, T. J. (2002). Acupuncture: Theory, efficacy, and practice. *Annals of Internal Medicine, 136*(5), 374–383.

Kessler, N. (2012). Chaos or relationalism? A pragmatist metaphysical foundation for human-nature relationships. *The Trumpeter, 28*(1), 43–75.

King, R. J. H. (1991). Caring about nature: Feminist ethics and the environment. *Hypatia, 6*(1), 75–89.

LeDoux, J. (2000). Emotion circuits in the brain. *Annual Review of Neuroscience, 23*, 155–184.

Merchant, C. (1980). *The death of nature: Women, ecology, and the scientific revolution*. San Francisco: Harper & Row.

Milton, K. (2002). *Loving nature: Towards an ecology of emotion*. London: Routledge.

Panksepp, J. (1998). *Affective neuroscience: The foundations of human and animal emotions*. Oxford: Oxford University Press.

Plumwood, V. (1993). *Feminism and the mastery of nature*. London: Routledge.

Warren, K. J. (1990). The power and the promise of ecological feminism. *Environmental Ethics, 12*(2), 125–146.

Warren, K. J. (2000). *Ecofeminist philosophy: A western perspective on what it is and why it matters*. Lanham, MD: Rowman & Littlefield.

Williams, D. R., & Vaske, J. J. (2003). The measurement of place attachment: Validity and generalizability of a psychometric approach. *Forest Science, 49*(6), 830–840.

Chapter 3
Posthumanism's Material Problem

The main arguments presented in this book fit comfortably within the boundaries of a basic definition of posthumanism: that humans are not, and ought not to be, taken as the center of consideration in human-nature relationships. I noted this in Chap. 1 while also noting my discomfort with what I see as posthumanism's over-reliance on a monist material ontology. Forms of posthumanism which argue that materialism presents the widest possible field of inclusiveness don't address directly that which materialism by definition eliminates—more-than-material elements of reality. In proceeding down a materialist path, theorists espousing such approaches risk basing their ontology on the dualist privileging of human projects and human qualities and capacities. Again, this is not in the Cartesian dualist sense, but the ecofeminist sense of human/nature dualism.

In addition, it is not my purpose in this chapter to exhaustively explore all aspects of posthumanist scholarship as it relates to the topics of this book. The literature is so vast and varied that to even speak of "posthumanism" as a monolithic literature is inaccurate and reductive. That said, there are some strong, central themes to posthumanism, and common ontological approaches exemplified in those whom I take to be the major voices working in the field—those of Barad, Braidotti, Coole and Frost, Haraway, and Bennett. It is mainly to their portrayals of posthumanism that my critiques in this chapter apply. In this chapter, I will also often move fluidly back and forth between the terms *new materialism* and *posthumanism*. I do this not because I take them to be identical, but because my discussion of posthumanism is an ontological one, and new materialism is the ontological underpinning one finds most often in the work of the authors I just noted. Thus, in this context at times the two are treated synonymously. I note that this sentiment is echoed by posthumanist theorists Cudworth and Hobden (2015), who say, "ideas about which ideas and positions might be 'posthumanist', and which might be 'new materialist' are often used interchangeably by many authors identifying with such positions, and also by their critics…" (p. 3).

© Springer Nature Switzerland AG 2019
N. H. Kessler, *Ontology and Closeness in Human-Nature Relationships*,
AESS Interdisciplinary Environmental Studies and Sciences Series,
https://doi.org/10.1007/978-3-319-99274-7_3

Finally, it is the main work of this chapter to trace the origins of monist materialism in posthumanist thought, explore the ways in which it operates dualistically, and
explore the limitations such thinking places on notions of human-nature relationships and the possibilities for closeness therein. It also focuses on the reductive
effect these authors' renderings of posthumanist theory have on notions of the
individual-in-relation, or the Self as I defined it in the first chapter.

3.1 Deleuzian *Becoming*

I begin my discussion of posthumanism by exploring a phenomenon upon which
many posthumanist theorists have come to rely heavily. It is the Deleuzian idea of
becoming as a means of understanding human-nature interchange in non-humanist/
non-anthropocentric terms. The notion of becoming is explored most thoroughly in
the work *A Thousand Plateaus* by Deleuze and Guattari (1987). Becoming is
described in that volume as a phenomenon where one being becomes like another
being—not in a literal sense, but neither as an imitation nor a resemblance. As the
authors describe it, becoming-animal, for instance means the taking on some set of
attributes of an animal while still not becoming any particular, "molar" animal as
the authors refer to it. To explain what is meant by the difference between the molar
and molecular the authors say this about becoming-woman:

> What we term a molar entity is, for example, the woman as defined by her form, endowed
> with organs and functions and assigned as a subject. Becoming-woman is not imitating this
> entity or even transforming oneself into it…we are saying…that these indissociable aspects
> of becoming-woman must first be understood as a function of something else: not imitating
> or assuming the female form, but emitting particles that enter the relation of movement and
> rest, or the zone of proximity, of a microfemininity, in other words, that produce in us a
> molecular woman, create the molecular woman. (p. 275)

As to the nature of this molecularity, the authors say,

> you become-animal only if, by whatever means or elements, you emit corpuscles that enter
> the relation of movement and rest of the animal particles, or what amounts to the same
> thing, that enter the zone of proximity of the animal molecule. You become animal only
> molecularly. You do not become a barking molar dog, but by barking, if it is done with
> enough feeling, with enough necessity and composition, you emit a molecular dog. (p. 275)

In the authors' discussion, the zone of proximity is key. As to how particles interact
in this zone and the reasons why they do, they say, "Becoming is to emit particles
that take on certain relations of movement and rest *because they enter a particular
zone of proximity*. Or, it is to emit particles that enter that zone *because they take on
those relations* [emphasis added]" (p. 273). As to whether the zone of proximity is
a spatial one, one end note that accompanies the discussion of the "zone of proximity" states that the authors' idea is rooted in "set theory." Set theory offers two types
of "near sets" (Peters, 2013, p. 3). The first are "spatially near sets" and the others
are "descriptively near sets" (p. 4). Since the latter concerns itself with "solving
classification and pattern recognition problems" (p. 6) then it must be the former to

which Deleuze and Guattari refer. Since Peters, as an example of a "spatially *non-near set*" (p. 5), offers a figure showing a knight being lowered onto his horse that hasn't yet reached the horse and been situated in the saddle, one can assume that the zone of proximity to which Deleuze and Guattari refer is one where physical proximity is a required characteristic.

Having suggested that physical proximity of particles is necessary for the becoming phenomenon, at other points it becomes less clear if this is the case. For example, the authors say,

> [Proust's] Albertine can always imitate a flower, but it is when she is sleeping and enters into composition with the particles of sleep that her beauty spot and the texture of her skin enter a relation of rest and movement that place her in the zone of a molecular vegetable: the becoming-plant of Albertine. And it is when she is held prisoner that she emits the particles of a bird. And it is when she flees, launches down a line of flight, that she becomes-horse, even if it is the horse of death. (p. 275)

In this instance, the authors seem to be demonstrating a way in which Albertine and the molecules of plant, bird or horse can be in a zone of proximity from anywhere—with no particles from a molar plant, bird or horse being necessary. Immediately, this begs the question of whose molecules or particles one's molecules and particles are occupying a zone of proximity with. In addition, the vast majority of Deleuze and Guattari's discussion seems to contradict this and speak to a specific, spatial zone of proximity. This makes sense since it is a material theory whose supposed material roots necessitate *spatial* proximity if becoming doesn't want to enter the realm of the fantastic, the ideal, or the more-than-material. Perhaps it is only in sleep like in the quotation above that such non-proximal becomings can occur, but whether that is the case remains an open question.

There is also the question of just how, if there is some entering into molecular relation of particular particles of a type of being that one is becoming, one takes up the molecules or particles such that one *becomes* the one whose molecular zone of proximity one has entered. To that end, the authors say that it is through the mechanism of "contagion." As to how this contagion operates, they say, "the demon functions as the borderline of an animal pack, into which the human being passes or in which his or her becoming takes place, by contagion" (p. 247). They describe the difference of contagion from heredity or some other biological mechanism by saying,

> The difference is that contagion...involves terms that are entirely heterogeneous: for example, a human being, an animal, and a bacterium, a virus, a molecule, a microorganism. Or in the case of the truffle, a tree, a fly, and a pig. These combinations are neither genetic nor structural; they are interkingdoms, unnatural participations. That is the only way Nature operates—against itself. This is a far cry from filiative production or hereditary reproduction, in which the only differences retained are...small modifications across generations. For us, on the other hand, there are...as many differences as elements contributing to a process of contagion... (p. 242)

While the authors have certainly just described contagion as the mechanism of becoming, they still have not explained *how* it functions. In a material reality, this must be known and if it is not, the confidence that it is an *explanatory* mechanism ought to come under scrutiny. That the material workings of contagion are poorly understood and articulated is echoed in Sampson's (2012) description of affective contagion:

> [A]ffective contagion is selfspreading, automatic, and involuntary and functions according to a hypnotic action-at-a-distance with no discernable medium of contact. Affective contagions are manifested entirely in the force of encounter with events, independent of physical contact or scale. (p. 57)

As one can see, contagion, the mechanism, remains opaque. One is seeing what one thinks is the effect of contagion and reverse engineering an explanation with little to show in terms of schematics. What one has above is more descriptive than it is explanatory. Contagion in that context is little more than a material label invented for what one knows nothing about, and in a material world, this will not do. If one were to include more-than-material elements in one's ontology, the way that this communication of particles occurs could be through some spiritual force. As I'll discuss in Chap. 10, this absence of particulars is also the case with theories of emergence that are often invoked by materialists to explain the appearance of things like consciousness from, it is supposed, physical structures like a brain. Deleuze and Guattari (1987), for their part, try to explain this kind of molecular communication by saying, "Under certain conditions, a virus can connect to germ cells and transmit itself as the cellular gene of a complex species; moreover, it can take flight, move into the cells of an entirely different species, but not without bringing with it 'genetic information' from the first host…" (p. 10). Sampson (2012) points out that some of the material explanation for phenomena like contagion in Deleuze and Guattari has to do with Tarde's notion of the social, which is

> not concerned with the individual person or its collective representation but rather with the networks or relational flows that spread out and connect everything to everything else. To be sure, Tarde's contagion theory is all about flows or vibratory events. This is what spreads—what he calls *microimitations*—a point Deleuze and Guattari stress in their homage to his project. (p. 7)

To quote directly from Deleuze and Guattari's "homage" to how Tarde explains the "similarity of millions of people" as a function of "microimitations," the authors say,

> Tarde was interested in the world of detail, or of the infinitesimal: the little imitations, oppositions, and inventions constituting an entire realm of subrepresentative matter…microimitation does seem to occur between two individuals. But at the same time, and at a deeper level, it has to do not with an individual but with a flow or a wave. (p. 8)

While again there is a lot of description here, it's still not clear how it actually works. Given the opacity of the mechanism by which the molecular interchange necessary for becoming takes place, and the possibility that without a definitive material explanation that more-than-material possibilities could fill the void, I'll note that many don't see Deleuze as purely a materialist in this thinking. For example, Hallward (2006) says that "rather than any sort of 'fleshy materialist', Deleuze is most appropriately read as a spiritual…thinker preoccupied with the mechanics of *dis*-embodiment and *de*-materialization" (p. 3). Others have argued that while his philosophy may not have begun in materialism, in his collaboration with Guattari, "he came to embrace fully his materialist…leanings" (Smith & Protevi, 2018, Sect. 3.2, last paragraph). Žižek (2004) also calls into question the materiality of the becoming phenomenon, identifying a deep ontological tension in the work of Deleuze and Guattari when suggesting that becoming can be interpreted both as the "EFFECT of bodily-material process-causes" and as immaterial, *"virtual spacetime* [that] progressively differentiates itself into actual discontinuous spatio-temporal structures" (p. 21) or bodied beings. Žižek suggests that the two poles of explanation are an example of ambiguity between materialism and idealism in Deleuze and Guattari's theory.

Žižek delves even further into these two possibilities, and shows how readings of Deleuze and Guattari could extend into an idealism that is the realm of the spiritual or G-d that could even be seen as "preontological," just as easily as it could fit into material notions of quantum physics. The important takeaway for the purposes of my discussion here is that becoming should not be uncritically accepted as a materialist phenomenon. The only way to do so is via the *assumption* of a materialist ontology. Deleuze and Guattari's explorations are not support for such an ontology.

The amplified uncertainty around the ontological roots of becoming ought to give pause to posthumanists who often rely on it to undergird their arguments for a monist material reality. If becoming is one of the pillars of their arguments, and it is in doubt as a material phenomenon, then posthumanists must either seek another means by which to bolster their materialist claims or, when deploying becoming in their own theories, must explore these areas of weakness in the underlying theory and offer more support for it than is found natively in Deleuze and Guattari.

To put Žižek's work in the relational context that is the subject of this book, one last aspect that is worth exploring is the ambiguity around just what—or *whose*—particles or molecules the one-becoming is becoming *with*. There are two main possibilities. The first is that the molecules, corpuscles, or particles are floating free everywhere in the universe, and that molar beings are constituted out of these particles as Žižek notes. In addition, as part of the process of becoming, these free-floating particles are the ones whose zone of proximity any number of molar beings may enter in order to become the type of being of which these particles are constitutive. In such a scenario, no molar being first owns or emits these particles, they are merely available under the right circumstances. Two circumstances cited in Deleuze and Guattari (1987) above are good examples. In the first, Deleuze and Guattari say, "You do not become a barking molar dog, but by barking, if it is done with enough feeling, with enough necessity and composition, you emit a molecular dog" (p. 275). Here the molecular dog must emerge from somewhere, and perhaps the authors even take this to be a situation where inside of every molar being are all possible molecule types such that, under the right circumstances, they are summoned and emitted. Either way, there does not appear to be, under this first scenario, the necessity of an actual molar dog to emit them such that the human can enter the zone of proximity with them to acquire them and emit them herself upon barking with the right level of gusto. The other example is that of Albertine's dream, that places her "in the zone of a molecular vegetable: the becoming-plant of Albertine. And it is when she is held prisoner that she emits the particles of a bird. And it is when she flees, launches down a line of flight, that she becomes-horse…" (p. 275). Here again one sees not the immersion within free-floating particles but the emission of them under the right circumstances—and those circumstances need not involve any proximity to another molar being for whom those particles are more" native" to the type of molar being they are What's interesting is that in both examples, the emission is from within a molar being. Given that there is no proximity to another molar being, one can assume that the particles that I envision to be free floating must float freely *within* molar beings—a veritable catalogue of all possible beings available to each individual under the right circumstances. Even though the means by which this might occur are not entirely clear, it strikes me as a nod in the direction of materialism that such molecules are contained within a spatially constituted being.

The second possibility, which is rooted in the other possibility offered by Žižek (2004), is that molecules are emitted by beings, and other beings in the zone of proximity have the opportunity to become the type of being that emitted them. This overlaps with the first scenario, but with one important distinction: instead of the becoming depending on a being summoning molecules of his own to facilitate the becoming of another type of being, here a molar instance of the other type of being emits the molecules, and the being in the zone of proximity with that emitting being's molecules then has the opportunity to become the emitting being. This second scenario is represented by the following example from Deleuze and Guattari (1987):

> Wasp and orchid, as heterogeneous elements, form a rhizome…[and] something…is going on: not imitation at all but a capture of code, surplus value of code, an increase in valence, a veritable becoming, a becoming-wasp of the orchid and a becoming-orchid of the wasp. Each of these becomings brings about the deterritorialization of one term and the reterritorialization of the other; the two becomings interlink and form relays in a circulation of intensities pushing the deterritorialization ever further. (p. 10)

Here, the wasp and orchid are clearly becoming one another through the sharing of what we must assume are emitted particles or molecules.

At this juncture, one thing that stands out about these potential arrangements is their lack of *relationality*. Of course at the molecular level, the relations run as deeply as is physiologically possible. There is the greatest of particulate intimacy. Yet, I must ask what becomes of the real relationships that we molar beings have, and that constitute the most important conjoinings of our lives? If all they are is the sum of the microstructural material interactions as embodied in becomings, then one is ordaining a very difficult to grasp, yet still deterministic form of materialism, the latter of which is generally seen as too reductive for most new materialists working in posthumanism. Such an implication from becomings makes sense given that the very notion of molecular interaction as the basis of larger and more complex relational interchange harkens back to Epicurean materialism and its swerving atoms (Foster, 2000). In Chap. 1, I discussed the experience of a German girl who recounts that she "would hug this old tree, and I always felt that it spoke to me…" (Hoffman, 1992, p. 24). Under Deleuze and Guattari's (1987) notion of becoming, this girl gets a "whiff" of the tree's molecules—or even less relationally, some *generic* tree molecules—such that the conditions for becoming-tree are met and the foundation at least, for intimacy with *that* tree is established. One issue with such an explanation of relationships is the presumption that it is *similarity,* embodied in becomings, that is the only vehicle of relational interchange. But, "opposites attract" is an aphorism because that is also a frequent vehicle for interchange in real relationships.

Another issue with this arrangement is that the structure and dynamics of relational interchange are undergirded by becomings that apparently do not require molecules from the *actual relational partners* in order for the becoming with those partners to take place. How then does becoming, one of the most transformative relational experiences possible, not require *something* of the partners in the relationship where this becoming actually occurs? Only molecules of the *type* of partner—woman, animal, mineral, etc.—are necessary. I reiterate a quotation from above to show that the authors affirm the lack of need for molecules from a molar relational partner when they say,

> Becoming-woman is not imitating this entity or even transforming oneself into it… these indissociable aspects of becoming-woman must first be understood as a function of emitting particles that enter the relation of movement and rest, or the zone of proximity, of a microfemininity, in other words, that produce in us a molecular woman, create the molecular woman. (p. 275)

Thus, the interaction with "this entity" is not a becoming, only the interaction between you and the molecules of her type: woman. This stands out as singularly odd given that certain posthumanist theorists believe that becoming is often the vehicle by which the deepest, most particular relational interchange is established (Coole & Frost, 2010).

In a way, becoming as Deleuze and Guattari (1987) have described it instantiates the ultimate and unbridgeable separateness that I suggest all materialism carries. In materialism the best one can expect is an asymptotic proximity, not overlap or merging, even if that merging can only be temporary. Neither can contagion be an existential bridge without any evidence whatsoever for it save a failure of available alternatives when a material ontology is deployed. The resonance between molecules that is the phenomenon of becoming can just as easily be explained more-than-materially—as love in operation—a wave of feeling that is not material at all, causing the microstructural material response of becoming in some top-down, from-the-molar-to-the-molecular, fashion. In the material reality that Deleuze and Guattari have constructed, however, Selves are just two ships passing in the night, barred ontologically from ever being allowed to truly intermix—their molecules so tantalizingly close, yet so unreachable no matter how much magnification is applied to the microscope. One partakes of the corpuscular mist, but what one is not partaking of is *the other.* Next, I'll explore two examples of the posthumanist use of becoming and the complications it creates for the authors deploying it.

In the first example, Watson and Huntington (2008) espouse a posthumanist approach to ontology by grappling with how different epistemic spaces influence portrayals of reality. Huntington, one of the authors, is an indigenous Koyukon man and a hunter who describes the capacity of a moose to communicate his desire to give himself to the hunters, and the hunters' capacity to receive this communication, thusly:

> I explained [to Watson] that people taught in the old ways can *sense the will of animals*; with many animals you see, you probably did not sense that they were there, and when you do see them 'they're just there for a second and they just go back into the woods.' Those ones, you usually won't even *feel them mentally or spiritually*. 'But the ones that are going to stand there and look at you,' *you're most likely going to feel them* way before you see them. 'You don't know what it is. . . . but when you come around that bend you just know something is waiting right there, and that that thing is not just going to run into the woods.' *You have intuited the will of the animal*, the gift from the Creator. '*Sometimes you can change it* a little by making calls, certain kind of calls to make an animal stop, but it doesn't always work.' Because ultimately *the animal is in control of the hunt*. (p. 263, emphasis added)

Postponing to later chapters a discussion of the potential mechanisms for Huntington to feel the moose mentally and spiritually as he describes it above, here it's possible to observe that Huntington is unambiguous in suggesting that such a thing is possible. In describing this communication, Huntington is positioning it as being as straightforward as if the moose had picked up a telephone and called Huntington to communicate his intentions. Watson, in her turn in the essay, attempts to understand what Huntington is describing by referring to Huntington's "intuition" as a "concept" instead of as a phenomenon. Doing so has the subtle yet powerful effect of transforming Huntington's intuition from a interbeing relational element or perceptual tool into a *post hoc* interpretation executed by his mind. This calls into question the legitimacy of the intuition *as an experiential phenomenon* and positions the interpretation of the experience by the human, rather than the relational experience between moose and human, as paramount. Then, when discussing how Huntington refers to using intuition to think "as" an animal, Watson says, "Intuition could result from an act of becoming: 'becoming-animal'...[where] Orville could feel the moose because he had to think as one; not a stable identity, not a literal becoming, but an identity erupting through the performance of the hunter's practice" (p. 265). One can see here that not only is the phenomenon of becoming suggested as the basis for the relationships between Huntington and the moose being hunted, but also as the basis of any intuition that Huntington might have about the will of the moose. Not only does the seeming suggestion by Watson, that by performing an identity he can intuit the moose's will, violate the very basic requirement of becoming that it not be imitation, but if Huntington did not actually enter the zone of proximity with the actual moose whose will he intuits instead of calling up some moose-type molecules, how could he possibly sense *that* moose's will? In addition, even if entering the zone of proximity with the particular moose's molecules would allow Huntington to intuit the moose's will, in the scenes the authors describe about the actual hunt, the hunters were not at all spatially close to the moose until the climax of the hunt when they finally find the moose in one of the drainages they are boating through and shoot him. Thus, no zone of proximity could have been entered in order for this intuition to operate.

One gets the sense from an application of becoming such as this that the concept is used rather loosely, as a catch-all explanatory framework of materialist posthumanists looking to bolster their suggestion that reality is at root material. Given becoming's inherent weaknesses as a material and also relationally particular explanation, and the lack of rigor at times used in its application, it strikes this author as failing to provide convincing reinforcement for posthumanist claims. In addition, by empowering an author like Watson to transform Huntington's intuition so radically—from an experienced form of more-than-material communication to a material, abstracted and generic form not even of interchange, but of humans generically accessing what it's like to be a moose—in at least this application, it is being used to abet dualistic marginalization of an incompatible ontology.

Another example of the loose application of becomings is from van Dooren (2014). He speaks poetically and rightly about how

> all subjects are formed in and through their interactions with others...enfolded in webs of intra-active becoming...[where] what we think of as "human nature" is, over both evolutionary and personal time frames, "an interspecies relationship." And so my comments about this entry of people into birds' lives are not premised on a refusal to accept the co-shaping of all subjects in and with members of other species, but on recognition of the *particular* modes of being and becoming with others that occur... (p. 102)

What one sees in van Dooren is exactly what I suggest in this volume is occurring in human-nature relationships: that humans are inextricably and foundationally embedded in, and deeply influenced by, their human-nature relationships. But, as with many other theorists who deploy a more shallow notion of becoming as only "interdependence" and "mutual influence," van Dooren seems not to see that his reading of "*particular* modes of being and becoming *with others* [emphasis added]" (p. 102) is not ontologically possible in the becoming phenomenon as described by Deleuze and Guattari. Since becoming is microstructurally material, the "others" are secondary to the process. As I'll explore later in the chapter, this loss of identity or Selves is a significant blow to posthumanist aspirations to de-centralize the human in human-nature relationships via subjugation of both human and more-than-human to non-identity based forms of being and relationship. In such a loss, relationships between *beings* as we and other species know them become ontologically inaccurate and of greatly reduced importance. Becoming, in a certain sense in Deleuze and Guattari, is deeply im-*person*-al, yet what van Dooren is so excellently describing is deeply intimate at the level of the Self for human and more-than-human both.

Becoming as a process of radical co-influence in a general sense is a perfect fit for what van Dooren seeks and, frankly, for what I seek in this book. Becoming as a quasi-material, depersonalized, and microrelationally driven and determined "reality" for relationships is not. As I mentioned above, the loose application of the phenomenon of becoming, while it may give theorists the imprimatur of legitimacy that is associated with Deleuze and Guattari's work, and seems to satisfy the desire authors rightly have to move the human out of the center of consideration in human-nature relationships, is ultimately not a good fit for any but the most radical ontologists who understand its limitations and can compensate for them. If these limitations are not accounted for, using becoming in one's theory may work directly against what authors like Watson and Huntington, and van Dooren, are attempting to enable. Some might counter my critique of this reliance on Deleuzian becoming by noting what Braidotti (2006) says, "There are no Deleuzeans...there are only people who engage with Deleuze's thought, figurations and intellectual intensities – in order to construct alternative ways of thinking" (p. 203). But, hers is more support for my suggestion that only meticulous ontologists can wield becoming in order to address its weaknesses and leverage its strengths. In the end, however, much of the posthumanist literature does not take this approach, and merely relies on the general notion of becoming as a stand-in for a materially based relationality.

3.2 Origins of Human Exceptionalism

Having discussed the limitations of becoming, I now move to investigate the roots of new materialism and posthumanism. The posthumanist project, at its most basic, is an attempt to respond to the "awareness [in human-nature relationships] of the limits of...hierarchical social constructs and human-centric assumptions" (Ferrando, 2013, p. 29). The main effort, therefore, is to "decenter" the human in the relationship. This is a laudable goal and, as I stated in Chap. 1, is one that this book shares. But, because of the close ties between posthumanism and materialist ontological approaches like *new materialism*, the identification of the origins of humanism are fundamentally different in posthumanism than they are in this volume. Again, it is for this reason that the present work should not be characterized as a work of posthumanism.

Posthumanist theories generally identify the source of humanism as the privileging of humans by ontologically separating them (or at least some of their qualities or capacities) from the rest of the more-than-human world. This is accomplished through theological or Cartesian dualism, the latter of which is something that Anderson (2014) notes. Depending upon the theorist, the qualities and capacities that emerge from such hyperseparation as human-only or nearly so are rationality, free will, cognition, emotions, consciousness, agency, teleology, self-awareness, and a soul (Anderson, 2014; Coole & Frost, 2010; Ferrando, 2013; Plumwood, 1993). In its most abbreviated form, the posthumanist literature describes this privileging of humans as human exceptionalism (e.g., Barad, 2007 or Twine, 2012).

Deconstructing and replacing this exceptionalism is the main work of the posthumanist project. In order to know how to replace it, though, one must first determine why it exists. The predominant analysis in posthumanist thinking is exemplified by Wilson (2017), who states that

> from the beginning, human *exceptionalism* has been the persistent motif of [materialism's] critics...As Diana Coole and Samantha Frost (2010) observe in their introduction to *New Materialisms: Ontology, Agency, and Politics,* philosophical reflection has favored the "immaterial things...language, consciousness, subjectivity, agency, mind, soul, imagination, emotions, values, meaning and so on. These have typically been...valorized as superior

to the baser desires of biological material or the inertia of physical stuff". The canon of philosophy, from Plato and the Stoics to Kant and beyond, is antimaterialist and exceptionalist… (p. 113)

Wilson (2017), like other posthumanists, correctly sees that human claims to sole possession of these immaterial qualities and capacities are part and parcel of human exceptionalism. But, she and other posthumanists trace the origin of that exceptionalism to the *invention* of the more-than-material category and the installation of humans in it. For example, Coole and Frost (2010) say that "there is an apparent paradox in thinking about matter: as soon as we do so, we seem to distance ourselves from it, and *within the space that opens up*, a host of immaterial things seems to emerge" (pp. 1–2). Thus, according to them, it is only the ontological vacuum that is created by denying the material that allows the immaterial to "appear" or, to more accurately reflect the posthumanist account, be *invented*. Figure 3.1 illustrates this posthumanist interpretation of Cartesian dualist events. To trace the history in this way is understandable, and is a natural outgrowth of their presupposition of a

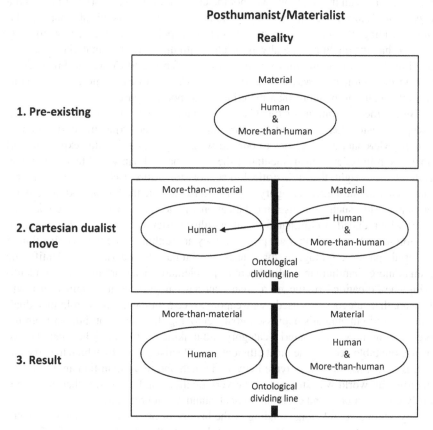

Fig. 3.1 Posthumanist/Materialist understanding of progression from existing/pre-Enlightenment reality through Cartesian dualism to resulting distortion of reality

material reality. What else could the more-than-material be but human fiction if reality *out there* and/or *in here* is material?

But, an alternative reading of these events is supported by Merchant (1980), whose excellent, in-depth treatment of the historical progression of humanism tells another story. Merchant notes that before the Scientific Revolution and Enlightenment periods, it was accepted that human *and* more-than-human had both more-than-material and material qualities and capacities. Thus, the materialist and posthumanist notion that the ancient Greeks, and later Cartesian dualists of the Enlightenment *created* a more-than-material category (Descartes' "mind") and installed humans in it, is incorrect. What Cartesian dualism did was to merely "harden" the divide between the material and more-than-material such that they were now to be seen as qualitatively, metaphysically different from one another. Midgley (2004) echoes this reading of Descartes' efforts by describing his actions as the placing of a "strange gap… between mind and matter" (p. 34) or elsewhere as "Descartes's violent separation of mind from body" (p. 57). If this reading of the history of the more-than-material ontological category is correct and it already existed—and what's more, is *real*— then the Cartesian dualist move was not the creation of this category but instead *the expulsion of more-than-human beings from it*. Figure 3.2 shows this alternative, relational reading of events. The result of such an expulsion was that afterward, *only humans* had the qualities typically associated with the more-than-material—the very qualities listed above and mirrored in Wilson's reference to Coole and Frost. Thus, human exceptionalism was a byproduct of this move to expel more-than-humans from the realm of the more-than-material. It was not the cause.

But, if the motivation for this Cartesian dualist move was not to create human exceptionalism, then why did the thinkers of that time expel more-than-human beings? Merchant again comes to our aid when suggesting that this expulsion was undertaken in order to instrumentalize the more-than-human world to serve human ends. If she is right, then it is critical to see that the desire was *not* to reify a more-than-material ontological category but instead to separate the human and more-than-human worlds such that instrumentalization of the more-than-human could take place with reduced or eliminated social or cultural friction. After all, what treatment of more-than-humans is there to protest if they are nothing more than a collection of material, passive, biological machines? Posthumanists are right to identify the human/more-than-human dualism here as problematic. They are mistaken in identifying the "creation" of the more-than-material category as its origination point. Instead, that category was used as a weapon. Since a human/more-than-human dualism was Enlightenment's expressed goal, removing the more-than-human from the pre-existing more-than-material category and making the boundary between the two insurmountable became the most efficient and effective means by which to carry this out. By creating this false divide between human and more-than-human, the more-than-human world was stripped of the very qualities and capacities that had previously given it moral, and even ontological, standing and relevance.

My alternative, relational reading of the history of human exceptionalism creates deep complications for the posthumanist solution to this problem: that of offering a monist materialism where "human and nonhuman actants can be regarded as co-existing on…a single material plane" (Anderson, 2014, p. 5). Because they read the

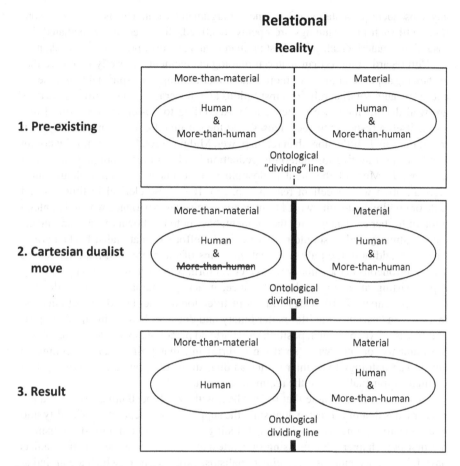

Fig. 3.2 Relational understanding of progression from existing/pre-Enlightenment reality through Cartesian dualism to resulting distortion of reality

existence of the more-than-material category as the means of dualist separation, post-humanists move to eliminate it. But if the ontological category is not the cause, an alternative solution presents itself. It is to simply undo the imposed division between human and more-than-human by conceptually reinstalling the more-than-human into the more-than-material category and once more softening the divide between the two. If a materialist balks at the suggestion of maintaining or reinstalling a more-than-material ontological category, I attribute such a response to the effectiveness of old materialist ecofeminist dualisms in hegemonically, and in self-reinforcing fashion, eradicating the more-than-material as deserving of serious consideration. As I'll address in discussing the work of Midgley (2004) in the "Midgley and Materialism" section below, such a pre-existing materialist commitment is one of the main reasons that a "new" (or really, any) materialism continues to be the choice of most theorists working in the ontology of human-nature relationships today. As I stated at the outset of this book, however, material reality has been *a priori* inserted to provide support for any subsequent presentation of evidence or arguments for it. This is circular reasoning. Until such underpinnings are openly identified, discussed and remediated for, assuming a material reality is at best premature and at worst, profoundly mistaken.

With regard to more-than-material ontological elements, I briefly note here that eschewing superficial forms of more-than-material things like spirituality is wise, as abuses by authoritarian religious institutions and theocracies down through the ages are well documented (see Foster, 2000). According to Foster, materialism of any sort is itself a response to such abuses, and the spiritual fictions concocted by such institutions to do little more than retain power. Midgley (2004) echoes this by noting that "a central motive for materialist reduction has been moral indignation against the Church. Most of the great philosophical reducers have been violently anti-clerical, often with excellent reason." (p. 58). But, those clerical institutions had little or nothing to do with what I term genuine spiritual experience which, I contend without further explanation for the moment, has yet to be shown to carry any inherent illegitimacy and so should not spark similar efforts of materialist-backed eradication. In addition, the only way that *all* forms of the more-than-material can be seen as erroneous in the same ways as clerical forms of spirituality are is by carrying a pre-existing atheistic ontological stance, which posthumanists largely do. For example, Bennett (2010) states without justification that her brand of posthumanist materialism "eschews the life-matter binary and does not believe in G[-]d or spiritual forces…" (p. 63). Such positions are part and parcel of any modernist ontology, posthumanist or otherwise, so it's not surprising that posthumanists seeking to develop an ontology to counter Cartesian dualism would gravitate toward one that eliminates spirituality from the realm of the existential.

I also pause here to note that the dualism with which posthumanists are mainly concerned is that of Cartesian dualism—the supposed false separation of reality into ontologically distinct categories that I distinguished from ecofeminist dualism in the first two chapters. By focusing on Cartesian dualism, however, posthumanists have failed to account for *ecofeminist* dualisms, explications of which are an indispensable tool for exploring the underpinnings of modern human-nature relationships. As I discussed earlier, in the ecofeminist sense of dualism there is either a bifurcation of reality into fictitious categories or the leveraging of existing and potentially accurate distinctions to install as superior one half of the created/recognized dyad. As in the previous chapter, I suggest that all forms of materialism are dualist in this ecofeminist sense, because they denigrate or ignore the possibility of more-than-material ontological elements without well-supported justification.

3.3 New Materialism's Old Roots

Because posthumanists often employ new materialism, no matter the degree to which they take this type of materialism to be inclusive, it fails to account for materialist dualism in the ecofeminist sense. Thus, there is the risk of perpetuating the latter types of dualisms in any posthumanist ontology that employs new materialism. This is precisely what has happened for posthumanists. Again, the materialist resolution for Cartesian dualism proposed in Fig. 3.1 is to move the supposed unique qualities of humans "back" into the material realm. In doing so, however, the supposedly unique human qualities and capacities (that became unique only because more-than-human beings were cast out of the more-than-material category) come along with them, maintaining their unique alliance with humans. All that is required for humans to maintain their claims of exclusivity to these qualities and capacities, then, is to fashion material explanations for them.

This is precisely what has occurred with "old" materialism historically. Midgley (2004) notes that "Descartes established the assumption that, since physical particles moved on the model of machines, the things made out of them, including human bodies, must do that too…[and t]hough Descartes himself exempted the mind from this machine, others quickly saw that it too could be reduced to fit the picture" (p. 49). Thus, as theories of materialism evolved, the explanations for qualities and capacities supposedly unique to humans shifted to the material, in the process either preserving human uniqueness or at least granting humans status as the most advanced or developed in these qualities and capacities. As can be seen, this is a form of human exceptionalism since it is sponsored by a precedent, historically lingering human exceptionalism. It is also dualist.

A first example of this shift of human exceptionalism to material explanations is in Damasio (1994), who provides an in-depth exploration demonstrating how the ability for rational thought is rooted in the functioning of human-like brains. Another capacity, free will, has been explained materially by Epicurean materialists, who root it in the indeterminacy of the swerve of atoms in the void (Foster, 2000). Brembs (2011), while noting that many neurobiologists have declared free will an illusion, suggests instead that it exists and is "a biological trait and not a metaphysical entity" (p. 930). He notes that humans and animals have this trait, and that for humans in particular, the "experience of freedom is an important characteristic of what it is like to be human" (p. 935). Additionally, many fields of study have emerged to take up consideration of the material bases for cognition. All of them root cognitive processing in the brain (e.g., Geary, 2005, and Springer & Deutsch, 1998) while some theorists lay human, social and cultural elements atop it (e.g., Tomasello, 2009, and Fiske & Taylor, 2013). Both Damasio (1994) and Panksepp (1998) root the capacity for human and animal emotion in the material functioning of neurobiological structures and processes. Leary and Buttermore (2003) are amongst innumerable theorists who trace the origins of the capacity for self-awareness to the evolution of uniquely human, brain-based cognitive abilities. In terms of agency, Bandura (2006) suggests that biological evolution is responsible for the human form of it, where

> The evolutionary emergence of language and abstract and deliberative cognitive capacities provided the neuronal structure for supplanting aimless environmental selection with cognitive agency. Human forebears evolved into a sentient agentic species. Their advanced symbolizing capacity enabled humans to transcend the dictates of their immediate environment and made them unique in their power to shape their life circumstances and the course of their lives. (p. 164)

Even human teleology and spiritual experiences have come to be explained not as existential ontological elements, but instead as products of human evolution explained materially. A good synopsis of these explanations is provided by Bulbulia (2004), who states that

> there are two dominant research strands in the naturalistic study of religion. In one camp are those that see religious cognition as a by-product of the evolved mind. For these spandrelists, religion has no adaptive value per se. The psychological architecture that produces god-related thought and activity has evolved for other purposes, and religion falls out of it as relatively harmless noise. On the other side are the adaptationists, who view religion as exquisitely functional, an elegant mechanism best explained as the target of natural selection, and best discovered by reverse-engineering its design. (p. 656)

In all of these examples, though some include animals, it is material structures and processes similar to humans that enable each capacity, and it is generally assumed that of these qualities and capacities, humans have them in their most advanced form. Thus, in old materialism, anthropocentrism and human exceptionalism are left intact.

Posthumanists, noting the problems with such an arrangement, have sought to strike out in a different direction. Critiquing the old materialist portrayal of matter as purely passive, they attempt to inject elements such as agency (Barad, 2007), what they call "lively" matter (Bennett, 2010; Coole & Frost, 2010; Cudworth & Hobden, 2015), and vitalism (Bennett, 2010; Lorimer, 2012) into materialist thinking. This is done in order to try to reconcile old materialism's reading of matter as passive and purely spatial with advances in physics and the recognition that *something* is animating this spatial matter. Not content to leave the animating force to more-than-material explanations, they attempt to construct ontologies where matter isn't just the inert bits awaiting some enlivening force, but are both the bits *and* the force in its dynamic, ongoing acts of self-creation. Barad's (2007) agential realism is a paradigmatic example of this move. She explains it thusly:

> Reality is composed not of things-in-themselves or things-behind-phenomena but of things-in-phenomena. The world is a dynamic process of intra-activity and materialization in the enactment of determinate causal structures with determinate boundaries, properties, meanings, and patterns of marks on bodies. This [is an] ongoing flow of agency through which part of the world makes itself differentially intelligible to another part of the world... (p. 140)

Subsuming these once more-than-material qualities under the rubric of materialism allows for agency to no longer be viewed as in possession of a being in any sort of individual or ontologically essential sense. What matters is matter. As Barad states, "The world is an open process of mattering through which mattering itself acquires meaning and form through the realization of different agential possibilities" (p. 141). Rather than abandon materialism, the posthumanist move is to divest humans of the qualities and capacities that supposedly make them unique. For example, Barad argues that agency is not *human*, it's a quality inherent in all matter. Therefore, humans must not be agents. There *are* no agents as ontological individuals (or "molar beings" as Deleuze and Guattari (1987) term it) according to her. Agency then coheres around individuals, or rather, individuals-in-relation cohere around agency or are its product. Ultimately, then, Barad (2007) is seeking to establish a compelling, inverted sense of agency. It's subtly brilliant in the way it incorporates all beings by making "all" secondary to the microstructural, self-perpetuating matter itself. The world is acting all the time, why not position all individual beings as being its recipients instead of its progenitors? This moves humans away from their exclusive claims and lets all manner of being into the frame of consideration.

If that is all there was to reconfiguring ontology to decenter humans, despite the determinism that lingers—that of individuals being determined by microstructural material components—there is an undeniable egalitarianism to such a theory. One side effect that may be intentional on the part of Barad and those who espouse her ontology but that I find problematic in ways that I will discuss later in the chapter, is that Selves as ontologically irreducible beings are either lost or their importance subordinated to disembodied—or rather, *pre*-embodying—matter. An acceptable cost, perhaps, at least to those working in new materialism and posthumanism. It is an acceptance I do not share, however, and is one that I'll argue next isn't necessary given the alternative interpretation of and response to Cartesian dualism that I defined in the previous section.

To return to Barad's agential realist ontology, another issue with it, and one that posthumanists correctly identify, is that not all beings receive equal shares of things like agency, consciousness, cognition and the other qualities and capacities heretofore attributed either exclusively, or most and best, to humans. This is the "asymmetry" of human-nature relationships routinely referenced in posthumanism and that I will examine in the next section. Given the posthumanist ontology that Barad advances, one must ask about which beings are the beneficiaries of which matter-underwritten qualities and capacities, and exactly how such determinations will be made. I suggest that problems will arise in such posthumanist determinations because divesting individuals of agency and other previously supposed uniquely human capacities will occur atop the old materialist, humanist explanations for those capacities as detailed in the previous section. If those old materialist determinations are left intact, and only the *origin* of the attributes moved out of the "individual," then it's likely that new material ontologies like Barad's will articulate beneficiaries of these qualities and capacities consistent with the pre-existing, old materialist hierarchies. Thus, while new materialism and posthumanism are sensible as attempts to contrapose the human exceptionalism underwritten by old materialist explanations of human qualities and capacities, if they fail to understand that the privileging hierarchy of old materialism is itself rooted in a misreading of the history and action of Cartesian dualism, then their ontologies may enable, instead of counteract, human/nature dualism.

One way to reveal whether old materialist dualisms are still lurking in new materialism is to take some posthumanist text and swap references to agential, self-directed and self-organizing matter for actual more-than-human beings. This should be effective for two reasons. First, the posthumanist literature is riddled with dyads set up not as human being/more-than-human being but as human being/matter. That is, instead of seeing the dyad as between Selves or at least potential Selves—instead of seeing it as relational in a non-reductive, non-microstructural material sense—

posthumanists tend to reduce the more-than-human half of the dyad to objects like microscopes and other "inert" constellations of matter (Barad, 2007; Pickering, 1993). But, as Midgley (2004) says, "reducing wholes to parts is always a good way to downgrade their value" (p. 53). Therefore, undertaking this text swapping will reveal whether the two types of dyad can or should be taken as interchangeable. The other reason for its effectiveness is that if I am right about the presence of these old materialist dualisms, by replacing such im-person-al matter with relational Selves, whatever dissonance the modified text creates with new materialist or posthumanist positions should point to where dualisms are in operation.

For this exercise I take the following passage from Coole and Frost (2010):

> Conceiving matter as possessing its own modes of self-transformation, self-organization, and directedness, and thus no longer as simply passive or inert, disturbs the conventional sense that agents are exclusively humans who possess the cognitive abilities, intentionality, and freedom to make autonomous decisions and the corollary presumption that humans have the right or ability to master nature. Instead, the human species is being relocated within a natural environment whose material forces themselves manifest certain agentic capacities. (p. 10)

Now let's replace the use of "matter" with various references to Selves as "beings" and see what is revealed:

> Conceiving [all beings] as possessing their own modes of self-transformation, self-organization, and directedness, and thus no longer as simply passive or inert, disturbs the conventional sense that agents are exclusively humans who possess the cognitive abilities, intentionality, and freedom to make autonomous decisions and the corollary presumption that humans have the right or ability to master [these other beings]. Instead, the human species is being relocated within a natural environment [where other beings] manifest certain agentic capacities. (p. 10)

Just as Coole and Frost point out about their conception of matter, I note that the passage resulting from converting "matter" to "beings" still maintains the "disturb[ance of] the conventional sense that agents are exclusively human." This hints at the fact that there are *two* paths out of human exceptionalism available when it comes to agency. The first, as I suggest, is to understand how all beings as onto-logically irreducible Selves could possess agency. This may take one down the more-than-material ontological path that I explore as part of the work of this book. The second is the posthumanist path: to locate agency not in beings, but in matter abstracted and reduced to being the true, relationally generative "substance" of the universe.

Recognizing agency's existence and importance, posthumanists are right to want to account for how more-than-human beings can exercise it. But, the genesis of their attempts to abstract agency from individuals-in-relation in order to allow more-than-human participation in agency is actually motivated by *two* things. First, by the need explain how that agency *can* appear to manifest itself as an essen-tial, *individual* trait in humans, while at the same time explaining how *it does not appear to manifest itself as an individual trait for most of the more-than-human world*. Thus, what posthumanist have apparently failed to recognize is that their solution reveals their *acceptance of the idealist and old materialist determinations* that *as individuals*, most more-than-human beings do not have agency—or con-sciousness, cognition, emotions, or anything of the other qualities and capacities under discussion in this chapter.

In some ways, it is astonishing how uncritically some of the distorting results of the old materialist approach to suppositions about humans, founded in "scientific" determinations of individual qualities and capacities, are accepted as "factual" in the posthumanist literature. All the more so given the hard work that theorists like Barad (2007) have put in to recognizing, for example, how a failure to account for human perspective, values, definitions of "equipment" and material effects of the human can greatly influences things like experimental results. In other words, in ontologies like Barad's, there is a clear recognition of the situatedness of the human in all her interactions with the more-than-human world. Yet, the near entirety of the posthumanist and new materialist project is predicated on the acceptance of the "fact" that most more-than-human beings do not as individuals possess agency, con-sciousness, etc. In response, posthumanists attempt to end-run this "fact" by decou-pling agency from individuals, even individuals-in-relation, and superadding it to matter. It is hoped by posthumanists, one must presume, that in doing so this agency-infused matter becomes an ontological tide rising to lift all boats—human and more-than-human alike. Ultimately, I neither believe such a tactic will work, nor do I think it necessary. I reject as dualist such materialist determinations about the qualities and capacities of more-than-human beings, and throughout the rest of this book will demonstrate the ways in which such "facts" are inaccurate and based almost entirely in the *a priori* installation of a material ontology.

Another example that carries this subtle yet powerful acceptance of old-materialist sponsored, human exceptionalism lies in how some posthumanist theo-rists seek not to eradicate consciousness, for example, as a unique human trait, but to downgrade its importance. For example, Snaza and Weaver (2014), say that "embodiment…proves to be of greater importance (ontologically) than conscious-ness" given the latter's identification with the seat of human identity within human-ist Western tradition (p.5). Hayles (1999) undertakes something similar when suggesting that consciousness is "an evolutionary upstart…that…is…a minor side-show" (p. 3). But again, such statements are only necessary when there already exists an unexamined acceptance of the old-materialist, humanist suggestion that consciousness can only be attributed to "higher" animals such as humans because they are the only ones with the material apparatus to support it.

As a last and detailed example, I'll consider Barad's (2007) discussion of a being called the Brittlestar. As with most erudite posthumanists, locating the humanism in Barad's theories and the assumptions underlying them is subtle work. Barad's dis-cussion of the Brittlestar is no exception. Yet, when examined closely enough, dual-ism shows itself. To begin, Barad takes great pains to demonstrate the ways in which the Brittlestar has neither consciousness nor cognition. She gets at this "fact" by making a multitude of assumptions about what the Brittlestar does or does not think—or more precisely, is or is not capable of thinking. For example, she says of the Brittlestar, "This is an animal without a brain. There is no *res cogitans* agonizing about the postulated gap (of its own making) between itself and *res extensa*"

(p. 375). In one interpretation of this statement, she could be taken as saying that with or without a brain, since *res cogitans* as a Cartesian creation doesn't actually exist, any mental agony over the gap between it and the body is pointless since there is no gap. It follows from this that the mental does not exist in *res cogitans*, but in *res extensa*. An alternative interpretation is more straightforward, and more humanist. That is, the Brittlestar does not have a brain, therefore it cannot think about anything, least of all such things as whether there exists a metaphysical gap between the mind and body. I tend to think, based on her other points about what the Brittlestar is or is not mentally capable of doing (that I'll discuss just below), that the latter is more likely her intent, though the former is not inconsistent with the tack of her ontology generally.

In another example of her comments about what the Brittlestar thinks or could think, she says,

> There is no optics of mediation, no noumena-phenomena distinction, no question of representation. Brittlestars are not fixated on the illusion of the fixity of 'their' bodily boundaries, and they wouldn't entertain the hypothesis of the immutability of matter for even a moment. (p. 375)

Here Barad is indicating that any of these "big" *human* thoughts are not the provenance of the Brittlestar because, it must be assumed she means, the Brittlestar does not have a brain, therefore doesn't have a mind. It's also interesting that she puts "their" in quotes. This could mean either that because they don't have a mind, Brittlestars don't have a stable identity in any sort of bounded, static or molar sense. Barad could also mean something related to her discussion of where an individual Brittlestar, who is able to shed body parts in response to predation, begins and ends. She notes this fluidity of the conception of an "individual" Brittlestar, thus what she could mean by "'their' bodily boundaries" is that this sense of boundaries is loose, and thus not a thing in strict possession of the Brittlestar as an ontological individual-in-relation. Regardless, the sense of identity is de-stabilized, a thing which I'll address later in this chapter, and the Brittlestar is still not thinking complex, supposedly human-like thoughts. In yet another example of this lack of thought, and especially these human-like thoughts, Barad says, "Brittlestars are not gripped by the idea of mirroring, imitation, reflection, or other modes of the tropology of Sameness. These echinoderms don't reflect on the world" (p. 382). Here, Barad is saying that Brittlestars have no capacity for mirroring, imitation, reflection, or representation. It's important to note at this point that without a humanist, materialist ontology, it's not a settled matter that a mind requires a brain, and better yet (so one can think the "big" thoughts) a human-like one. Yet Barad accepts this "fact" of minds requiring brains, and requiring human-like ones for certain kinds of cognitive and consciousness-based capacities, without question.

What I find particularly interesting in her discussion of the Brittlestar is that she appears to be hedging her ontological bets with regard to whether a Brittlestar has a mind. As I already mentioned above, when discussing the mental capacities of the Brittlestar or their lack, her invocation of *res cogitans* to stand in for "the mind" creates ambiguity around what she thinks *res cogitans* means. If she is defining *res cogitans* as the fictional mental substance Descartes invented as a separate meta-physical category, then neither the Brittlestar *nor* the human have it. But, if she means that *res cogitans* is a mind more generally, then the human *does* have one while the Brittlestar does not.

Since in some ways Barad is making the case that agency and thoughts etc., originate in matter, it's of even more interest, and creates even more ambiguity, when she makes the following statement: "'mind' is a specific material configuration of the world, not necessarily coincident with a brain. Brain cells are not the only ones that hold memories, respond to stimuli, or think thoughts" (p. 379). One element of the ambiguity results from her placement of quotes around "mind." Usually, such quotes qualify the use of a term and convey that it's not to be taken as literal truth. Thus, Barad might mean that *minds* don't exist as we understand them: as attributable to individuals in any molar way. In this sense of it, neither humans nor Brittlestars have minds. Another possible meaning to the quotes is that minds *do* exist, but they are based in mattered agency, with the quotes indicating that she is radically reconstructing the notion of mind in a posthumanist light. Therefore, the mind becomes fairly foundationally different from the old materialist, humanist view of it being in the possession of an individual, molar being. In this case, both human and Brittlestar have it.

But, there is yet a third meaning possible that I think is the one she most likely intends. That is, in humans it's *minds* without quotes because they occur with a human brain. In Brittlestars, what she is calling a 'mind' does not occur with a human-like brain. The quotes, therefore, are needed specifically for the Brittlestar because, without a brain, in Brittlestars *it's not really a mind.* This reading of her statement is reinforced by the second sentence where she suggests that in some beings, brain cells hold memories and have thoughts, whereas in other beings, those memories and thoughts exist, just not in brain cells and therefore, not in an existential mind like humans have. Of course this begs the question of where thoughts and memories do occur for the Brittlestar if not in brain cells and thus an individual mind. Barad's answer appears to be that for the Brittlestar, they are "*materially enacted*" (p. 375). That is, they occur directly in matter without the "middle man" of a human-like mind and its humanistically, old-materially derived, unique capacities for reflection, representation, etc., that, by their presence *within* human minds, tend to muddy the waters of Barad's agency originating in the microstructurally material.

Leaving that avenue aside for now, the main point I make here is that with *mind* in quotes, but thoughts and memories outside of them, the best reading of Barad we can have here is that thoughts and memories are real for the Brittlestar, but the mind is not. The converse implication is that for humans, all three—thoughts, memories, and a mind for them to call home—are real. This brings me back to my original suspicion that Barad is hedging her bets on whether the Brittlestar has a mind. She recognizes that what we generally characterize as mind-based elements in humans—memories, agency, knowledge, thoughts, etc.—are also occurring, or at least a subset of them, in Brittlestars. But, because of her allegiance to the humanist, old materialist privileging of these elements via insistence that they are only produced by a human-like brain, for the Brittlestar her only recourse is to disperse them across the mattered universe.

For her ontology to be consistent, however, this scattering must also underpin human thoughts, memories, agency, etc. She explains this very thing throughout her work, where even in places in her treatment of the Brittlestar, human similarity in this regard garners mention: "These [Brittlestars] don't reflect on the world; they are engaged in making a difference in the world as part of the world in its differential becoming, and *so are we* [emphasis added]" (p. 382). Even though she clearly positions humans as being the product of agential matter here, and thus having these scattered material elements, she also still maintains a differential, humanist privilege by allowing humans to have an internal-to-the-individual, brain-based mind in which the more advanced thoughts of reflection, mirroring, representation and others take place. Thus, human exceptionalism is intact. Humans have minds that think thoughts. Brittlestars have thoughts, but they aren't really "theirs."

As a final point about Barad's lurking humanism, I'll point out that if she were to just dispose of the material contingency and let the Brittlestar have a mind, how much more consistent, and how much more sense would it make for us to have a thinking, remembering, intentional, agential—that is, *minded*—Brittlestar? Alas, Barad is a committed materialist, and the old dualist materialism says that if you don't have a brain, you at best have a "mind" instead of a mind. Just remember that this refusal to consider more-than-material explanations and instead to advance monist material ones relies wholly on an *assumption* of a material ontology and the dualisms that underpin that assumption.

As a final note, my discussion of Barad here is certainly not meant to be an exhaustive treatment of her ontology, nor even her discussion of the Brittlestar. It only serves as an example of how dualist, humanist elements can both linger in and potentially distort even magnificently wrought, de-anthropocentrizing ontologies such as hers.

3.4   Asymmetry

In some ways, new materialist and posthumanist theorists are openly content with the old materialist determinations about which individuals-in-relation mattered agency, consciousness, etc., have chosen to manifest themselves in or to produce. Posthumanists seem undisturbed by the striking similarity of that list to the one produced more openly by reductive, human/nature dualist, old material ontologies—a similarity which should be a strong indication that dualisms are at work. Posthumanists routinely talk about this as an "asymmetry" between human and more-than-human in human-nature relationships, and take those differences generally as both a factual backdrop to the action of mattered agency, cognition, self-awareness, consciousness, etc., and also as a positive step toward abandoning the need to anthropomorphize more-than-human beings into a false symmetry with humans so that the former can be admitted into ontological, epistemological and ethical realms previously reserved for humans alone. Unfortunately, the posthumanist theorist's seeming unawareness of, or lack of interest in, interrogating these assumptions to reveal any dualisms tends to undercut posthumanism's goals of moving the human out of the center of consideration. As feminist theorist Jenkins (2012) is correct to note about human-animal dichotomies (though it could just as easily be applied to any human/more-than-human one), "Although…posthuman theories reject the ontological dualism between human and animal, hierarchical dichotomies reside within these theories' normative presuppositions" (p. 505).

As an example of this acceptance of dualistically defined difference in posthumanism, in the posthuman field of common worlds pedagogy, Taylor and Giugni (2012) see the role of their field as one that takes up "the ethical and political challenge of learning how to live well together and to flourish with [human/more-than-human] difference" (p. 109). In their essay, the dualism percolates to the surface in their consideration of a whale in a children's story. Noting the differences between human and whale, the authors warn us off of treating the whale as another human. I explore extensively the reasons they give in Chap. 5, so here only point out that they do so because of what they take as the "fact" that the whale does not have the relational capacity to interrelate with the human in the story. Thus, treating the whale as having the human-like qualities of relational reciprocity would be a mistake. They further reveal a dualist hierarchy hidden within their supposedly value-neutral asymmetry when suggesting that "in a common worlds relational ethics…all [the] unlikely intimacies, predictable asymmetries and radical differences pose significant challenges for living together" (p. 113). Again, I will discuss their article in much greater depth in Chap. 5, but here note that intimacies are only "unlikely" if one sets out with the assumption that intimacy with a more-than-human being *is* unlikely, and thus when it occurs, should surprise the human participant. That is a dualist assumption rooted in the materialist suggestion that more-than-human beings don't generally have the material structures and processes in place to engage in intimacy with humans. Another dualist assumption is that the asymmetries are "predictable." If one is really delving to the roots of human-nature relational ontology, as common worlds pedagogy is claiming to do, then examining what is predictable and unpredictable about human-nature relational asymmetry ought to form a central component of the project. Here, it is assumed that the authors and readers are "on the same page" about what is and is not predictable with regard to the qualities and capacities that both humans and more-than-humans contribute to any human-nature interchange. This assumption is dualist and should be set aside.

While author Donna Haraway (2008) claims not to be a posthumanist (p. 19) it appears that she makes this claim (at least in part) to underscore the fact that posthumanism is less a rigid theory and more an ongoing, situated, human-nature relational becoming. Ferrando (2013) echoes the general openness of posthuman theories when stating that "The posthuman overcoming of human primacy…is not to be replaced with other types of primacies" and "does not employ any frontal dualism or antithesis" (p. 29). Thus, for Haraway (2008) to identify as posthumanist betrays, in a way, what it is to *be* posthumanist. Having qualified Haraway's posthumanism thusly, it's essential to note that she is one of the great forebears of posthumanism, with her theories around companion species and cyborgs deeply influencing posthumanist ontology. That's why posthumanists like Braidotti (2006) say that she would "classify Haraway's work as 'high post-humanism'" (p. 197). Because of this, Haraway's discussion of human-nature relational asymmetry is essential to consider.

At first glance, Haraway (2008) appears to be espousing an asymmetry that is somewhat liberatory, for animals at least. She argues that this asymmetry does not preclude things like communication, response, and mutual responsibility between human and animal. For example, when discussing dog agility training, she says, "Humans in agility are not handlers (nor are they guardians); they are members of a cross-species team of skilled adults. With an ear to the tones of asymmetrical but often directionally surprising authority in contact zones" (p. 225). Here she positions the dogs as "adults" who at times appear to be the authorities in their relationship with humans. This means that the asymmetry she describes carries some promise that, at times, the nonhuman animal can escape the inferior position to which he is relegated in the old materialist, or Cartesian dualist hierarchy.

In Haraway's discussion of laboratory animals, she notes the asymmetry by saying, ""[T]he capacity [of people and animals in labs] to respond, and so to be responsible, should not be expected to take on symmetrical shapes and textures for all the parties" (p. 71). But responding and responsibility are still within the matrix of human-nature relationships. Weisberg (2009) notes this, saying that Haraway "emphasizes the centrality of response and responsibility in companion species relationships" (p. 41). Further, Haraway (2008) suggests that laboratory animals display agency even when they are part of laboratory experiments. For example, she notes that lab animals "have many degrees of freedom...including the inability of experiments to work if animals and other organisms do not cooperate" (pp. 72–73). Here she suggests that they have the agency, apparently, to intend to die rather than continue on in the lab. In another example referencing factory farms, Haraway again expresses this sentiment, saying that animals have agency in their "refusal to live" (p. 73) under the conditions in which they are placed by humans.

If these last two examples strike the reader as odd, it's likely because of what Weisberg (2009), who has been invaluable in my analysis of Haraway here, points out about Haraway's suggestions. She calls Haraway's "freedom" a "perverse conception" (p. 35), noting that what is really happening in the lab and the factory farm is a brutal subordination of more-than-human beings where humans leverage "inequality, asymmetry and calculability [to perpetuate] unfreedom and violation" (p. 29). As Weisberg notes, "Haraway attributes to [nonhuman animals] an agency which they are denied in reality" (p. 35).

Weisberg's discussion of Haraway's subversion of the notion of agency reminds me of the article by Watson and Huntington (2008) discussed above in the "Deleuzian Becoming" section of this chapter. Using the moose hunt scenario in it, we can portray a clearer sense of the difference between true more-than-human agency and that which is characterized by Haraway. In the article, Huntington speaks of the moose somehow communicating his will to give himself to the hunters, and through human intuition connected to the moose (and the moose's actions when found—he won't turn and flee), the moose's will to give himself to the hunt is understood. In other words, the moose's true agency is exercised. As Weisberg (2009) notes, to truly exercise agency means to do so in the context of a "freedom *from* exploitation and violence and freedom *to* fulfill one's potentialities" (p. 35). Applying *Haraway's* construction of agency to the moose hunt, Huntington would have had to lock the moose in a cage and then approach him to shoot him. According to Haraway, if the moose didn't will himself to die before being shot, then the moose would be exercising his agency to be gunned down in a cage. This just cannot be taken seriously as an example of agency. What it can be taken as is an example of a human being seeking to, for whatever reason, reinforce human superiority and the largely unchecked ability to instrumentalize more-than-human beings in the peculiar ways that modern societies do. Granted, Haraway does attempt to differentiate between outright domination and that which results when taking animals more seriously as individuals with Levinasian "face" (see Weisberg again). But if, in the end, the result is the same, then such explanations of more egalitarian interchange really only serve to mask the same old humanism, making them it that much more insidious when employed in earnest posthumanist attempts to de-anthropocentrize modern human-nature relationships.

But that is not all. There is something even more deeply seated and subtle as it relates to the asymmetry that Haraway describes, and that betrays the humanist, hierarchical asymmetry from which Haraway unwittingly works to advance her views. As stated above, what is clear from Weisberg's damning critique of Haraway is that Haraway has advanced a portrait of animal agency that equates the tortured reactions of nonhuman animals under near total control of humans with the life choices of an animal free to live as he will. Three possible explanations emerge from this misrepresentation of agency. The first is that Haraway does not understand what agency is. I sincerely doubt this. The second is that she sees more-than-human agency as so limited that acts of true agency and reactions to torment *can* be equated since nothing more of agency can be displayed *by animals*. The third is that nonhuman animals have no agency at all, at which point Haraway is free to project any construal of agency at all onto the blank slate of more-than-humans acting in relation with human beings. In that third sense, she would be taking advantage of their lack of agency to advance one that fulfills what appear to be her own ends: that of justifying animal experimentation and other dualist, humanist acts of instrumentalizing subordination of animals. In any of these three scenarios, Haraway's perspective on the agential capacity of animals is deeply humanist.

Braidotti (2006) does not read Haraway in this way, however. Instead, she says,

> Both Deleuze and Haraway refuse to underplay the contradictions and discontinuities between the human and the non-human environment. They also refuse to romanticize the interaction between them. This sentimental glorification of the humans' proximity with animals is especially problematic in contemporary culture... (p. 200)

Here, Braidotti takes Haraway's "realistic" consideration of animals—we use them in labs as much as we keep them as affectionate pets that we love—as evidence for the argument that we should encourage the former and discourage or abandon the latter. It's worse than that, however, because in Braidotti suggesting that closeness or a bond between human and animal is sentimental, she is saying that the feeling is an amplification of an existentially less powerful or fully developed emotional bond. This means that the strength of the feeling is not in the relationship, nor is any resultant strong feeling inside the human rooted in that relationship. One must assume that for Braidotti, this is because more-than-human animals are thought to be incapable of such a contribution. That is a dualist conclusion to reach, and thoroughly, modernly humanist. It is rooted in the same belief structure I accuse Haraway and other posthumanists of participating in. That is, that animals have more limited abilities to enter into close relationships. Therefore, to feel very close with, or a strong affinity for, animals is to delude oneself. Thus, asymmetry in these posthumanist construals becomes a vehicle for oppression and an impediment to relational possibility. No matter how accepted as "fact" one finds Haraway's or Braidotti's characterization of the qualities and capacities of more-than-human beings to enter into relationships with humans, such acceptance is undergirded by a dualist assumption that places humans at the apex of any relational hierarchy. Because we use animals in labs does not mean that we *should*. It is certainly not *evidence* of its rightness. As the US Forest Service interpreter, Ernestine, also a native Tlingit woman, once told me on a ferry headed north to Alaska long ago (personal communication), "Just because you do a thing doesn't mean you have *the right* to do it."

In a way, Haraway and Braidotti have taken the humanist definition of animals as passive material objects and merely made them active material objects even if the activity is matter's, and not truly theirs. With such a transformation, I suppose that the place of the more-than-human being in relational interchange becomes more essential, and certain allowances for largely material, relational qualities and capacities previously denied them come along. But, as I've argued thus far, they are still dualistically constrained to being lower down in the hierarchy. Ultimately, I'm unsure if this is an improvement, as all it's seemed to do is position animals to be *blamed* for the various predicaments in which humans have placed them without a thought to their genuine welfare and without any mutuality to the decision or the relationship overall. They've animated the passive yet left the circumstances that created it justified. Thus, posthumanists like Haraway say, *if* the animal is animated with agency and finds him or herself in this predicament, since they are animated with this agency, the predicament must be *of their own doing*. It's almost incredible that they don't assign blame to humans for this—that in seeing the suffering of this now active more-than-human Other, that instead of moving to alleviate it, they conceptualize the circumstances of the suffering as if it were an inevitability whose origins are mysterious and whose effects are intractable—like a rainstorm. This, instead of it being *us* who do this, and what's worse, to a now active other with desires of her own!

While posthumanists hold out hope that their ontology will move beyond anthropocentric humanism, if the mechanism by which this is achieved carries the vestiges of dualisms, then such efforts will fail. In some ways, this indicates that posthumanists believe, as I do under different circumstances than these, in what Warren (1990) suggests. Warren says,

> there may be nothing *inherently* problematic about 'hierarchical thinking' or even 'value-hierarchical thinking' in contexts other than contexts of oppression...The problem is not

simply *that* value-hierarchical thinking…[is] used, but *the way* in which each has been used *in oppressive conceptual frameworks* to establish inferiority and to justify subordination. (pp. 128–129)

Let me pause to say that I deeply respect the posthumanist impulse to move humans out of the center of consideration, and clearly recognize that posthumanist thinkers are not intending it, but they have imported from materialism's history an asymmetry, or hierarchy, that is a direct product of an oppressive conceptual framework. They think they have escaped it or neutralized it by suggesting that certain key elements are not traits of beings in any essential sense but inhere in matter—and as matter forms and reforms itself, all beings are the beneficiaries. But, the old dualisms still linger in dictating the asymmetries—in dictating which kinds of being either have access to it or are manifested with it by matter's ongoing dynamism. So yes, asymmetry or hierarchy is possible without the logic of domination, but when the hierarchy itself is *the product of* the logic of domination—of the Enlightenment and then old materialist installation of human superiority—then the hierarchy and the logic of domination cannot be separated. In that case, the hierarchy *is itself a human/nature dualism.* In such an instance—an instance which fits the assumptions underpinning posthumanist asymmetry perfectly—the hierarchy is inaccurate and must either be looked at with an eye toward complete overhaul or be discarded.

One final note here on asymmetry is that I also recognize difference and asymmetry in human-nature relationships. It is rather self-evident that a bird can fly and I cannot. But, I refuse to accept that agency, consciousness, cognition, feelings and other attributes are unique to humans and animals with neural and other physiological structures and processes similar to humans. Even if one were a materialist, I note recent research that shows that pigeons have a sense of space and time (De Corte, Navarro, & Wasserman, 2017). It's unknown where, materially, such a capacity might be located since their neurological structure is so radically different from that of humans. Of course, with dualisms pushed aside, one might suggest that the sense of space and time are not material ontological elements at all. Regardless, with research like this appearing all the time, I am not sanguine with the old, materialist hierarchy—the same hierarchy which undergirds the asymmetry that posthumanism has embraced. Because I espouse use of both material and more-than-material ontological elements to understand what capacities humans and more-than-humans have, and thus what is possible in human-nature relationships, I imagine that any differences, asymmetry, or hierarchy produced by this materialist and more-than-materialist approach would be substantively different from that which is produced by posthumanism.

One last point. Posthumanists at times equate suggestions of symmetry or sameness with anthropomorphism. They don't necessarily employ the term anthropomorphism directly, but they routinely, and without justification or even an attempt to explain it, scoff at suggestions of symmetry or sameness. A good example is from common worlds pedagogy, where Taylor and Giugni (2012) speak denigratingly about notions of symmetry when saying that as theorists, they "are not reiterating harmonious Disney worlds – worlds of happy (cross-species) pluralism, in which cute and innocent children and animals only ever frolic together as equals" (p. 113).

As I discuss later in this volume, anthropomorphism is only inaccurate when more-than-human beings *don't* have the qualities one believes them to have. Letting them in because they are made of matter while embracing asymmetry is saying you accept the assumption that more-than-human beings largely can't have continuity with us in our special, unique, human-only qualities and capacities like consciousness, agency, etc. This only reinforces my suspicions that posthumanists and their new material ontology are, perhaps unwittingly, harboring the same old humanist dualisms.

3.5 Justifications for Materialism

My suggestion in the sections above is that posthumanism has two options to neutralize human exceptionalism—to eradicate the more-than-material half of the Cartesian metaphysical dyad, or to eradicate the differential apportionment of certain qualities and capacities, some material and others more-than-material, to humans as opposed to more-than-humans. As discussed above, the vast majority of posthumanists have chosen the former. Thus, it's worth exploring some of the arguments, beyond assumption, that posthumanists have offered for this choice—and whether such arguments have merit.

I'll first consider Bennett's (2010) take on a vitalist new materialism. As I've already quoted Bennett, when arguing for a vitalist materialism, she states simply that her materialism "eschews the life-matter binary and does not believe in G[−]d or spiritual forces" (p. 63). Here we find the rejection of Cartesian dualism (the "life-matter binary") along with anything more-than-material ("G[-]d or spiritual forces"). As I noted earlier in this chapter, rejection of spiritual ontological elements is often accomplished by equating any spiritual experience with religious totalitarianism. Bennett seems to hold a similar perspective when applauding Hans Driesch for "distinguish[ing] his vital principle from the idea of a disembodied spirit [and in so doing]…explicitly eschewed religious dogmatism" (p. 57). In discussing Driesch, she notes his having stopped short of embracing a materialist ontology, however. She says, "I applaud the way Driesch yokes his vital principle to experiential activities in the lab. This helps him to ward off the temptation within vitalism to *spiritualize* the vital agent" (p. 56). What's fascinating about such a statement is the assumption carried within it that science and scientific ways of knowing are more accurate and thus, superior to some other means of knowing, such as relational ones that I explore in Chap. 7. To Bennett, scientific perspectives act as some "cure" for the weakness of appeals to more-than-material ontology. Again, this flows from the anti-clerical tendencies of materialists overall. But, as I've already pointed out, one cannot easily equate religious totalitarianism with spiritual experience of all kinds. To do so, as Bennett seems to, is foundationally dualist, and only reinforces my suggestion that materialist positions are arrived at not by evidence, but by a self-reinforcing ontology and epistemology that accepts no other forms of knowledge but their own, therefore hegemonically clearing the field of any competing perspectives.

With Bennett having rejected any sort of spirituality, one might expect her next move to be explaining to the reader why Driesch is mistaken in not "taking the plunge" (p. 49) into monist materialism. Strangely, she never does. She details abuses of spiritual or more-than-material elements in the hands of authoritarian-minded humans. She notes some weaknesses of Driesch's perspectives on *entelechy*, the indeterminacy of life that he felt could not be reconciled with purely material explanations. But, when she stands face-to-face with his explanations of why he rejects materialism, she can only say, "If we think of the term *entelechy* as an attempt to name a force or an agency that is naturalistic but never fully spatialized, actualized, or calculable then this vitalist gesture is not inimical to the materialism I seek" (p. 63). To translate, she says that if we simply *imagine* that this indeterminacy is naturalistic instead of more-than-material, then it could be called material. That is not an argument, but a thought-experiment that could be countered by any number of contradictory ones. I repeat a quote of hers from earlier to show it in the context of her inability to overcome Driesch's main objections to materialism. In the face of his objections and her suggestion that one might imagine his *entelechy* as material, she presses on by saying,

> This materialism, which eschews the life-matter binary and does not believe in G[–]d or spiritual forces, nevertheless also acknowledges the presence of an indeterminate vitality-albeit one that *resists* confinement to a stable hierarchy in the world. It affirms a cosmos of a lively materiality that *is* my body and which also operates outside it to sometimes join forces with it and sometimes to vie against it. (p. 63)

In such a statement, she simply transmogrifies this indeterminacy into a cosmos of lively materiality even though Driesch refused to do the same, as Bennett notes when saying, "Despite his great admiration for the wondrous complexity of nature, Driesch could not quite imagine a 'materialism' adequate to it" (pp. 63–64). While I would have to agree with Driesch, Bennett seems ultimately unfazed by his reluctance when she says,

> Nevertheless, I now locate my "vital materialism" in Driesch's wake. Emerson wrote in his journal: "I have no longer any taste for these refinements you call life, but shall dive again into brute matter?" I too go diving there, and find matter not so brute at all. (p. 64)

Thus ends her article, and thus her arguments for a vital materialist ontology. As a set of claims for a materialist ontology, these are very weak. Essentially, what she is saying is that even though she cannot reconcile the foundational difference between her material vitalism and Driesch's immaterial one, she will still assume that materialism can explain what precedes it and so feels justified in making use of the rest of his point of view. This is untenable, at the very least because his entire argument for vitalism leads him to the conclusion that vitality is not material. Inserting her conclusion where his was doesn't just change the conclusion, it destabilizes the entirety of his philosophical project. The sum total of her argumentation distills to:

1. Vital forces exist.
2. I believe in a material reality.
3. Therefore, vital forces are material.

This is not a viable argument. Thus, while Bennett is free to suggest that vitality is a material phenomenon, she's failed to add to the corpus of support for its possibility. That this sort of argumentation is largely unquestioned in posthumanist circles is testament to the hegemonic effectiveness that materialist ontology has had in modern, scientific and academic thought. So effective has it been that it forms a kind of half-seen backdrop, where simply wanting materialism, as is the case with Bennett, is enough to give it argumentary weight. In this book, it is not enough, as I again point out that the acceptance of such desires, where questions about their validity no longer need to be answered, are predicated on a dualist ontological mechanism that self-perpetuatingly works to reinforce its own position while eroding the basis of any that contradict it. This cannot be accepted without critical examination of its origins and a correcting of any mistakes caused by its application.

The other essay I'll consider in order to examine the posthumanist rationale for espousing a materialist ontology is Coole and Frost's (2010) oft-cited *Introducing the New Materialisms,* the introduction to their edited volume, *New Materialisms: Ontology, Agency, and Politics.* As a first step, it's worth exploring features of their opening statement, in particular because its overall tenor stands in notable contrast to the rest of the essay. In most of the rest of the chapter, the authors paint material reality in the more positive and benevolent light to which most new materialist posthumanists are accustomed, but the opening passage veritably spills over with equal parts excitement and dread at staring into the maw of an unforgiving material world. Such a characterization, which they offer as the genesis of their materialist approach, invokes many dualist tropes about the impersonal power of nature as some kind of overwhelming foe. For instance, they speak of matter as "restless" and "intransigent." They speak to humans as having to "endure" matter's "imperatives." They speak of the totality of human action in the material world as either attempting to "reconfigure" or "consume" matter. They speak of humans as being somehow passively buffeted by their own bodily and cellular reactions and also by "pitiless cosmic motions." What's striking about their description is the undeniably ominous quality that the material world takes on within it. Ultimately, it bears no small resemblance to characterizations of life given by Hobbes' (2004/1651) as "nasty, brutish and short" (p. 92) or by Spencer, whose philosophy of a struggle for existence made its way into Darwin's natural selection when the latter defined natural selection as "the preservation of favoured races in the struggle for life" (Gerhardt, 2002, p. 15). After Coole and Frost (2010) finish portraying the impersonal and mindless machinations of material reality thusly, they dramatically ask, "In light of this massive materiality, how could we be anything other than materialist?"

Indeed, if it were only so easy to establish the basis of a monist material ontology. As a proof, it is a weak one. Suggesting that the existence of something is sufficient to establish it as the only thing that exists simply doesn't bear weight. Just like with Bennett (2010) above, I take the absence of the authors' need to provide any more of a proof as, itself, evidence of how thoroughly effective the dualist wielding of materialism has been down through the centuries. When it has been so well established that the material world is the origin of all things, what more proof is needed? Theirs is more a nod to the awesome power of matter than it is a justification for using it as an ontological inception point.

In addition, Coole and Frost's characterization of the world as, at best, a heedless enemy (or unavoidable bedfellow) also harkens back to the human/more-than-human bifurcation that dualism has wrought, and that most posthumanists generally seek to counteract. In contrast, the renowned anarchist, Kropotkin (1902) does not see the world as a pitiless, intransigent foe to be endured, reconfigured, or consumed. Instead, he sees the fundamental fitness of those who survive in an evolutionary sense as based on mutual aid. I've noticed this awe and fascination with the power of material reality in much of the new material posthumanist literature (Barad, 2007; Braidotti, 2006; Coole & Frost, 2010). While science can certainly deliver experiences of the potency of materiality, accepting such potency as sufficient proof of its primacy or its role as ontological progenitor is unjustified. But, when more-than-material ontological elements are assumed not to exist, what choice do posthumanists have? Better to find the roots of what they are right to acknowledge as non-anthropocentric, agential, discursive, generative, relational life within whatever is left. But that is not necessary if one questions the eradication of these more-than-material elements. That such elements are taken not to exist by dualist assumption and not by evidence, however, should give anyone seeking the roots of these forces significant pause. Therefore, even though we've followed first Bennett (2010) and now Coole and Frost (2010) this far, we still have mostly assumption, and very little argument, for a materialist interpretation of reality.

Coole and Frost appear in general to have this tendency, that of offering more of a rationale than a well-constructed philosophical argument to advocate for their materialist position. There is no inherent harm in this, one could suppose, but when matters of ontology are being openly discussed, and one is attempting to justify one's reliance on a certain set of foundational facts about reality in opposition to other possible sets of facts, such an approach is generally an insufficient one. Given the authors' influence in new materialist and posthumanist fields, however, it is nevertheless worth exploring more of their arguments and analyzing the strengths of their claims *either* as philosophical argument or more general rationale.

A first justification they use for adopting a materialist ontology is by suggesting that advances in natural sciences should spur us to use a new materialist approach. Of this they say,

> While Newtonian mechanics was especially important for…older materialisms, for post-classical physics matter has become considerably more elusive (one might even say more immaterial) and complex, suggesting that the ways we understand and interact with nature are in need of a commensurate updating. (p. 5)

This statement indicates that to them, the ontological question before them is not a choice between the material and more-than-material/ideal. Instead, they treat that question as settled in favor of the material, and so move on to suggest that there needs to be a move from the old materialism sponsored by classical science to new materialism. Within a materialist ontology where the natural sciences and physics are considered the best and most accurate means of perceiving and interpreting reality, this is a reasonable enough suggestion. However, it does little to undergird the merits of materialism, old or new, as an ontological approach. What's more, suggesting that the primary means by which we "understand and interact with nature" is through physics speaks to what they see as the foundation of human-nature relationships—matter and the mattered forces that govern it. That is, the foundation is *not* the meaningful, substantive interchange possible between relational Selves that are ontologically irreducible to constituent particles, matter, or forces operating at a microstructural material level. If physics is taken to be that which gives access to the most ontologically basic, then one has already assumed that this is the best and most accurate way to understand human-nature relationships. Recognition of advances in physics does not support this choice, it is the product of its *a priori* installation as the point of reference to use when grappling with the ontological underpinnings of human-nature relationships. If one instead holds human and more-than-human Selves, and their interchanges at the level of the Self, as ontologically basic, physics would either be irrelevant, or merely a secondary, correlated material manifestation of a primary and precedent set of relational elements. It is only when posthumanists accept the dualist suggestion that more-than-human beings cannot interrelate with the human world first at this relational level—they cannot be symmetrical with humans in the sense of also being Selves, even in their differences—does a focus on the material, and advances in physics, come to be central. In this book no such acceptance of dualist negation of more-than-human Selves is assumed or overlooked, therefore no appeal to physics or anything overtly material is necessary.

Another justification the authors offer for a turn to materialism is "the emergence of pressing ethical and political concerns that accompany the scientific and technological advances predicated on new scientific models of matter and, in particular, of living matter…" (p. 5). They note advances such as "biotechnological engineering of genetically modified organisms, or the saturation of our intimate and physical lives by digital, wireless, and virtual technologies" as just two of these technological/scientific advances. Essentially, what the authors are saying is that because science and technology are impacting our lives in materially troubling ways, that we ought to pay exclusive attention to the material in our lives. While I would agree that we should pay attention to these concerning material developments, if the *origin* of the problem is not *assumed* to be material, then turning to a materialist ontology to grapple with them would be a mistake, and potentially a profound one.

The authors go on to explain their recommendation for a material focus by saying,

> [W]e are finding our environment materially and conceptually reconstituted in ways that pose profound and unprecedented normative questions. In addressing them, we unavoidably find ourselves having to think in new ways about the nature of matter and the matter of nature… (p. 6)

But again, what if the root cause of this reconstitution is not material? Or, if one insists on a materialist view, what if the root cause is not most helpfully understood as a matter of physics instead of say, a matter of individual-level human-nature relationships? If that's the case, then thinking in new ways about matter as such is essentially conflating the effects with the causes. It's akin to suggesting that because a beach is being bombed in advance of an amphibious landing, that we ought to turn our attention to the nature of sand. This might aid us in the most superficial sense of finding some way to prevent the sand from getting blown out into the surf, but to get at the root of the problem, it'd be best to understand the nature of bombs, or the planes overhead that drop them, or the geopolitics that spurred the conflict which led to the bombers, and so on. Because techno-scientific advances affect matter, the solution to any issues caused by this need not also be rooted in matter. The only way that they are is if one has already assumed a materialist ontology.

In the context of my discussion of Coole and Frost here, I note that one of new materialism's hallmarks is converting ontological elements previously held to be more-than-material to the material realm. An example of this in the authors' article is when they say that "materiality is always something more than 'mere' matter: an excess, force, vitality, relationality, or difference that renders matter active, self-creative, productive, unpredictable" (p. 9). Usually, and as is the case here, no explanation for the ways in which the physical world produces such phenomena is offered. It's hard to accept as material a thing which cannot be explained materially. Yet this is routinely undertaken within posthumanist thought. Again, the reason that new materialist and posthumanist authors don't feel the need to explain this practice is because of their reliance, unwitting or otherwise, on the hegemonic work of dualisms. Thus, materialist explanations are the only possibility left.

So thoroughly eradicated has been more-than-materialism that the authors can even refer to it by name and know that their readers won't take them literally. For example, they say about matter that it "has become considerably more elusive (one might even say more *immaterial* [emphasis added])" (p. 5). A similar pattern can be seen in my discussion of Bennett (2010) above, where she rejects spiritual forces in favor of a materialism that "nevertheless also acknowledges the presence of an *indeterminate vitality* [emphasis added]" (p. 63). Such a vitality is seen by her as a "force or an agency that is naturalistic [because it is assumed to be so]..." (p. 63). Thus and again, this force or agency is naturalistic by assumption, even though it does not obey material laws such as being spatial or actualized. From whence does such posthumanist confidence in materiality originate? It's from unadulterated assumption. The dualist sponsoring of the subsumption of the more-than-material is so strong within posthumanism and new materialism that even including spirituality itself under the new material umbrella can be proffered without explanation. For example, Coole and Frost (2010) state, "In this monolithic but multiply tiered ontology, there is no definitive break between sentient and nonsentient entities or between material and *spiritual* [emphasis added] phenomena" (p. 10). One other question that comes to mind in reading this is why such a statement recommends a *materialist* ontology? If the authors see no definitive break between material and spiritual phenomena, the explanation *could be* that the material should be subsumed under a purely spiritualist ontology. The only reason this is not what they conclude is, again, that a materialist ontology is assumed. Ultimately, arguments such as these, that position material assertions and assumptions about reality as evidence for materialism are unconvincing. Until new materialists grapple openly with the possibility of more-than-material ontological elements, any dismissal of them or failure to acknowledge them is both humanist and dualist.

3.6 Midgley and Materialism

In the previous section. I discussed the ways in which some posthumanist authors argue for and justify their materialist ontologies. While I've demonstrated the ways in which they lack the proper support to overwhelm the possibility of more-than-material ontological elements, Mary Midgley suggests some other, more general reasons that are worth exploring briefly. Midgley, in her important book *The Myths We Live By* (2004), states that one of the reasons materialists pursue their monist ontology is that for many theorists, developing a monist ontology is a "Serious and reputable wish to bring explanations of mind and matter together in some sort of intelligible relation" (p. 54). She notes that the unbridgeability of the gap between "the two superpowers" of mind and body in Cartesian dualism is "most unsatisfactory" (p. 54) and therefore some way of bringing the two together, through "conquest" (p. 54) if necessary, is appealing to many philosophers. According to Midgley, "Materialists who reduced mind to matter certainly did think they were simply following the example of physics. But that example cannot decide *which* of the superpowers is to prevail" (p. 54). What she is saying is that it is just as easy to "absorb matter into mind" (p. 54) through an idealist ontology. That said, she notes that in modern societies there is an "asymmetry between our attitudes to mind and body" with a heavy emphasis on the body (i.e., the material). Ultimately, she says that reducing one to the other—matter to mind or mind to matter—"spring[s] from the strong demand for unity, from the conviction that reality simply cannot be arbitrarily split in two down the middle. But [again] this formal demand for unity cannot help us to pick sides" (p. 55).

She goes on to say that the attempt to reduce mind to matter is not because *reality* is that way, or because it's the simpler path since all manner of paradox arises from trying to create a monist ontology. Midgley suggests that those espousing one or the other monist position are instead fueled by "the pursuit of an ideal" (p. 55). For example, the materialist notion that "only the physical body is actually *real,* while talk of the mind or soul is mere superstition, can have a startling influence..." (p. 56). What's particularly interesting in Midgley's discussion is her point that the "atheism and anti-clericalism" that permeate modern attempts at monist material ontology "do not actually require materialism" (p. 59). What she means is that in an idealist view, where an immaterial mind is real, does not require anything *spiritual* at all. Therefore, one can be an atheist, disavowing the reality of the spiritual, but not be forced by doing so to choose materialism. Having said that, she notes that "At present...materialism strikes most atheists as a more straightforward path..." (p. 59). I note that atheism is pervasive in almost all modernist scientific and academic circles, and is expressed routinely as a basis for asserting that reality is at root material (see my reference to Bennett (2010) above about not believing in spiritual forces).

Given Midgley's perspective, my earlier suggestion that it is erroneous to equate the shallow spirituality of clericalism and religious authoritarianism with more legitimate forms of spiritual experience becomes more well-supported. I note here that the argument many posthumanists (and modern scholars in general) make to equate these two things is to create a straw man argument against the more-than-material, articulated as follows:

1. Clericalism and religious authoritarianisms are at root humanist, self-serving and have nothing to do with G-d, the soul or other spiritual elements.
2. Clericalism and religious authoritarianisms are the only kinds of spirituality human beings can have.
3. Therefore, all spirituality is erroneous.

In particular, it is claim 2 that is problematic. The entirety of indigenous spirituality the world over, Eastern religions, and small, highly personal forms of modern, spiritual experience stand as counterargument to such a claim. It is a shallow, facile claim and cannot be uncritically accepted.

Further, while Midgley (2004) and many posthumanists like Barad (2007) see the poles of ontology as between the material and the ideal, the spiritual is yet a third alternative. Midgley (2004) does consider this when discussing the possibility of the soul, but in that discussion even she seems to eventually succumb to the possibility that spirituality is not real in the sense of "spiritual forces acting on human beings from the outside" (p. 61). She doesn't reject spirituality outright, but when entertaining the possibility of spiritual forces, she says,

> If indeed it is true that there are no spiritual forces acting on human beings from the outside – if our spiritual experiences are entirely generated from within – then what goes on in our spiritual life is unavoidably psychological business. (p. 61)

In other words, *if* there are no real spiritual forces, then humans are making it up. But, instead of questioning the basis for the conclusion that there are no spiritual forces, she appears to accept that there are not by following this statement with a discussion of the ways in which its important to deal with spirituality as a "psychological business" and what that means for human beings having such experiences. As I've stated in the context of my discussion of posthumanism and new materialism in this chapter, the equation of a false, clerical spirituality with real spiritual experience is dualist, serving mostly to reinforce the dominant, atheistic and materialist position in modern scientific and academic discourse. Thus Midgley, while asking excellent and challenging questions regarding the basis of materialism, abandons her position when faced with a choice between existential, more-than-material, and potentially spiritual ontological elements and those created by humans. This is dualist as well. As with posthumanists, Midgley establishes no basis for this turn away from the potential for legitimate more-than-material experiences. She has either so accepted it as reality that it needs no further discussion, or she's concerned about being thought of as unserious in academic circles if she supports such a possibility more explicitly.

3.7   Relations, *Relata*, and the Loss of the Self

Earlier in this chapter, I made mention of how microstructural relational considerations work against notions of Selves-in-relation that I provisionally take to be ontologically basic. In this and the next two sections, I discuss the consequence that accepting a new materialist, posthumanist ontology has for Selves and their importance in those relations. In this section, when setting out to explore this first topic of relations and *relata* in Barad's (2007) work, I desire again to note the compelling power of her ontology. The egalitarianism is absolutely foundational. That she has made agency available to all beings is clear. Her ontology requires us to grapple with relations as ontologically foundational and generative of what it means to exist. For example, she is precisely right when noting that

> the agential realist ontology that I propose does not take separateness to be an inherent feature of how the world is. But neither does it denigrate separateness as mere illusion, an artifact of human consciousness led astray. Difference cannot be taken for granted; it matters—indeed, it is what matters. (p. 136)

She is also correct in stating that, "It is through specific agential intra-actions that the boundaries and properties of the components of phenomena become determinate" (p. 139) where the arrangement of the elements of phenomena, which she suggests are material but which I think require no such constraint, "effects an agential cut between 'subject' and 'object'" (pp. 139–140). She concludes this description by stating that "*relata* do not preexist relations; rather, *relata*-within-phenomena emerge through specific intra-actions" (p. 140). The baseline ontological notions here, of the rejection of metaphysical separation while acknowledging the importance of difference; of the understanding that things come to be known as *things* through the phenomena of relations; and of the rejection of the notion that *relata* exist prior to relations are all correct and, really, foundational to any relational ontology. I'll discuss these features more in Chap. 6 when I take up and reject the Aristotelian view of substantivism that posthumanists, including Barad, also reject.

Having noted the strengths of Barad's ontology, there are some subtle weaknesses that when teased apart, threaten the totality of the application of her ontology to understandings of what's possible in human-nature relationships. This is especially so when considering whether and in what ways closeness and care are possible. The concern under specific consideration is that her ontology cannot help us get to the Selves, both human and more-than-human, which figure so crucially in the establishment and maintenance of close human-nature relationships. The first set of weaknesses important to contend with are her claims regarding *relata*.

At some point in articulating her ontology, Barad says, "Phenomena are ontologically primitive relations—relations without pre-existing *relata*" (p. 137). My work in this book also advances a baseline relational ontology, especially in Part III, and her statement is totally consistent with it. The key to this consistency is her stating that *relata* cannot exist before relations. The trouble comes, however, in her end-note for this statement. In it, she explains that "the mutual ontological dependence of *relata*—the relation—is the ontological primitive" (p. 429). Here, it becomes clearer that not only is she rejecting the notion of *things* existing prior to relations, she is pressing her case even further by stating that *relations* alone are the ontological primitive. If that's the case, then relations are more ontologically basic than *relata*. Such a suggestion can only be interpreted in two ways. First, that relations originate the *relata* spatially, as if relations are the soil in which plants as *relata* grow. The other interpretation is of relations preceding *relata* temporally. That is, relations exist first, and from the phenomena of relations, *relata* subsequently emerge.

The first issue with either possible arrangement is that the ontological dependency between *relata* and relations is, as Barad herself notes, "mutual." If it is a mutual dependence, then the dependence *must* run both ways—from relata to relations and vice versa. That's inherent in the definition of mutuality. And while it may not be a symmetrical mutuality in the sense of what each contributes to the interchange, if it is *mutual* dependence, then to say that one precedes the other is problematic. Of course there could be some period in which only relations exist, and once they produce *relata*, they *become* mutually dependent, but that contains a paradox which I'll discuss in a moment below. If we take them to be totally mutually interdependent here, her conclusion that relations are *the* ontological primitive is incorrect by her own characterization of the nature of their intertwinement. Further, if Barad is right that they are mutually dependent *and* that relations are ontologically primitive or basic, then so must *relata* be ontologically primitive or basic. In the statement that follows the one I just referenced she states something slightly different, a thing that is more accurate about the mutuality when saying, "i.e., there are no independent *relata*, only *relata*-within-relations" (p. 465). This is exactly right. But it's not what she said in the preceding sentences.

Elsewhere within this portion of her discussion, the dependence could be read differently, not as between the *relata* and relations (and through mutuality, vice versa), but instead as a situation where the *relata* have dependence on multiple intra-actions or agencies. For example, in her end-note she says that "the term 'intra-action' signifies the *mutual constitution of relata within phenomena*...In particular the different agencies remain entangled" (p. 429). This could be understood as there being multiple agencies inherent in relations that "triangulate" to create a *relatum* in space and time—to make her "agential cut." (p. 140). But if that was the case, the word "mutual" would not be applicable, since there is nothing traveling causatively back to the intra-actions' agencies from the *relatum*. Therefore, such an explanation is not viable.

With regard to the potential paradox I made mention of above, the more obviously problematic element in her statements on relations preceding *relata* is that in order for "relations" to be a coherent term, there must be *relata* upon which they operate. Definitionally, relations and *relata* are a mutually interdependent, inseparable dyad, each depending on the other for sensibility as both concept and phenomenon-in-the-world. To reference Taoism, Yin is not Yin without Yang, and vice versa. A curious thing takes place if one tries to establish one as ontologically precedent to the other either spatially, temporally or both. That is, the ontological element that is reified while the other is negated, even if only momentarily, becomes a monad or monolith. In the case of a reified relations, not only do they lose definability *as* relations, but if relations are phenomena that pre-exist *relata*, they suddenly become a sea of indistinguishability—a pure, monotonous flux without difference or separation of any kind. This is something that Barad most clearly does not desire as an outcome. I quote her above speaking to difference as real and as what matters. In addition, if in this moment of pre-existence, relations cease to be relations and become a monad or monolith, then they become the very thing against which Barad argues most vigorously—that is, a *"thing"* that is metaphysically distinct, without relation to anything else. They become a metaphysical individual isolated as all that exists. They are relations without relationship to anything and containing nothing around which to cohere.

If one were to object to my suggestion by proposing that Barad doesn't mean that relations could exist without *relata*, she offers statements that clearly contradict such an objection. At one point she says, "Relations do not follow *relata*, but the other way around" (pp. 136–137). Elsewhere she says, "what is at issue is the *primacy of relations* [emphasis added] over *relata* and the intra-active emergence of 'cause' and 'effect' as enacted by the agential practices [of relations] that cut things together and apart" (p. 389). If one were to argue that Barad only means that there are always relations producing *relata*, that may be true. But without deploying the notion that there is a moment in time, perhaps *the* first moment in time, when relations pre-exist *relata* (which takes us down the monadic road described above) then *relata* must be seen also to pre-exist relations in precisely the same way. If one were to counter *that* suggestion by arguing that only relations have causative power, Midgley (2004) makes a relevant and powerful statement in the context of discussing the causative power of consciousness, but that she intends as generally applicable. She says, "[T]he idea of anything occurring without having effects is an extremely strange one (p. 57). If one were to respond to this by suggesting that *relata* don't have causative power because they don't really exist—that they are only relations congealed with the agential cut at a moment in time—then Barad's (2007) suggestion about the essential reality of difference and separation is lost. It is also at this point that we arrive back at the insensibility of relations without *relata*. That, or we must embrace some form of idealism, radical constructivism, or relativism, all of which Barad and other posthumanists flatly reject. Thus, while one can always look back and find a set of relations preceding any *relata*, in each and every case one can accomplish the inverse task.

To get out of this chicken-and-egg conundrum, one must realize that the relations and *relata* co-occur in time and space. One cannot find one without the other. They make no definitional, logical, or existential sense otherwise. Therefore, a mutualistic middle ground is the only way. This does not mean that *relata* are primary, it just means that neither and both are. They are totally, mutually interdependent. For a scientific thinker seeking a monist ontology, as Barad is, this is not good news. It is a difficult juxtaposition to leave teetering there in the middle. Yet it is the only way if one is not to go to the extreme and either reject all *relata* as a relativist would, or reject all primary relations, as Aristotle, Democritus and other atomistic philosophers would.

Thus and in conclusion, *relata* are to be taken as ontologically basic and primitive in the same way relations are. Again, Chap. 6 contains more of this discussion. Who and what are these *relata*, one might ask? In this book, they are Selves. They are embedded forever and always in relations, but they matter as much as relations do and not just because they are made out of matter, which is yet to be firmly established in materialist ontologies according to my research. These Selves have causative power. They have it because of the qualities and capacities that, yes, are dynamically generated and altered over time by the relations in which Selves are embedded, but that also persist in oftentimes remarkably stable ways that seem at odds with the ephemerality and dynamism of matter to which many posthumanists respond with nearly rapturous reverence when trying to describe it (Barad, 2007; Braidotti, 2006; Coole & Frost, 2010). These stable qualities and capacities are things that one comes to know through relations. In doing so, one comes to know that very thing so particular and unique to every individual Self: her *identity*.

3.8   Relations of Matter or Relations of Selves?

In the previous section I spoke to the "nearly rapturous reverence" that new materialist/posthumanist ontologists seem to hold for the ephemerality and dynamism of matter.

Barad (2007) speaks extensively to this when describing "matter as a dynamic and shifting entanglement of relations" (p. 35), as reality containing "the dynamic and contingent materialization of space, time, and bodies" (p. 35) She also speaks centrally about

> [t]he world [being] a dynamic process of intra-activity and materialization in the enactment of determinate causal structures…This ongoing flow of agency through which part of the world makes itself differentially intelligible to another part of the world…does not take place in space and time but happens in the making of spacetime itself. (p. 140)

At times in posthumanism, even dynamism is dynamic, as when Barad states,

> The changing topologies of the world entail an ongoing reworking of the notion of dynamics itself. Dynamics are a matter not merely of properties changing in time but of what matters in the ongoing materializing of different spacetime topologies. (p. 141)

Braidotti (2006) echoes this when saying that within her notion of "nomadic thought, a radically immanent intensive body is an assemblage of forces, or flows, intensities and passions that solidify in space, and consolidate in time, within the singular configuration commonly known as an 'individual' self" (p. 201). She calls this self an "intensive and dynamic entity" (p. 201). Bennett (2010), for her part, speaks of "materiality as a continuum of becomings, of extensive and intensive forms in various states of congealment and dissolution" (p. 63). She also says of Driesch's *entelechy*, that for him it is the thing which animates matter but for her is a material "force or an agency that is naturalistic but never fully spatialized, actualized, or calculable" (p. 63). Coole and Frost (2010) are also effusive about the dynamism of matter when saying that

the physical world is a mercurial stabilization of dynamic processes. Rather than tending toward inertia or a state of equilibrium, matter is recognized here as exhibiting immanently self-organizing properties subtended by an intricate filigree of relationships... phenomena are now understood as emergent systems that move with a superficially chaotic randomness that is underlain by patterns of complex organization Such systems are marked by considerable instability and volatility since their repetition is never perfect; there is a continuous redefining and reassembling of key elements that results in systems' capacities to evolve into new and unexpected forms. (pp. 13–14)

One can see here a description of the basis of reality as a thing which is changing from moment-to-moment, a veritable welter of material shifting, forming and dissolving. I can understand the fascination with what pragmatists call this "flux" of primary experience. Yet, when it becomes the primary vehicle by which to understand human-nature relationships, several relational consequences follow that work against both understanding and improving those relationships.

The first effect I note is that in posthumanist or new materialist theories, there is a superficiality to the understandings and descriptions of the kinds of relations under consideration in this volume and that I described in Chap. 1. This manifests itself in positioning material interchange at the root of even the deepest relational experiences, which I find to be counterintuitive. As a first example, I'll contrast a human/more-than-human relational encounter described by Buber (1947/2003) with one described by Haraway (2008). Of his childhood encounter with a horse in a stable, Buber (1947/2003) says,

I must say that what I experienced in touch with the animal was the Other...which, however, did not remain strange...but rather let me draw near and touch it. When I stroked the mighty mane...it was as though the element of vitality itself bordered on my skin, something that was not I, was certainly not akin to me, palpably the other...and yet it let me approach, confided itself to me, placed itself elementally in the relation of Thou and Thou with me. The horse, even when I had not begun by pouring oats for him into the manger, very gently raised his massive head, ears flicking, then snorted quietly, as a conspirator given a signal meant to be recognizable only by his fellow-conspirator; and I was approved. (pp. 26–27)

In contrast, Haraway's (2008) description of the qualities of interchange between her and her dog are as follows:

We have had forbidden conversation; we have had oral intercourse; we are bound in telling story on story with nothing but the facts. We are training each other in acts of communication we barely understand. We are, constitutively, companion species. We make each other up, in the flesh. Significantly other to each other, in specific difference, we signify in the flesh a nasty developmental infection called love. (p. 16)

Both of these descriptions characterize the perceived foundation for human-nature closeness. Buber's resonates like the beginnings of a love affair—of intimacy. And while Haraway's also contains elements of intimacy, because of her materialist ontological commitments, it also necessarily features the material elements of their interchange. One issue that arises in Haraway's account is that the material descriptions sound, at least to these ears, a discordant note when interleaved with elements that are more descriptive of a more-than-physical connection. For example, "We have had forbidden conversation" sounds like intimacy, but "we have had oral intercourse" sounds like the byproduct of intimacy and not its foundation. It sounds carnal. While these material elements do go along, at times, with intimacy, I do not believe they are at the core of its formation—at least not initially. One could just as easily say that the materialist descriptions work *against* understanding her interchanges as closeness between her and her dog. While one might counter my analysis of these two portrayals by noting that the physical touching in the Buber account is also a material act. I'd respond by noting that while this is true, it is not the touching that is the *basis* of the sense of I and Thou, it is its physical complement, and has no inherent claim to preceding it.

This "fleshy" and almost hedonistic characterization of relational elements appears in the work of Braidotti (2009) as well. For example, when exploring how Deleuzian becoming influences thought, and how such thought does not or should not conform to normative images of it, she says, "Thinking is life lived at the highest possible power, both creative and critical, enfleshed, erotic, and pleasure-driven" (p. 527). While thinking is pleasurable, it is a fine line to walk between deriving enjoyment from thought and thinking with *the goal* of pleasure that at times becomes erotic. Braidotti also notes that when becoming is used to "explode...the skin of humanism" to create a "'body without organs'...[i]t also introduces a joyful insurrection of the senses, a vitalist and panerotic approach to the body" (p. 527). Braidotti appears to take the entering into Deleuzian "zone of proximity" as a melding of flesh across the boundaries that humanism would suggest are real (e.g., between "human" and "animal"). Again, while such interchanges can be and are a means of interrelating, to position them as *the* means by which to enter into intimacy or closeness of a less superficial kind is only possible if a materialist ontology is assumed such that no other avenues exist.

Another interesting effect of this bodily focus in Braidotti's essay it that, even though the title of the essay from which I quote above is "Animals, Anomalies, and Inorganic Others" it's dedicated almost entirely to discussion of humans, and how their relationships with a vaguely referenced and generic "animal" might change for the better should the supposed barriers erected by human essentialism and exceptionalism be broken down. By focusing only there, what becomes clearer is that in this hedonism and focus on dynamic material processes that propel *us humans* further with animals, little is asked of the animals and how they might feel about such an altered arrangement. The animal recedes into the background, and what comes to the fore is something that begins to approximate human self-indulgence, not relationships of any sort of genuinely non-humanist, reciprocating kind. In fact, to have this focus seems directly at odds with the whole posthumanist project of moving humans out of the center of consideration in human-nature relationships. Perhaps in this essay Braidotti doesn't think animals have anything else to give but their skin to contact. In the end, this is not an acceptable position to take without justification. Ultimately, this is my concern with focusing on material interchange, or suggesting that it is all there is and thus the basis of closeness and the other elements of substantive relationships. It encourages a view of those relationships as shallow.

Perhaps this is due to the lurking humanism that says that more-than-human beings don't have the qualities and capacities as individuals-in-relation to reciprocate or interact in any other way. I reject such characterizations as dualist. The remainder of this book will present arguments against such dualistically sourced assumptions.

Let me pause here to note that my exploration of eroticism in Braidotti's and Haraway's posthumanism is not meant as a critique of eroticism *per se* or the elevation of the erotic to a proper topic of consideration in the philosophy of relationships. But, to *originate* intimacy or closeness in eroticism strikes me as unfounded. It's rare indeed that a "one-night stand" in the human world trends to a long-term, stable relationship. It's not unheard of, but when it does work it's because of something else—loyalty, appreciation of intangible qualities, and hard work. All these must come to the fore to shore up the stability of the interchange. Therefore, while eroticism has its place in close relationships, it is usually as a means of expressing something that is not of the flesh directly. Taking it to be so is the product of the *a priori* installation of a materialist ontology.

This hedonism and ephemerality of relations also seems to appeal to those who equate the material with the less tangible elements of close relationships—as if the erotic *is* love. I, on the other hand, draw a stark line between the two—a line that approximates the demarcation between love and lust. To take lust and couple it with the microstructurally material as a basis for explaining things like love, cognition, etc., seems counterintuitive and achieved more via assumption of materialism than anything about deeper relational elements pointing back to the material as an origin. That's why in my discussion of the two modes in Haraway's description of love with her dog above, I note that the material and the intimate seem to exist in tension with one another. I suggest it is because the latter is not experienced initially or wholly as being material.

The notion that new materialism promotes, instead of discourages, relational detachment is further evidenced by the egalitarianism of *all* relations in this ontology. In reading some descriptions, like Coole and Frost's (2010), one gets the sense almost of glee at the rapidity with which specific interchanges grow and die. Of course material interchange certainly is part of all relations, and yes, relationships wax and wane every day and in every moment. But to take such attributes of relationships as their most basic, detaches them from what makes relationships strong and valuable. In fact, the two kinds of relationships are so different—the ephemeral, microstructurally agential, and material ones and the close, caring and intimate ones between Selves—that I'm not sure we're even speaking of the same kinds of relationships. Wilson (2017) gets at this point to some degree by saying that "the problem of moral normativity is seriously undertheorized within new materialism… New materialist are prone to declare, rather than argue explicitly for, an intrinsic connection between…[its attributes like] indeterminate spontaneity and…animal welfare" (p. 114). Her point is that new materialists have not made a robust connection between their ontology, with its micromaterial energetics, and the effects that construing relationships thusly have on human-nature relational problems like animal welfare.

In the first chapter, I differentiated relations of matter from relations of Selves, suggesting that the latter are ontologically basic, and critical to identifying what is not functioning well in modern human relationships with the more-than-human world. Relations as relations of matter, on the other hand, seem to be the detached, "10,000 foot view" of relationships which lacks investment in any particular one of them, or any particular kind. Where is the real anguish at the loss of some relationship, or the extreme joy of participation in a valued one persisting over time? Some might counter this question by noting that relations of matter are detached because they cover all possible relationships, and only serve as the foundation of these more developed, inter-"personal" ones. But, in Chap. 6 I will present an argument that such a conclusion is itself undergirded by a materialist ontology, therefore it cannot be accepted uncritically. If one were to argue that my sorts of relations don't cover all possible relations as agential realism does, then one can see we have returned to dualist assumptions about what kinds of human-nature relationships are possible based on dualist, hierarchical assessment of the qualities and capacities of certain beings to enter into and participate in them substantively. This book embodies a challenge to the assumption that there is any limit at all on who can participate in close human-nature relationships. Thus, it is my contention here that substantive, reciprocal relations are *also* universal and therefore should displace reductive, material versions of them.

In some respects, for all the argumentation around this dynamism of the physical world, what I believe posthumanists are really trying to say is that we are all connected and that humanism's bestowal of unique, isolating and superiorizing qualities and capacities upon the human is wrong. This is an excellent position to take. The difference between posthumanism and the approach in this book is that here, a dynamic physics is unnecessary to counter humanist arguments. Such materialist arguments have their own, internal and eventually *fatal* flaws. As a standalone constellation of ontological elements, humanism is rife with assumption and self/human-serving "facts" that requires no external lens or alternative view of a dynamic physical reality to debunk. My suggestion here is that both material and immaterial humanism have the same origins—ecofeminist dualisms. Posthumanists have only managed to identify the deep flaws in the Cartesian form of immaterialism and move against it. By failing to trace the precedent effect of ecofeminist dualisms there, and also in older forms of materialism, they've left these dualisms intact in the foundations of their own theories. Thus, humanism is once more allowed to creep back in, significantly weakening the posthumanist position and greatly reducing the chances that their prescriptions to counteract humanism will succeed.

One final note. Throughout this book I refer to the dualistic characterization of more-than-human beings as passive, material objects. One logical extension or byproduct of locating matter in agency instead of in Selves is that the Self, or the individual-in-relation remains passive. The active portion, or agency, is not the individual's, but instead that of microstructural matter. As Coole and Frost (2010) put it, "Our existence depends from one moment to the next on myriad micro-organisms and diverse higher species, on our own hazily understood bodily and cellular reactions and on pitiless cosmic motions" (p. 1). Therefore, *as an individual-in-relation* a being is still passively receiving the activity and dynamism—she is still an *object* of it—it just boils up from both within the individual because the being itself is an intra-action, or from an exteriority within the larger intra-action within which the individual is embedded. But when one is an identity and/or a Self, while one is affected by the workings of the universe and matter, one also affects others. Posthumanism wouldn't deny this, necessarily, but in that family of theories at best you *are* the action, it's just not *your* intent, therefore it really is not *you.* There is something intuitively unsatisfying about such an explanation that is counter-relational in anything but a brute, microstructurally material sense.

3.9   Ecofeminist Selves vs. Posthumanist "Selves"

As support for the contention discussed in the previous section that all human-nature relations carry the potential for closeness, I turn to ecofeminism and the esteemed Val Plumwood (1993). As part of the effort to define the features of the "relational self" that Plumwood constructs, and that I leverage for my notion of a Self in the first chapter, Plumwood says the following:

> The relational self gives an account of the non-instrumental mode, which includes respect, benevolence, care, friendship and solidarity, where we…treat at least the general goals of the other's wellbeing, ends or *telos* as among our own primary ends…The relational self delineates the general structure of a relationship of respect, friendship, or care for the other as a variant on Aristotle's account of friendship: wishing for the other's good for their own sake. (p. 155)

One can see here that elements of care, friendship, respect, and reciprocity between humans and the more-than-human world form the core of Plumwood's ecofeminist ontological commitments. Another example that carries much the same tenor of description, as well as contents of relationship, is from Warren (1990), who says about climbing a mountain that

> ecofeminism makes a central place for values of care, love, friendship, trust, and appropriate reciprocity—values that presuppose that our relationships to others are central to our understanding of who we are. It thereby gives voice to the sensitivity that in climbing a mountain, one is doing something in relationship with an "other," an "other" whom one can come to care about and treat respectfully. (p. 143)

What can be seen here is that selves in ecofeminism have these relational qualities and capacities, and the relationships within which they are embedded have close relational elements. This stands in stark contrast to descriptions of selves in posthumanism, and the mattered relationships from which they supposedly emerge or that they really should be taken to *be.*

I find the contrast between ecofeminism and posthumanism particularly interesting given that they have some common origins in feminist thought. Without probing too deeply into the causes of their divergence, I'll note that posthumanism did not emerge from ecofeminism, and seems instead to have come forward from more scientifically-oriented feminist theories. This is not surprising given that some of the most influential posthumanist ontologists have a scientific background. Barad (2007), for example, started her academic career as a physicist. Andrew Pickering (1993) also has a doctorate in physics. Braidotti (2006) notes that "Haraway's training in biology and sociology of science are very useful here" (p. 198), and goes on to say of Haraway's perspective that "in [her] line of thinking the practice of science is not seen as narrowly rationalistic, but rather allows for a broadened definition of the term, to include the play of the unconscious, dreams and the imagination in the production of scientific discourse" (p. 198). Braidotti thus intimates that science broadened beyond the rational helps form the basis of posthumanism. This reliance on science is also seen in Coole and Frost (2010), who note that for new materialism, "Of great significance here are, firstly, twentieth-century advances in the natural sciences" (p. 5). Barad (2007) states plainly that her "approach is to place the understandings that are generated [from natural and human sciences] in conversation with one another" (pp. 92–93). Wilson (2017), in her discussion of the history of materialism, suggests that "like the old materialism…the new materialism demands philosophy's engagement with the natural sciences" (p. 111). As I'll explore throughout this volume, such a reliance can introduce dualisms that carry over from the historical forms of science, and often go unrecognized. I've already articulated some of them in this chapter as working directly against a more fulsome understanding of the relational qualities and capacities of humans and more-than-humans, and thus the possibilities for relationship between them.

In contrast to the scientific grounding of the understanding of relationships is the ecofeminist approach I just described above. It is "less scientific" in the sense that it comports poorly with modern scientific judgements about the relational qualities and capacities of more-than-human beings and how one might arrive at them. But, since I take those judgments to be dualistically grounded, and thus sometimes inaccurate, a truly broadened science may yet determine that ecofeminists are more accurate in their assessments. In another sense, the less scientific tenor of ecofeminist theory overall works to counter the dualisms inherent in the sciences, and put the discussion on firmly relational grounds—originating in Selves in interbeing relations instead of in particle or quantum physics. What's more, I see ecofeminist thinking as more aligned with the basis of care ethics, which is less about scientific accuracy and more about reciprocity and care in a relational matrix. This view is exemplified by Noddings (1984), where one passage of hers in particular comes to mind as it relates to this discussion. When talking about the basis of reciprocity in her care ethics, Noddings says,

> It is not at bottom a matter of knowledge but one of feeling and sensitivity…When I receive the other, I am totally with the other…I cannot be wrong in responding…It is not a matter of knowledge at all…[I]n the receptive mode itself, I am not thinking the other as object. I am not making claims to knowledge…What is offered is not a set of knowledge claims to be tested but an invitation to see things from an alternative perspective… (p. 32)

Essentially, what she is saying is that the critical part of the reciprocity is not having knowledge about the other, but in responding to the other *as an engaged relational partner*. It is not so important to be accurate in reading the other, but instead to accept the invitation from the other to see from his or her point of view. This is precisely the foundational relationalism—between *Selves*—that I suggest is missing from posthumanism, for all its general claims that mattered ontology will eventually produce this, or at least something that moves the human out of the center of consideration. As I've stated above, I'm not as confident in this eventuality, especially given the lack of robust explanation by posthumanists on how the bridge from agential matter to relational Selves will be crossed. Haraway (2008), for her part, does attempt to provide some of this, but as I noted when discussing her take on asymmetry, there is either something missing from her prescriptions, or some dualistic tendencies operating within her work such that more-than-human beings are not accurately represented as Selves.

3.10   Agency Without Agents

As a last topic of consideration in my critique of posthumanism, I take up the agential realist notion of agency without an identifiable agent. Barad and others' attempts to dislocate agency from agents as individual beings-in-relation means that the traditional philosophical definition of agency is at risk of having to be abandoned entirely. One of those definitions states, "In very general terms, an agent is a being with the capacity to act, and 'agency' denotes the exercise or manifestation of this capacity" (Schlosser, 2015, paragraph 1). Here, clearly, agency requires an agent.

Clearly, posthumanists such as Barad (2007) are looking at the issue from a different perspective, because to them agency is not a "trait" and should not be associated with individual mental states, intentions or individuals at all. For example, Barad notes that

> there is no determinate fact of the matter about both our thoughts and our intentions concerning the object of our thoughts…We are used to thinking that there are determinate intentional states of mind that exist "somewhere" in people's brains…But…[i]n the absence of [the conditions under which we could measure these intentional states of mind]…the notion of an "intentional state of mind" [is] meaningless… (pp. 21–22)

Barad is essentially making the point that if intentionality cannot be measured, it does not exist as a distinct and real mental phenomenon. She goes on to note that in order to "speak in a meaningful way about an intentional state of mind, we first need to say what material conditions exist that give it meaning and some definite sense of existence" (p. 22). Given the incredible complexity of doing so, she suggests that "we are led to question whether it makes sense to talk about an intentional state of mind as if it were a property of an individual" (p. 22).

This proof must be unpacked to test the validity of her argument, as her conclusion is not self-evident. Barad's first suggestion is presented as an assumption—that is, intentionality can only be meaningful or real if it can be measured objectively in some scientific and material way. Even in scientific circles, this requirement is unfounded. The entirety of qualitative data collection speaks against such a requirement. While there has been a long-running and lively debate about just how scientific qualitative data collection is, that it is not materially measurable is certain. The next requirement she proposes is that we must be able to articulate all the material conditions that give individual intentionality meaning and existence. If we can't, she suggests, intentionality likely does not exist as the property of an *individual*. And, if it doesn't, then it must inhere in matter itself. The problem with this second requirement relates to her theory of agential realism in general. That is, the material conditions that relate to *any* phenomenon in her ontology are so complex, and in such a constant state of flux, that to describe "what material conditions exist" surrounding anything at all is impossible. By extension, *no phenomena* can meet her requirement, therefore nothing can be taken to have meaning or exist, not only as the property of an individual, but really to exist at all. This outcome is certainly not her intention, but that she requires this of individual intentionality and not of other phenomena may speak as much to her "intent" to dislocate intent from individuals as to any impartial assessment of the nature of intentionality. One other weakness to her argument is that even if *we can't* articulate the material conditions that exist it does not follow that they don't exist. It may simply be beyond our capacity as humans to inventory and articulate them.

Her argument is not without some valuable elements, however. Specifically, her suggestion that an individual's intention may be informed and shaped by, though perhaps not wholly determined by, a host of external-to-the-individual but internal-to-the-intra-action elements is a worthy perspective. It helps situate the individual-in-relations. This is a useful perspective for any analysis of human actions in encounters with the more-than-human world.

Those valuable elements aside, her argument that intentionality cannot reasonably be attributed to an ontologically isolated individual has merit only because an ontologically isolated individual doesn't exist.

The rationale by which she denies intentionality to an individual-in-relation, however, is far weaker. As it relates to individual *agency*, the importance of her exploration of intentionality lies in her understanding that it forms the basis of that agency such that, without this basis, humans can no longer be said to possess individual agency either. As an example, Barad says, "In an agential realist account, agency is cut loose from its traditional humanist orbit. Agency is not aligned with human intentionality or subjectivity" (p. 178). I suggest that one of the central reasons for denying the reality of individual intentionality, even for acknowledged individuals-in-relation, is to free agency to be a force acting in the world. For example, Barad says,

> Perhaps intentionality might better be understood as attributable to a complex network of human and nonhuman agents, including historically specific sets of material conditions that exceed the traditional notion of the individual. Or perhaps it is less that there is an assemblage of agents than there is an entangled state of agencies. (p. 23)

As I've noted elsewhere in this chapter, a baseline motivation in posthumanist attempts to divorce agency from agents is to allow agency to flow to more-than-human beings that, as individuals, may not be seen to have it. This is not to downplay another motivation Barad might have to simply argue for what physics tells her about the dynamism of reality. But, that posthumanists seek a non-humanist avenue for ontological understandings of human-nature relationship is not particularly controversial. The strategy of positioning agency in matter is deployed to combat the Cartesian dualism that isolates agency in humans alone, either as immaterial quality in the Cartesian sense or as an old material one inhering in human or human-like minds. That Barad also seeks the displacement of agency from individuals for these reasons—because she subtly yet clearly accepts the humanist account that most individual more-than-human beings cannot have agency—can be seen in the following statement:

> From a humanist perspective, the question of nonhuman agency may seem a bit queer, since agency is generally associated with issues of subjectivity and intentionality. However, if agency is understood as an enactment and not something someone has, then it seems not only appropriate but important to consider agency as distributed over nonhuman as well as human forms. (p. 214)

Here, she appears to be stating that agency can be distributed to more-than-humans only via this de-individualized agency. This is reinforced when she says, "the space of agency is not restricted to the possibilities for human action. But neither is it simply the case that agency should be granted to more-than-human beings as well as humans" (p. 178). Given her position, one must assume that, were she forced to assign agency to individual beings, she is poised to assign it only to humans. This is dualist and ought to be rejected.

I see agency and intentionality as inherent in both human and more-than-human Selves. Because Selves are individuals-in-relations, if agency and intentionality in relations is lost, one also loses all meaning of close relationship. With whom am in love if not the Self into whose eyes, or leaves, I stare? I understand that Barad may not worry over such a loss because individuals still get to be and act like individuals—but the logical extension of what she is saying is that the situatedness of beings-in-relation is *so* situated that to speak of agency or intent in those embedded beings is the stuff of illusion. It's one thing to ignore the "external" influences on molar beings and treat those beings as ontologically distinct. She is right to critique this. It's quite another to suggest that the external influences *are* the relationship, and that the individuals embedded in them aren't "really" acting, caring, reciprocating, intending, exercising agency or performing any other foundationally relational activities. Never mind how agency or other supposed material "forces" come to exist (that they are taken as inherent in matter is wholly sponsored by dualist assumption), here we have a clear indication that we are mere actors in matter's play of agency. And in a way, one can accept that we are embedded in this way while still wondering how it's possible that we as "we's"—as we know our Selves and other Selves—don't have an influence in our capacity *as Selves* on the course of our relationships.

As a simple way to underscore the difficulty of her contention that intent is not in the possession of an individual-in-relation, I note that in several places in her book Barad speaks of various humans' intent. For example, she says, "It is not my *intention* [emphasis added] to contribute to the romanticizing or mysticizing of quantum theory" (p. 68). Here, clearly she is referring to intention that she, as an individual-in-relation, is capable of having. She does this again when saying, "Although I will ultimately add substantive qualifications to this definition, I *do not intend* [emphasis added] to weaken what I take to be the spirit of Cushing's demand" (p. 44). She also makes such references when discussing others' work, for example when she says, "despite *Hacking's best intentions* [emphasis added] to leave representationalist beliefs behind..." (p. 55). I point out these examples not as a way to "catch her" in a contradiction, but to reinforce the point that it is exceedingly difficult to divest agents of agency, and doing so may be an inaccurate reading of how agency actually works. It may be the product of innumerable relations, even material ones, but that it is gathered in some coherent way and deployed by an individual-in-relation into the world—even if its effects are never wholly controllable or predictable—is undeniable. Even if one were to want to situate the individual in relations, as is right and proper, to deny that individual all efficacy *as* an individual, especially when speaking to matters of relationship between individuals, just doesn't ring true. Does such a move enable the inclusion of more-than-human beings in agency's play? At some level, I suppose the answer is yes. But in the end, it actually doesn't, because what's lost in the process is the sense of a more-than-human being as a Self. The same occurs for human beings in such an ontology. Because of materialist commitments, in order for theorists like Barad to de-anthropocentrize their ontologies, they are forced to de-individualize it to the point where it simply fails to comport with common sense or any common ability to *make* sense of the world and the relationships within which one is inextricably embedded. Because I reject any Cartesian or materialist humanist supposition that more-than-human beings can't have agency as individuals, I need no matter-based workaround to explain reality.

Not all posthumanist theorists are as accepting of the Baradian move to position agency as inherent in matter. For example, posthumanist theorists Cudworth and Hobden (2015) note that "there have been strong criticisms of such a position of distributed agency" (pp. 2–3). In the context of consideration of *human* agency, the authors ask, "If we humans are simply another node in the relational net of lively matter then how exactly can we be seen to act in and on the world..." (p. 3). Again in the context of human agency, the authors address critiques that "new materialism ignores the unique specificity of human agency..." (p. 3). Leaving the potential dualism in such a statement aside for a moment, the authors raise an excellent point about issues with mattered agency that appears both Bennett and Latour's work. They note Bennett's position that "non-human assemblages can act" (p. 7). In response they say,

> However, what she actually seems to mean is that assemblages can have an impact or effect on humans and non-humans. Here, Bennett is conflating the idea of the properties and powers of beings and things, and the notion of action and the idea of agency. (p. 7)

The authors are getting at my earlier point that the very notion of agency is bound up with an agent. *Action* as matter in motion can have a causative effect in the world, but that is not synonymous with agency. Cudworth and Hobden underscore this point when considering Latour's Action Network Theory. In this theory, they say that

> agency is inflated conceptually (so that it becomes simply a capacity for action) and exten-
> sively (so that anything that has an effect on something else is seen as an actant, from fisher-
> men to scallops). However, the difficulty with Latour is that in his broad sweep, all agency
> is understood as of the same quality. (p. 8)

Cudworth and Hobden also reference Chandler (2013) in their discussion, and it is worth considering some of the key points he raises directly and in a bit more depth. Chandler's critique of distributed agency echoes some of the points I make above but amplifies them. He says that these processes of material agency

> create contingent and fleeting chains of *causation* that are not in themselves traceable as
> causal relations…Our new attachments therefore mean that we can no longer act as subjects
> in the world. We can speak, have understandings and views…but we can never be *human*
> subjects, collectively understanding, constituting and transforming our world. (p. 528)

As I will discuss below, both Chandler and Cudworth and Hobden are still mainly concerned with *human* agency, but when more-than-human beings are released from the materialist strictures that assume they do not and cannot have such rela-tional qualities and capacities as agency, Chandler's statement could also apply to them. Like these authors, the concern centers around the loss of identity or Self-ness. Of this, Chandler says that with distributed agency "we can never be human subjects, collectively understanding, constituting and transforming our world" (p. 528). Cudworth and Hobden point out that Chandler's concern rests in the fact that, with an emphasis on process and becoming, distributed agency transforms all beings into "objects transforming objects – rather than subjects transforming objects" (p. 529).

In attempting to address this concern about an agent-less agency, Cudworth and Hobden (2015) propose separating agency into categories. One is "transformative agency," where "humans and possibly some other creatures engage in a struggle over resources and social organization to effect differences in that distribution" (p. 9). This is the human agency that both Chandler and Cudworth and Hobden appear most concerned with preserving in the face of distributed agency. Another is "affective agency" which the authors introduce thusly: "The human world overlaps with innumerable non-human systems, both animate and inanimate, which can impact and influence, and indeed radically change the structures of the human world. We describe this as 'affective agency'" (p. 9). As one can see, this affective agency is meant to align with the authors' notion that "things" can have actions that "affect" other things. The clear distinction between the two is that affective agency is not in possession of an agent, especially a human one.

To expand upon who has only this "affective agency," the authors say, "[W]e consider the agency of non-human species to be constricted in the extreme and that privileged groups of humans exercise considerable power over the lives of human and non-human animals and intervene dramatically and often disastrously in non-human lifeworlds" (p. 9). In first reading this statement from the authors, I thought I had misjudged them, and that they must mean that the agency that more-than-human beings *do have* is constricted because of the considerable power exercised over them by humans. In other words, more-than-human beings were not free to exercise their agency because of external human impingements. However, that is not what they mean. What they mean is what I feared they meant initially, that most more-than-human beings have little to no transformative agency at all, and instead only affective agency. Two elements indicate the presence of this dualist assumption on their part. First, in the quote above they don't include dominated *humans* as having their agency "constricted in the extreme" even though in the political realms to which they partially direct their discussion, humans are also impinged upon by other humans. Thus, it must only be more-than-human beings for which this occurs. Second, the authors follow the statement above by saying, "We have argued for a conception of differentiated agency in which the agential being of non-human animals, particularly mammals is countenanced, and the possibilities for agency very much depends on the relational systems which produce such being" (p. 9).

What one ultimately finds here is the garden-variety dualism of privileging humans and human-like animals as identity-based or Self-ed agents and the rest as mere material forces with effects, but no agency as Chandler's (2013) subjects "collectively understanding, constituting and transforming [their] world" (p. 528). Cudworth and Hobden (2015) confirm this dualistically marginalized status for more-than-human beings when saying, "We would use affective agency to discuss the significant effects of natural systems and the beings and things caught up in them and in their relations with other systems" (p. 9). It's interesting that, though the authors do note "humans and possibly some other creatures" have subject-based agency, they don't really go on to discuss which creatures these are, nor their basis for thinking these few more-than-human beings have this capacity. Instead, when giving examples of more-than-human beings who only possess affective agency, they choose generally inanimate beings, or those for whom few would ascribe any other kind of agency. For example, they note affective agency in forces such as "The impact of global warming, or the effects of a viral pandemic" (p. 10).

Ultimately, Cudworth and Hobden seek a middle ground between agential realism where only de-individuated matter has agency, and the humanist view that ontologically isolated human individuals have it. To this point, they say,

> our conception of agency incorporates the idea that non-human life and non-human animals are social actors able to exercise agency without seeing agency simply as a capacity that material beings can exercise. We need a situated and differentiated notion of agency that understands the ability of creatures and things to 'make a difference in the world' as a question of *situated relations* rather than *intrinsic capacity* alone. (p. 10)

While I agree wholeheartedly with the project as described, their unwitting inclusion of dualist views of more-than-human beings work directly against the achievement of what they seek. To suggest such an egalitarian and open view of agency as associated with beings and then couple it with the suggestion that more-than-human agency is "constricted in the extreme" seems to leave behind nothing more than human exceptionalism, or the privilege of transformative agency given only to those whom the old materialist privileging of human-like brains allows.

To conclude, Cudworth and Hobden, along with Chandler, are exactly right to decry the loss of agency and identity for situated individuals. Just as they do, I also see it as an untenable position to take, ignoring the power of embedded individual cognition, consciousness, intent and all other things that we use to make sense of the world beyond it being a chaotic welter of material interactions. Reality has the dynamism that posthumanists like Barad and others point out, but that dynamism produces Selves with many relational qualities and capacities, and that form the irreducible basis of intelligible human-nature relationships. As Midgley (2004) notes, "reducing wholes to parts is always a good way to downgrade their value" (p. 53). I believe that in denying agency to agents, this is exactly what has happened to agents—to Selves—in the mainstream of posthumanist and new materialist theory.

3.11   Conclusion

In this chapter. I've discussed some of what I take to be the foundational weaknesses of posthumanism when it depends on a materialist ontology, which it most often does. Given such weaknesses, I do not feel that posthumanism does enough to displace humans from the center of consideration, or to allow for a more accurate assessment of what more-than-human beings are actually like. At best, posthumanists have installed the important notion that relations are essential, and that humans cannot make exclusive claim to having the only substantive forms of them. Moving forward in this volume I'll refer periodically to facets of posthumanist theory when they apply. But, I won't uncritically accept any objection that begins "But in posthumanism…" I certainly won't accept it as a challenge to such things as correctly identifying that modern theorists still relationally position more-than-human beings as passive, material objects. And I won't accept it when I correctly identify human-nature relational symmetry when explorations of specific human-nature relational interchanges uncover it.

References

Anderson, K. (2014). Mind over matter? On decentering the human in Human Geography. *Cultural Geographies, 21*(1), 3–18.

Bandura, A. (2006). Toward a psychology of human agency. *Perspectives on Psychological Science, 1*(2), 164–180.

Barad, K. (2007). *Meeting the universe halfway: Quantum physics and the entanglement of matter and meaning*. Durham, NC: Duke University Press.

Bennett, J. (2010). A vitalist stopover on the way to a new materialism. In D. Coole & S. Frost (Eds.), *New materialisms: Ontology, agency, and politics* (pp. 47–69). Durham, NC: Duke University Press.

Braidotti, R. (2006). Posthuman, all too human: Towards a new process ontology. *Theory, Culture & Society, 23*(7–8), 197–208.

Braidotti, R. (2009). Animals, anomalies, and inorganic others. *PMLA, 124*(2), 526–532.

Brembs, B. (2011). Towards a scientific concept of free will as a biological trait: Spontaneous actions and decision-making in invertebrates. *Proceedings of the Royal Society of London B: Biological Sciences, 278*(1707), 930–939.

Buber, M. (2003). *Between man and man.* (R. Gregor-Smith Trans.). London: Routledge. (Original work published 1947).

Bulbulia, J. (2004). The cognitive and evolutionary psychology of religion. *Biology and Philosophy, 19*(5), 655–686.

Chandler, D. (2013). The world of attachment? The post-humanist challenge to freedom and necessity. *Millennium, 41*(3), 516–534.

Coole, D., & Frost, S. (2010). Introducing the new materialisms. In D. Coole & S. Frost (Eds.), *New materialisms: Ontology, agency, and politics* (pp. 1–43). Durham, NC: Duke University Press.

Cudworth, E., & Hobden, S. (2015). Liberation for straw dogs? Old materialism, new materialism, and the challenge of an emancipatory posthumanism. *Globalizations, 12*(1), 134–148.

Damasio, A. R. (1994). *Descartes' error: Emotion, rationality and the human brain.* New York: Putnam.

De Corte, B. J., Navarro, V. M., & Wasserman, E. A. (2017). Non-cortical magnitude coding of space and time by pigeons. *Current Biology, 27*(23), R1264–R1265.

Deleuze, G., & Guattari, F. (1987). *A thousand plateaus: Capitalism and schizophrenia.* Minneapolis, MN: University of Minnesota Press.

Ferrando, F. (2013). Posthumanism, transhumanism, antihumanism, metahumanism, and new materialisms. *Existenz, 8*(2), 26–32.

Fiske, S. T., & Taylor, S. E. (2013). *Social cognition: From brains to culture.* Thousand Oaks, CA: Sage.

Foster, J. B. (2000). *Marx's ecology: Materialism and nature.* New York: Monthly Review Press.

Geary, D. C. (2005). *The origin of mind: Evolution of brain, cognition, and general intelligence.* Washington, DC: American Psychological Association.

Gerhardt, U. (2002). *Talcott parsons: An intellectual biography.* London: Cambridge University Press.

Hallward, P. (2006). *Out of this world: Deleuze and the philosophy of creation.* London: Verso.

Haraway, D. J. (2008). *When species meet.* Minneapolis, MN: University of Minnesota Press.

Hayles, N. K. (1999). *How we became posthuman: Virtual bodies in cybernetics, literature, and informatics.* Chicago: University of Chicago Press.

Hobbes, T. (2004/1651). *Leviathan, or, the matter, forme, & power of a common-wealth ecclesiasticall and civill.* New York: Barnes and Noble Books.

Hoffman, E. (1992). *Visions of innocence: Spiritual and inspirational experiences of childhood.* Boston: Shambhala.

Jenkins, S. (2012). Returning the ethical and political to animal studies. *Hypatia, 27*(3), 504–510.

Kropotkin, P. A. (1902). *Mutual aid, a factor of evolution.* New York: McClure Phillips & Co.

Leary, M. R., & Buttermore, N. R. (2003). The evolution of the human self: Tracing the natural history of self-awareness. *Journal for the Theory of Social Behaviour, 33*(4), 365–404.

Lorimer, J. (2012). Multinatural geographies for the Anthropocene. *Progress in Human Geography, 36*(5), 593–612.

Merchant, C. (1980). *The death of nature: Women, ecology, and the scientific revolution.* San Francisco: Harper & Row.

Midgley, M. (2004). *The myths we live by.* London: Routledge.

Noddings, N. (1984). *Caring: A feminine approach to ethics & moral education.* Berkeley: University of California Press.

Panksepp, J. (1998). *Affective neuroscience: The foundations of human and animal emotions.* Oxford: Oxford University Press.

Peters, J. F. (2013). Near sets: An introduction. *Mathematics in Computer Science, 7*(1), 3–9.

Pickering, A. (1993). The mangle of practice: Agency and emergence in the sociology of science. *American Journal of Sociology, 99*(3), 559–589.

Plumwood, V. (1993). *Feminism and the mastery of nature.* London: Routledge.

Sampson, T. D. (2012). *Virality: Contagion theory in the age of networks.* Minneapolis, MN: University of Minnesota Press.

Schlosser, M. (2015). Agency. *The Stanford Encyclopedia of philosophy* (Fall 2015 Edition). Retrieved from https://plato.stanford.edu/archives/fall2015/entries/agency

Smith, D., & Protevi, J. (2018). Gilles Deleuze. *The Stanford Encyclopedia of Philosophy* (Spring 2018 Edition). Retrieved from https://plato.stanford.edu/archives/spr2018/entries/deleuze

Snaza, N., & Weaver, J. (2014). Introduction: Education and the posthumanist turn. In N. Snaza & J. Weaver (Eds.), *Posthumanism and educational research* (pp. 1–16). London: Routledge.

Springer, S. P., & Deutsch, G. (1998). *Left brain, right brain: Perspectives from cognitive neuroscience.* San Francisco: W. H. Freeman and Company.

Taylor, A., & Giugni, M. (2012). Common worlds: Reconceptualising inclusion in early childhood communities. *Contemporary Issues in Early Childhood, 13*(2), 108–119.

Tomasello, M. (2009). *The cultural origins of human cognition.* Cambridge, MA: Harvard University Press.

Twine, R. (2012). Revealing the animal-industrial complex: A concept and method for critical animal studies. *Journal for Critical Animal Studies, 10*(1), 12–39.

Van Dooren, T. (2014). *Flight ways: Life and loss at the edge of extinction.* New York: Columbia University Press.

Warren, K. J. (1990). The power and the promise of ecological feminism. *Environmental Ethics, 12*(2), 125–146.

Watson, A., & Huntington, O. H. (2008). They're here—I can feel them: The epistemic spaces of indigenous and western knowledges. *Social & Cultural Geography, 9*(3), 257–281.

Weisberg, Z. (2009). The broken promises of monsters: Haraway, animals and the humanist legacy. *Journal for Critical Animal Studies, 7*(2), 22–62.

Wilson, C. (2017). Materialism, old and new, and the Party of Humanity. In S. Ellenzweig & J. H. Zammito (Eds.), *The new politics of materialism* (pp. 111–130). London: Routledge.

Žižek, S. (2004). *Organs without bodies: On Deleuze and consequences.* London: Routledge.

Part II
Dualism and Relational Structure

Chapter 4
Human-Nature Relationship Model

4.1 Parallels Between Interhuman and Human-Nature Relationships

Theorists contributing to the human-nature relationship literature attempt to both explain the structure and dynamics of human-nature relationships and to offer prescriptions for the rehabilitation of poorly functioning ones. For example, Brooks, Wallace, and Williams (2006) seek to understand how visitors to wildlands settings build relationships with those places over time so that they can offer guidance to public lands managers seeking to provide visitors a good experience with the more-than-human beings there. This research is indicative of a larger trend as well—that of theorists attempting to draw parallels between human-nature relationships and interhuman ones (Martin, 2007; Mayer & Frantz, 2004; Nisbet, Zelenski, & Murphy, 2008; Sanders, 2003). For example, Martin (2007) suggests that there is a possibility of "people's relationships with entities in the environment hav[ing] structural similarities to interpersonal relationships" (p. 60).

Many theorists have also noted that in well-functioning human-nature relationships, human participants are often behaving more benignly toward the environment. What theorists have noticed in particular is that these human beings are having relational experiences with more-than-human beings that mirror the kinds of close, personal relationships that human beings have with each other. For example, Mayer and Frantz (2004) attempt to adapt social psychological concepts of "interpersonal closeness, perspective taking, and altruism" (p. 504). Sanders (2003), in researching close relationships between humans and their companion animals, characterizes his work as an attempt to "reexamine close relationships [in order] to move beyond the limiting anthropocentric orthodoxy that presents the bonds and interactions between humans and nonhuman animals as qualitatively different from—and, by implication, inferior to—those between humans" (p. 406). In discussing the roots of pro-environmental behavior, Nisbet et al. (2008) say that "[p]ersonal relationships with

© Springer Nature Switzerland AG 2019   91
N. H. Kessler, *Ontology and Closeness in Human-Nature Relationships*,
AESS Interdisciplinary Environmental Studies and Sciences Series,
https://doi.org/10.1007/978-3-319-99274-7_4

nature may provide some insight into the way people treat the environment" (p. 2). Brooks et al. (2006) ask, "Do visitors develop deep lasting committed relationships with [Rocky Mountain National Park]? How are these bonds similar to relationships people have with human partners…?" (p. 332).

Still other researchers, while perhaps not speaking to the concept of "closeness" specifically, have focused on understanding human-nature relational elements that are identical to ones that figure prominently in interhuman closeness. These include relational "bonds" (Williams, Patterson, Roggenbuck, & Watson, 1992, p. 30), "care" (Saunders, 2003, p. 138), "love" (Haraway, 2008, p. 16; Kals, Shumacher, & Montada, 1999, p. 180), "oneness" (Dutcher, Finley, Luloff, & Johnson, 2007, p. 479), "kinship" (Warren, 1990, p. 143), and "intimacy" (Rogan, O'Connor, & Horwitz, 2005, p. 156). Thus, whether explicit or not, a portion of those studying the human-nature relationship attempt to understand and prescribe remedies for poorly functioning human-nature relationships that move them *toward* closeness in an interhuman sense.

That being the case, posthumanist theorists reject the anthropocentrism inherent in holding interhuman relationships as any kind of standard against which to measure the structure and dynamics of human-nature relationships. As discussed at length in the previous chapter, their emphasis is on the asymmetry of these relations, where they caution us not to anthropomorphically ascribe to more-than-humans the capacity to participate in closeness in this way. What's more, posthumanists such as Barad (2007) reject the metaphysical individualism and essentialism inherent in attempts to measure the qualities of relationships based on their participants. In posthumanist views such as these, all relations tend to be ontologically equivalent, at least at their inception point. Barad uses the term "intra-action" to describe such relational units. Deleuze and Guattari (1987) use the term "assemblage." As I stated in the previous chapter, while one need not take ontological individualism as a starting point in relations, rejecting or reducing the importance of the influence that persistent Selves have on such relations tends to artificially limit or distort the understanding one can gain of the structure and dynamics of those relationships. While my houseplant will be different tomorrow than she is today, she will not be a Mastodon. She will be a houseplant. Therefore, I can count on being able to relate with her in certain ways, and for there to be a unique and persistent (if not unchanging) structure and dynamics to any relationship that develops between us. That's not essentialism, that's just certain qualities and capacities tending to inhere in certain kinds of beings. It's another way of speaking to a being's identity.

The logical extension of posthumanist ontological intra-actions or assemblages is not inconsistent with the notions of stability and persistence that I find essential to exploring and understanding human-nature relationships. But, as Wilson (2017) says, relatively little posthumanist work has been carried out in order to connect the material interchange at this microstructural level to how real, functioning relationships tend to develop and persist, and what their features are. While some of this may be a simple lack of effort in a rapidly expanding and evolving theoretical framework, the tenor of posthumanist explorations tends to be one of de-emphasizing the relevance of this kind of exploration. What comes across is that the dynamic, material force at work in relations is the thing of primary importance, and takes precedence over the influence of, or effects upon, *relata*. In fact, as I explored in the last chapter, relations are portrayed as so dynamic and variable—and based on such specific instances of relationship at a moment in time—that there doesn't seem to be much room for relational stability and persistence either in relations or in *relata*. I qualify such observations by noting that no posthumanist theorist is saying this directly, but the strong emphasis on the intra-action, and the seeming unassailability of the completeness of its characterization of relationships, has a dampening effect nonetheless.

As I noted in the last chapter, it's interesting that much of new materialist theory emerges from those with experimental scientific backgrounds who are working in science studies. In such contexts, the "nonhuman" element involved in such intra-actions are as much microscopes or other mechanical apparati as they are the animals, plants and other beings with whom such apparati are mostly involved. Critical animal studies fares better at putting more-than-human beings in the relational frame, but as I will argue in the next chapter when discussing critical animal studies (and as discussed in the previous chapter) the dualism of materialism and anthropocentrism has significantly limited just which entities in the more-than-human world are thought to deserve significant relational consideration.

Ultimately, while posthumanist ontological suggestions about the agency of matter have merit, so do explorations that treat the Self as having ineradicable agency. That I intend to write this book as I am doing right now is not something I am willing to relegate to the accretion of *micro*agential activity I see Barad's (2007) model suggesting. Just as emergence theorists recognize that things like consciousness are qualitatively different than the material microstructures and processes of the brain (as I discuss in Chap. 10), here I suggest that the agency that I, as a Self-in-relation, or my dog, also as a Self-in-relation, exhibit, cannot be uncritically reduced to the sum of our agential parts. Besides, if those microagential moments themselves can be reduced to ever more minute microstructures with their own agencies, where does it end? And how quickly do such microagencies lose their explanatory efficacy as they relate to possibilities for closeness in human-nature relationships? My intent—the dog's intent—are things of value and relational essentiality unto themselves. They help form the "glue" of our interactions. What we, as imperfectly definable, dynamic, and ever-embedded Selves bring to our relations is worthy of study *as they are*.

4.2   Existing Human-Nature Relationship Theories

Given the work of human-nature relationship theorists to extrapolate the experience of interhuman closeness to human-nature relationships, one might expect there to have been developed some clear models to explain the structure and dynamics of the human-nature relationship itself. Yet, this is generally not the case. If articulated at all, theorists tend to be vague about these elements, and also to inject a heavily anthropocentric orientation. For example, Mayer and Frantz (2004) loosely adapt an interhuman relationship model from Aron, Aron, Tudor, and Nelson (1991), but do so without acknowledging that Aron et al. position their work outside the mainstream of interhuman relational theory. Instead of being focused on the "mutual influence…[and] interdependence" (p. 241) of relationships, Aron et al.'s work revolves around the "cognitive significance…for each person" (p. 241). This and other aspects of Aron et al.'s work applied to human-nature relationships puts the individual human at the center of consideration with the more-than-human being either at the periphery or out of the frame entirely. Again, while my anti-anthropocentric orientation aligns well with posthumanism, here I'll only suggest that the "cure" for this anthropocentrism isn't the re-characterization of the human as "merely" a material intra-action embedded in a larger one along with all other elements of existence. Instead, I suggest that we reassess the qualities and capacities of beings in the more-than-human world without dualisms like monist materialism, at which point the possibilities for human-nature relationships may radically change.

While use of work such as Aron et al.'s to model human-nature relationships provides safe harbor to theorists seeking (consciously or otherwise) to dualistically focus their attention on the human side of the relational experience, it cannot be used to model the *reality* of the relevant relational elements without justification. One could only do this if the more-than-human contribution was already assumed to be relationally less relevant or entirely superfluous. Many other examples of anthropocentric modeling like this can be found, such as in Saunders' (2003) view of the field of Conservation Psychology where reciprocity of human-nature relationship is conceived of as what humans receive and what they give, instead of an actual reciprocal exchange where both human and more-than-human give and receive. When Brooks et al. (2006) characterize "relationship to place as the active [human] construction and accumulation of place meanings" they center the definition of the relationship in the action of the humans within it. Van den Born, Lenders, Groot, and Huijsman (2001) offer support for my suggestion that more-than-human beings are improperly dislocated from the relational epicenter when they say, "An extensive body of research focuses on environmental behaviour, attitudes and values…[but] the problem with this body of research is that [more-than-human] nature is only a marginal element in the approach" (p. 66).

The marginalization or omission of more-than-human beings from close human-nature relational theory could be characterized as simple oversight, but even if this is unconsciously the case, I believe it to be indicative of a deeper, unexamined assumption at work. That is, that more-than-human beings *ought to be* thought of as peripheral, if thought of at all. Posthumanist theorists recognize this issue, but again as I argued in the previous chapter, I don't think that their predominant materialist ontology establishes any basis to supplant it. Though in materialist posthumanist ontology such as Barad's (2007), there is the suggestion that agency is where you find it as a product of the agential intra-actions, the decoupling of that agency from Selves so that closeness between individuals-in-relation becomes secondary to the flux of primary experience, we are little closer to an analytical apparatus that allows us to determine whether the intra-action that is the family dog, embedded in the larger intra-action of relational interchange with the intra-action that is me, can produce closeness.

The byproduct of materialist posthumanism seems instead to be a vague, flattened biophysical egalitarianism that admits the rest of the more-than-human world but does little to displace the human from her seat atop the relational hierarchy to which old material dualisms have elevated her (discussed in the "Asymmetry" section of the previous chapter). As I will discuss in the next chapter, one never hears of critical plant studies because plants are assumed not to have the material capacity for close relationships with humans. But, as I'll also explore in later chapters, any evidence offered for plants not having such capacities is itself rooted in dualist assumption, and thus should not to be taken as definitive evidence to support the claim of the plant's relational incapacity. In posthumanism, plants are admitted as part of assemblages, and one might suppose that if a posthumanist were feeling saucy, he might talk about taking Deleuzian becoming out for a spin by *becoming-vegetal*. Sarcasm aside, Deleuze and Guattari (1987) explicitly suggested that there was no problem with this very possibility. But even my suggestion of the possibility here likely evokes an almost reflexive dismissal (I have to fight it in my own thinking) of what that means or what it might look like. The belief that nothing is there relationally for a human and plant is overpowering in modern discourse, which reveals the hidden hand of a deeply entrenched, hegemonic dualism taking hold of our thinking about this. Though posthumanism has provided theorists an avenue to admitting other beings in the more-than-human world into the relational conversation, by using materialism as the foundation for understanding human-nature relations, old materialist dualisms still exert their influences against beings such as plants.

Also, while the posthumanist literature addresses anthropocentrism and the need for a more inclusive and reciprocal portrait of human-nature relationships, less work has been done within this and other literatures to understand specifically the structure and dynamics of the relationship itself. For example, Morris (2014) mentions close relationships when discussing posthumanism and educational research. But, while endorsing their possibility, there is still a lurking dualism as to what role human and more-than-human play in such closeness. Specifically, she says that "we must…build ethical relationships with nonhuman animals…Posthuman education also means that we must remember our own animality, *but we must move beyond this too* [emphasis added] in order to embrace and even love nonhuman animals" (p. 53). She takes an odd turn when saying that we must remember our animality but discard it in order to embrace and love species that are not our own. If love, the act of embracing, and its byproduct of closeness are common to human and nonhuman animal alike, then isn't it the very animality that she suggests we discard that which is the matrix within which the love develops? To suggest otherwise is to allow a dualist assumption to slip in: that animals, or at least the basic animality that we share with them, itself cannot sponsor the embrace or the love as humans know and experience it. Haraway (2008) also speaks to love when describing how she and her dog share it. But, by referring to it as a "contagion," (p. 16) she puts it in a passive and material light, as if it's something that happens to us instead of it being an active relational element that yes, we are at times subsumed by, but at other times try consciously and actively to administer as relational Selves.

Ultimately, when the more-than-human relational partner is conceptually diminished, or made less "weighty" through dualist mechanisms, whether posthumanist or from some other school of thought, it makes sense that theorists would leave relational structure and dynamics largely untended when building theories of close human-nature relationships. In such views, there is very little relationship in a realistic, reciprocal sense, *to* model.

4.3   Interhuman Closeness and Interdependence Theory

Since I've argued that one cannot uncritically assume a peripheral more-than-human partner, and since so far one cannot merely draw a dividing line between those with whom humans can and cannot be close based on monist materialism or any other dualistically influenced ontological position, a relational model that can be applied to all scenarios/beings is sorely needed. Such a model will form the context within which I will openly examine not only the qualities and capacities of human and more-than-human in relationship, but any *a priori* dualisms in ontological operation therein. To allow for the possibility that I assert in this book: that close human-nature relationships are akin to close interhuman ones, then choosing a model from the literature on close interhuman relationships seems sensible. Here I'll briefly explore and make use of one of the more widely known of these theories—*interdependence theory* (Kelley et al., 1983). I pause here to recognize that there is a vast and diverse literature on close human relationships. My goal is not to undertake an extensive examination of this literature, but instead to simply make use of one of its well-established models to "snap a line" as it were, against which one can hold the capacities of humans and more-than-human beings to engage in reciprocally relational closeness. I choose interdependence theory over, for example, two other well-established close relationship theories: Communal Theory (e.g., Clark & Mills, 1993) and Attachment Theory (e.g., Hazan & Shaver, 1994), due to its clear articulation of how thoughts, feelings and actions interoperate to form closeness.

Kelley et al. (1983) suggest that a relationship between two persons can be understood to exist most basically "when their two chains of events (of thoughts, feelings, actions, and other attributes) are *inter*connected" (p. 36). Fig. 4.1 depicts the interaction of these elements of thought, feeling and action between two persons. As can be seen, each person's thoughts, feelings and actions can affect any and all of the same three elements in the other person. Also, while in interdependence theory one does not see direct feeling-to-feeling interaction, as I'll explain in places in Parts III and IV, I believe that this vector of interaction is also possible.

Building on this basic notion of interdependence, Kelley et al. define the kind of interdependence in a *close* relationship as being "strong, frequent, and diverse... [and lasting] over a considerable period of time" (p. 38). In a *positive* close relationship such as those between close friends or spousal partners, the authors say that

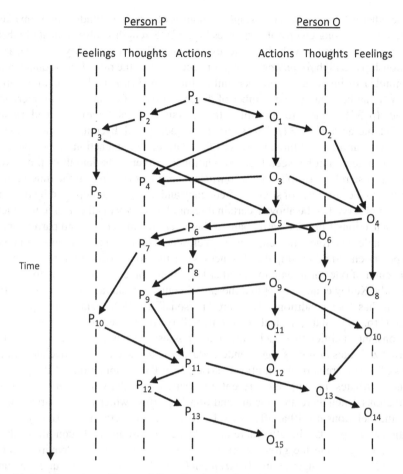

Fig. 4.1 Interdependence Theory of Close Interhuman Relationships showing interaction of feelings, thoughts, and actions. (Adapted from Kelley et al. (1983))

their interdependence has "symmetry; and high mutual facilitation" (p. 37). "Symmetry" is loosely meant as a reciprocity where each contributes elements contributing significantly to the closeness. This, as opposed to other positive close relationships Kelley et al. note such as parent-child relations, which can be quite "asymmetrical." It's interesting to note that in posthumanism, symmetry of human-nature relations is generally looked down upon. Granted, when relations are between human and more-than-human, there is an in-built difference, but if the relational capacities for thoughts and feelings are present in both partners, there ought to be nothing problematic in discussing the symmetry that exists in this relational context. Any persistence in denial of such symmetry is more likely a result of dualism than any evidence that the symmetry is not possible.

To explain each of these close relationship elements in a bit more detail, the authors say that interdependence is *strong* when a change produced in one participant by the other is "great, involv[es] single responses of large amplitude...[and] invokes numerous" or "long chains of responses" (p. 33). Strength is also indicated when changes are produced "with short latency or with high _dependability" (p. 33). The relationship has high *frequency* of interconnection when the rate of interconnection, or "number of interconnections per unit of time" (p. 34), is high. A relationship's *diversity* can be seen in "the number of different kinds of events that are interconnected" (p. 34). In close relationships, the diversity is greater, with a "broad, richly textured interaction" (p. 34). This strength, frequency and diversity of interconnection can occur along all three vectors of thought, feeling and action in the chain of events of interaction between the two relational partners. The length of time over which a close relationship lasts—its *duration*—can be measured in the time period over which the attributes of strength, frequency and diversity are reported by participants, or observed, to be above a certain threshold. *Symmetry* is the degree to which interconnections from one participant to the other are similar for both participants. For example, in a romantic couple there is closeness based, in theory, on symmetry. In a parent-child relationship the closeness is asymmetrical. *Facilitation* is the interconnection of one participant to the other helping the other reach certain goals.

Lastly, Kelley et al. suggest that the causal context of interaction in a close relationship has four conditions. They are "personal" (p. 96), "relational" (p. 103), "social" (p. 105), and "physical environmental" (p. 106). The "personal conditions" are constant qualities that each participant possesses such as genetic propensity, internalized rules or norms, age, gender, education, and social, cognitive and affective skills. The "relational conditions" arise out of the combinations of the participants' attributes in both the present interaction and deriving from their past interactions. Then there are "social conditions" (p. 105), which are the surrounding or "other relationships [that affect] the frequency and patterning of...[the present] relationship" (p. 105). Finally, there are "physical environmental" conditions that, because I am laying the groundwork for a relationship with that "environment," means that it is no longer applicable *as a backdrop*. In the context of human-nature relationships, I suggest that personal or social conditions have a great influence on whether a human, for example, enters into a close relationship with a more-than-human being or believes she is able to do so. Imagine how odd or even frightening it would be to feel love toward an animal or piece of land if there is little experience of having done so in the past—or if other humans within one's social context call such love a form of either self-indulgent anthropomorphism or worse, mental imbalance. In the incisive sarcasm that is the hallmark of Turner's (1989) essay on human relationships with wild nature, he suggests that most modern attitudes toward feelings of closeness like I describe are thought of by modern societies as "an emotional mistake—like being in love with the number 2" (p. 85). From the more-than-human side, past relational conditions can also have an influence over a willingness to form close relationships with humans. Most people have heard stories of a shelter animal having a fear of humans wearing certain clothing, like a baseball cap, because of past association with an abusive owner who wore one.

Casting such impediments and their underlying assumptions aside, we'll allow interdependence theory's reliance on discernible events and their interplay to facilitate an accurate measure of relationships and their closeness. In attempting to determine whether human-nature relationships are close, establishing that both human and more-than-human have the capacity to think, feel and act in relationally reciprocating ways is crucial. Parts III and IV of this book will focus primarily on establishing that there is nothing ontologically problematic with all manner of beings in the more-than-human world having the capacity for the kinds of thoughts and feelings that support closeness as defined here in interdependence theory. That is the main work of this book. And while personal, relational and social conditions, as well as strength, frequency, diversity, duration, and the tendency toward facilitation are important to examine as a next step, an exhaustive treatment of these elements is beyond the scope of this volume.

4.4   Interdependence Theory Applied to a Human-Nature Relational Encounter

I've chosen a close relationship model containing elements of thought, feeling and action to apply to more-than-human beings as well as humans. As I'll argue in Part IV, the capacity for thoughts and feelings is ontologically unproblematic even for "inanimate" more-than-human beings, thus this model is applicable in all human-nature relationships.

How such a model would work to explain human-nature interactions is best served with an example. I begin by quoting the illustrious Alice Walker (1989) at length as she describes an encounter she had with a horse named Blue:

It was quite wonderful to pick a few apples, or collect those that had fallen on the ground overnight and patiently hold them...up to [the horse, Blue's] large, toothy mouth...I had forgotten the depth of feeling one could see in a horse's eyes.

I was therefore unprepared for the expression in Blue's. Blue was lonely. Blue was horribly lonely and bored. I was not shocked that this should be the case; five acres to tramp by yourself, endlessly, even in the most beautiful of meadows—and his was—cannot provide many interesting events...No I was shocked that I had forgotten that human animals and nonhuman animals can communicate quite well; if we are brought up around animals as children we take this for granted...

But then, in our second year, something happened in Blue's life. One morning...I saw another horse, a brown one, at the other end of Blue's field. Blue appeared to be afraid of it, and for several days made no attempt to go near. We went away for a week. When we returned, Blue had decided to make friends and the two horses ambled or galloped along together, and Blue did not come nearly as often to the fence underneath the apple tree.

When he did, bringing his new friend with him, there was a different look in his eyes. A look of independence, of self-possession, of inalienable *horse*ness. His friend eventually became pregnant. For months and months there was, it seemed to me, a mutual feeling between me and the horses of justice, of peace. I fed apples to them both. The look in Blue's eyes was one of unabashed "this is *it*ness."

It did not, however, last forever. One day, after a visit to the city, I went out to give Blue some apples. He stood waiting, or so I thought, though not beneath the tree. When I shook the tree and jumped back from the shower of apples, he made no move. I carried some over to him. He managed to half-crunch one. The rest he let fall to the ground. I dreaded looking into his eyes—because I had of course noticed that Brown, his partner, had gone—but I did look…The children next door explained that Blue's partner had been 'put with him'…so that they could mate and she conceive. Since that was accomplished, she had been taken back by her owner, who lived somewhere else.

Will she be back? I asked.

They didn't know…

Blue was like a crazed person. Blue *was*, to me, a crazed person. He galloped furiously, as if he were being ridden, around and around his five beautiful acres. He whinnied until he couldn't. He tore at the ground with his hooves. He butted himself against his single shade tree. He looked always and always toward the road down which his partner had gone. And then, occasionally, when he came up for apples, or I took apples to him, he looked at me. It was a look so piercing, so full of grief, a look so *human*, I almost laughed (I felt too sad to cry) to think that there are people who do not know that animals suffer…But most disturbing of all, in Blue's large brown eyes was a new look, more painful than the look of despair: the look of disgust with human beings, with life; the look of hatred. (pp. 4–8)

Figure 4.2 depicts the initial part of Walker's interaction with Blue, showing a link between Blue's feelings of loneliness and boredom and Walker's perception and thoughts about them. Within this interchange one sees the interaction of feelings that I will claim in Chap. 9 is direct and unmediated by thought or material intermediaries. In Fig. 4.3, one sees the interdependence of Walker and Blue when Blue is in love with his partner. Here, one can see Walker's perception of Blue's arrival with his mate leading to her emotional response of happiness. One can also see that I've depicted Walker's statement regarding a "mutual feeling between me and the horses of justice, of peace" (depicted as a gray background) as a suffusing phenomenon whose existentiality I'll also argue for in Part III. Because I don't take feelings to be purely materially sourced, they can exist as phenomena within which an individual relational participant can exist. See Chap. 7 for more discussion of this possibility. In Walker's description above, both she and Blue do exist within it over a period of time, the duration of which contributes to a feeling of closeness between Walker and Blue. One can see also that there is a strength of response between the two, Blue with his love for his mate intermingling with Walker's witnessing of this to create powerful feelings of peace and justice.

Lastly, Fig. 4.4 depicts the interaction of Walker and Blue once Blue's mate is taken away from him. One can see that the removal of Blue's mate sparks a cascade of thoughts and feelings between Walker and Blue. Something about Blue's action makes her think that something is wrong. In trying to give Blue an apple, she sees his fury, moving directly from feeling to actions such as tearing at the ground with his hooves. The grief, then disgust and hatred that Blue feels, are all communicated through his eyes to Walker. One can also see a decrease in the frequency of interaction over time due to Blue's grief. Thus, what might've been a developing close relationship between Walker and Blue has stopped. What's more, the element of mutual facilitation that is so important in positive close relationships is absent from Walker's response to Blue's fury.

Putting interactions like this into a different context, imagine a U.S. Forest Service District Ranger developing a land management plan for a national forest. The Forest Service's imperative to base plans on the principle of Multiple Use Sustained Yield (MUSY) (Coggins & Evans, 1982), attends very little, if at all, to

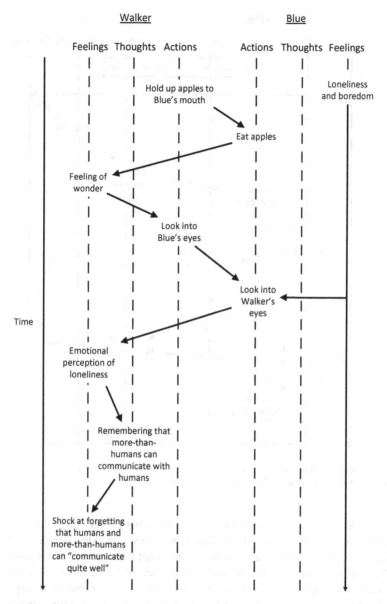

Fig. 4.2 Initial interaction between Walker and Blue. Shows link between Blue's feelings and Walker's thoughts about them

anything relationally mutual between human and more-than-human. The very notion of human "use" of more-than-human beings implies this relational imbalance. That's why I see Walker's essay less about what happens between Blue and his mate and more about what the dynamic is or could be between the human Walker, and the more-than-human Blue.

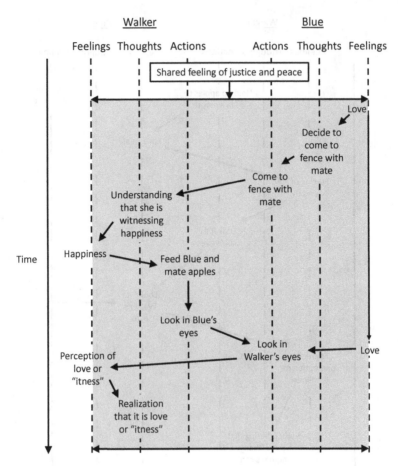

Fig. 4.3 Interdependence of Walker and Blue. Shows comingling of thoughts and feelings between Walker and Blue when Blue is in love with his mate

Walker's description of this relationship shows that the closeness that might've developed between them did not, or at least did not over an extended period of time as she's described it to her readers. And while land management plans guided by MUSY may not ensure a failure of this kind of potential for closeness, they do nothing intentional to allow for it unless the human consciously or unconsciously translates her desire into a "use" of a national forest to get close to the more-than-human beings living there.

Further, my reading of the events between Walker and Blue suggest that the failure of close relations was not because Blue didn't have the qualities and capacities to support closeness, just as the failure of the Forest Service to plan for humans and more-than-human beings to become close has nothing to do with whether the more-than-human beings living in a national forest have the qualities and capacities to engage in this activity either. Neither does Walker, for her part, suggest that more-than-human beings are missing these things. Instead she attributes the failure of

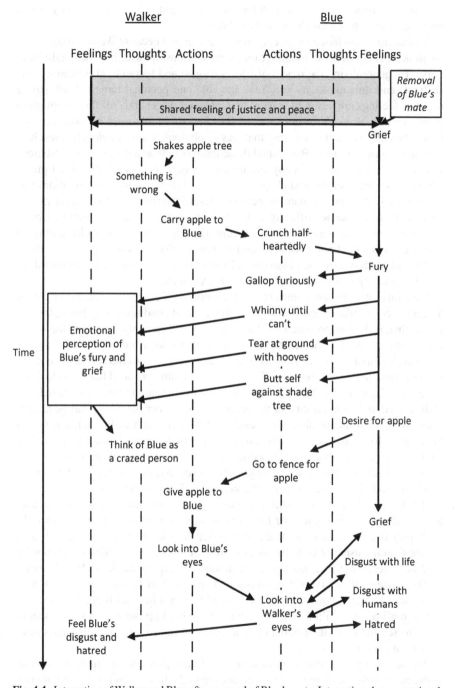

Fig. 4.4 Interaction of Walker and Blue after removal of Blue's mate. Interaction shows emotional and cognitive interdependence leading to disgust and hatred

closeness to Blue's grief turned to hatred and disgust for all of humanity. But I respond by asserting something slightly different.

I think closeness failed to develop in this case in part because Walker *failed to act on Blue's behalf* as an individual human in relationship with him, the individual horse. By generalizing the object of Blue's disgust and hatred to all humanity, she abstracts and thus masks its very real, tangible and personal targets. Such strong negative feelings are never, at least at first, abstract or general. All three humans— Blue's owner, Brown's owner, and *Walker herself*—fail to act substantively on Blue's behalf, to ease the suffering that was so obvious, and so obviously a result of their actions or their lack. Blue hated those three people and they failed to relationally respond to that hatred in any ameliorative way. This is precisely what I meant when, at the outset of the first chapter, I suggested that human-nature relationships are flawed first at an intimate or personal level. Rainforest deforestation occurs because that very same suffering is ignored on an individual-by-individual basis until it reaches some massive, biome-disrupting scale. Just the same, clearcutting in national forests continues because land managers charged with the care of those places fail to recognize the possibility of these individual-level close relationships and the suffering caused by their fracture or prevention.

Certainly, other factors contribute to this ignoring of close relational possibilities. To reference interdependence theory, Walker's social conditions may have played a role in that, if she *had* protested to Blue's owner, she'd certainly be risking the owner responding to her as if she had fallen in love with Turner's "number 2." But, wasn't it Walker's responsibility *to Blue as a relational* partner to overcome such an internalized obstacle and register her objection anyway? Blue included her in his feelings for Brown. He waited and then brought his mate to the fence to meet her. He deliberately waited and then did this, in the process showing her his "independence...self-possession...[and] inalienable horseness." And Walker was a willing witness to that blossoming love, sharing in and benefitting from the feelings of peace and justice it created. Then, when someone stepped in and took that all away, Walker did nothing, save a tepid inquiry into Brown's fate posed not even to Blue's owner, but the owner's children—individuals hardly in a position to bring Blue's mate back.

The lessons she chose to distill from the experience are indeed powerful, but they are not allowed to influence her interactions with Blue as a relational partner. As such, they become a *using* of Blue and his grief that may be to a much less abusive degree than the using of Blue by his or Brown's owner—or the using of a forest by land managers—but it seems to me a difference of degree, not kind. What's more, Walker's use of Blue in this way misses perhaps a deeper lesson: that the potential for a close human-nature relationship existed between her and Blue, and that it is her (and all of our) failure as modern humans to hold up our end in such encounters—to see the more-than-human self as a Self—that so often derails such closeness from coming to fruition.

Walker recognizes that communication of thoughts and feelings is possible between animals and humans. That she does not take that a step further and act in a way that would benefit Blue—and that Blue would be able to identify her as having undertaken—is the real failure. Let me pause to say that this isn't meant as a real criticism of Walker as a person, as an author or as a great humanitarian. I have the utmost respect for her, and for her courage in this very essay to address human-nature relations in the deeply thoughtful ways she has. I wanted to use the reaction of someone who *is* so tuned to these interactions that she's willing to write about them to show how deeply embedded in all of our thoughts as modern human beings is that we cannot act—and that the response of our more-than-human relational partner will only take us so far.

In the end, I don't know if Walker could've brought Blue's mate back. The modern world is constructed to treat more-than-human beings such as horses as slaves, as Walker is right to point out. But Walker could have given Blue's owner an earful anyway. Even if she had no confidence that it would work, the possibility would be registered with the owner and the wider world—a world that might've included Blue if she'd done it out in the open near his pasture. Perhaps Blue would have seen her yelling at the owner and pointing at him. He might've *felt* her care. And perhaps in response to witnessing this, instead of Blue arriving at the fence by the apple tree the next time with hatred and disgust for a humanity that included her, he might've come there to rub that great viola-case of a grief-stricken head against her shoulder to salve his broken heart with one who cared enough for him to protest on his behalf—to feel *with* him instead of *about* him. In interdependence theory, all the thoughts and feelings in the world won't amount to a hill of beans without the third vector of *action*. Walker appears to have chosen not to act in her relationship with him at that time.

Every day we modern humans engage in a multitude of such failures to act—most of it sponsored by a failure to think and feel, at least consciously, with our more-than-human relational partners. The filling in of a wetland to build a road that I drive down every day puts the count of lost possibilities for closeness in the thousands. The twinge of remorse or the notion that "this is wrong" felt by a logger clearcutting old growth in a Southeast Alaska national forest may not be the logger's feeling, but his internal response to the beating of the feelings of the trees against the dualistically hardened exterior of his consciousness. If he doesn't respond with action, the relationship remains as both material and as the worst sort of "use." In such a view, even the smallest act of weeding one's garden appears to be a relational catastrophe. I suggest that it may very well be that! Such characterizations of course require justification—and so it is the business of this work to offer them—to pry open the steel jaws of self-reinforcing dualisms that deny this possibility so that some other ontological position may enter. This is not my way of saying that we shouldn't alter land or that we should become Jainist Buddhists sweeping the bugs from our path as we walk (unless we feel compelled, by paying attention to our human-nature relationships, to do so). But, it is to say that as we move through the world, if we do so with the conscious knowledge that thoughts, feelings, and actions are being exchanged between humans and more-than-human beings, a radical alteration in how modern human-nature relationships unfold will be the inevitable result. I'm afraid to say that MUSY will cease to exist as a tenable relational directive. Throughout the rest of the book I will use views of human-nature relationships modeled on elements of the interdependence theory I've explored here both as a framework to consider the workings of close human-nature relationships and as a critical lens through which to examine the human-nature relationship literature as it currently exists.

References

Aron, A., Aron, E. N., Tudor, M., & Nelson, G. (1991). Close relationships as including other in the self. *Journal of Personality and Social Psychology, 60*(2), 241–253.

Barad, K. (2007). *Meeting the universe halfway: Quantum physics and the entanglement of matter and meaning*. Durham, NC: Duke University Press.

Brooks, J. J., Wallace, G. N., & Williams, D. R. (2006). Place as relationship partner: An alternative metaphor for understanding the quality of visitor experience in a backcountry setting. *Leisure Sciences, 28*(4), 331–349.

Clark, M. S., & Mills, J. (1993). The difference between communal and exchange relationships: What it is and is not. *Personality and Social Psychology Bulletin, 19*(6), 684–691.

Coggins, G. C., & Evans, P. B. (1982). Multiple use, sustained yield planning on the public lands. *University of Colorado Law Review, 53*, 424–429.

Deleuze, G., & Guattari, F. (1987). *A thousand plateaus: Capitalism and schizophrenia*. Minneapolis, MN: University of Minnesota Press.

Dutcher, D. D., Finley, J. C., Luloff, A. E., & Johnson, J. B. (2007). Connectivity with nature as a measure of environmental values. *Environment and Behavior, 39*(4), 474–493.

Haraway, D. J. (2008). *When species meet*. Minneapolis, MN: University of Minnesota Press.

Hazan, C., & Shaver, P. R. (1994). Attachment as an organizational framework for research on close relationships. *Psychological Inquiry, 5*(1), 1–22.

Kals, E., Schumacher, D., & Montada, L. (1999). Emotional affinity toward nature as a motivational basis to protect nature. *Environment and Behavior, 31*(2), 178–202.

Kelley, H. H., Berscheid, E., Christensen, A., Harvey, J. H., Huston, T. L., Levinger, G., et al. (1983). *Close relationships*. San Francisco: W. H. Freeman and Company.

Martin, P. (2007). Caring for the environment: Challenges from notions of caring. *Australian Journal of Environmental Education, 23*, 57–64.

Mayer, F. S., & Frantz, C. M. (2004). The connectedness to nature scale: A measure of individuals' feeling in community with nature. *Journal of Environmental Psychology, 24*(4), 503–515.

Morris, M. (2014). Posthumanist education and animal interiority. In T. Snaza & J. Weaver (Eds.), *Posthumanism and educational research* (pp. 43–55). London: Routledge.

Nisbet, E. K., Zelenski, J. M., & Murphy, S. A. (2008). The nature relatedness scale: Linking individuals' connection with nature to environmental concern and behavior. *Environment & Behavior, 41*(5), 715–740.

Rogan, R., O'Connor, M., & Horwitz, P. (2005). Nowhere to hide: Awareness and perceptions of environmental change, and their influence on relationships with place. *Journal of Environmental Psychology, 25*(2), 147–158.

Sanders, C. R. (2003). Actions speak louder than words: Close relationships between humans and nonhuman animals. *Symbolic Interaction, 26*(3), 405–426.

Saunders, C. E. (2003). The emerging field of conservation psychology. *Human Ecology Review, 10*(2), 137–149.

Turner, J. (1989). The abstract wild. *Witness, 3*(4), 81–95.

Van den Born, R. J., Lenders, R. H., De Groot, W. T., & Huijsman, E. (2001). The new biophilia: An exploration of visions of nature in Western countries. *Environmental Conservation, 28*(1), 65–75.

Walker, A. (1989). *Living by the word: Selected writings, 1973-1987*. San Diego, CA: Harcourt.

Warren, K. J. (1990). The power and the promise of ecological feminism. *Environmental Ethics, 12*(2), 125–146.

Williams, D. R., Patterson, M. E., Roggenbuck, J. W., & Watson, A. E. (1992). Beyond the commodity metaphor: Examining emotional and symbolic attachment to place. *Leisure Sciences, 14*(1), 29–46.

Wilson, C. (2017). Materialism, old and new, and the party of humanity. In S. Ellenzweig & J. H. Zammito (Eds.), *The new politics of materialism* (pp. 111–130). London: Routledge.

Chapter 5
Dualist Effects on Structure and Dynamics

The close human-nature relational model I articulated in the previous chapter can help identify where dualisms have their reductive or negating effects in conceptualizations of the structure and dynamics of human-nature relationships. In this chapter I will explore various fields, subdisciplines and schools of thought to illustrate how this occurs.

5.1 Conservation Psychology

Saunders' (2003) essay on the field of conservation psychology is a good place to begin this chapter's exploration. It is an example of how even fields built up around better understanding, and through that improvement of, human-nature relationships, can allow anthropocentrism and other dualisms to destroy the structure and dynamics of close human-nature relationships as I've articulated them in the previous chapter. Conservation psychology is defined by Saunders as "the scientific study of the reciprocal relationships between humans and the rest of nature" (p. 138). Because it is an attempt to speak first to human-nature relationships as reciprocal, if one finds dualisms at work this gives some support to the notion that dualisms work to *post hoc* conceptually distort the reciprocity of these relationships.

The first place we can see dualisms at work is in Saunders' portrayal of the qualities and capacities of beings other than humans. In her essay, there are only four brief mentions of them. The first is in regard to getting a better understanding of "the restorative and healing aspects of nature" (p. 139), the second in the "ability to perceive beauty in nature" (p. 144), the third in how "interactions with nature positively affect multiple dimensions of human health" (p. 143), and the fourth in "the effect of nature on [human] spiritual well-being" (p. 143). Of these four mentions of nature's conditions, three of them position nature as only having instrumental value. The "restorative and healing aspects of nature" are restorative and healing *for*

© Springer Nature Switzerland AG 2019
N. H. Kessler, *Ontology and Closeness in Human-Nature Relationships*,
AESS Interdisciplinary Environmental Studies and Sciences Series,
https://doi.org/10.1007/978-3-319-99274-7_5

humans. The qualities of nature (themselves not described) that positively affect human health and human spiritual well-being are also only noted for their instrumental value. 'Beauty' is the only non-instrumental quality that nature has in Saunders' discussion, though if it's mentioned because of the purpose it serves in human recognition of it, or whether or not that beauty is material, goes unspecified. Given this scant articulation of nature's Self-related qualities, what results in her characterization of conservation psychology is both an instrumental and reduced material view of more-than-human beings, and also a potentially anthropocentric view of the reciprocity dynamic of the relationship.

One place where the latter appears is when Saunders, in order to emphasize this very "reciprocal quality of relationships between humans and the rest of nature" (p. 141), recommends two main areas of study. The first is "how humans behave toward nature" (p. 141). If one believed that a real relational reciprocity with more-than-human Selves was possible, then one might expect the second major area to be "how nature behaves toward humans." But this is not the case. Instead Saunders suggests that the second area of study ought to be "how humans care about/value nature" (p. 141). By stating the other half of the reciprocity in this way, Saunders "deactivates" more-than-human beings as substantive participants with contributory qualities and capacities, and transforms them into unknown entities to which value is attached or onto which it's projected by the human. This move undercuts the mutuality implied by the notion of reciprocity and relocates its substance to the human's internal value assignations projecting outward from a relational center that only the human occupies. It also subverts the dynamics of close relationships portrayed in my interdependence model. The reciprocity becomes instead a consideration of human actions alone: either as actor *toward* nature, or as assigner of value *to* nature's (otherwise irrelevantly valued, it must be assumed) workings. This instantiates a dualist anthropocentrism predicated on an unspoken assumption about nature's qualities and capacities, or a lack thereof.

The effect this has on the structure and dynamics of my close relationship model is shown in Fig. 5.1, which I've modified from the original Fig. 4.1 in the interdependence theory of Chap. 4. As one can see, the thoughts, feelings and actions of beings in the more-than-human world have been eliminated as having influence over the human, and the thoughts, feelings, and the actions that emerge from the human to influence the more-than-human do so from some unspecified place within the human (not in *response* to more-than-human beings' thoughts, feelings or actions, which is the very definition of interdependence and reciprocity). I acknowledge that some in the conservation psychology field might disagree with my interpretation of their framework. Specifically, they might claim that theirs does not *eliminate* more-than-human beings' influence over humans since, in humans assigning value to nature, nature is having an effect that elicits such a valuing. In response, though, I'd point out that in Saunders' article there is no discussion of how the valuing is to be conducted. Thus, at least in her interpretation, nowhere is there an assurance that the valuing of more-than-human beings that humans undertake is tied to the thoughts, feelings or actions of more-than-human

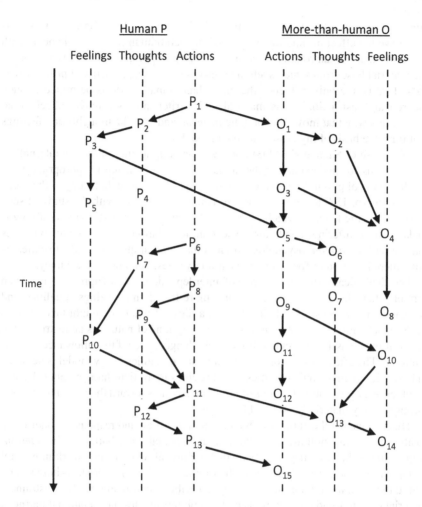

Fig. 5.1 Modified interdependence diagram from Fig. 4.1 with all thought and feeling-based, more-than-human reciprocity removed

beings. In fact, from the tenor of her discussion, it seems that such a valuing process is sourced somewhere wholly within the human, and thus outside the reach of more-than-human beings' influence. One is forced, therefore, to assume that it is not an imperative that human valuing of more-than-human beings come in response to being in relationship with more-than-human beings at all. The relationship could instead be the *post hoc* result of an ontologically distinct set of human circumstances. Figure 5.1 above shows this loss of reciprocity, and thus of true interdependence, via this unidirectional dynamic.

Posthumanist theory addresses this kind of issue by suggesting that the more-than-human world is inextricably caught up in any qualities and capacities the human is supposed to possess as an individual-in-relation (e.g., Barad, 2007). This does have the effect of binding the genesis of such human relational elements with the more-than-human relational partners, but as I've pointed out above, because of the material base from which such theories operate, human relational primacy persists. That is, the only influence the more-than-human world is permitted in such theoretical constructions is as material underwriter, not as a universal relational partner in a way that moves beyond the material and into the thoughts and feelings that form the basis of my close relational model.

In defense of Saunders' (2003) portrayal, one might suggest that an internal-to-human focus for understanding the human-nature relationship is appropriate for a subdiscipline of psychology that this field is. But, I'd suggest that study of the "self-in-relation" (p. 138) as Saunders puts it, is not synonymous with the study of self-*as*-relation—the latter of which Saunders' portrayal of conservation psychology is in danger of proffering. The failure to account for relational structure and dynamics that, in part, *determine* psychological and other human qualities and capacities, is not a mere isolation of focus along disciplinary lines, but a lack of accuracy.

The dualistically driven collapse of interdependence or reciprocity goes even further when, by failing to include any more-than-human beings' qualities and capacities in the model of relationship, regardless of what one might take them to be, Saunders allows dualistic ontological conceptions of nature as an instrumental, material-only, passive object to intrude unchallenged, even if this allowance is unintentional. The effect this has on the structure and dynamics of my model is depicted in Fig. 5.2, where the only qualities and capacities more-than-human beings have is to either receive the output of more-than-human beings toward them, and to respond through purely materially observable actions.

This means that the entirety of the human interaction and response to nature can easily have almost nothing to do with nature's actual contribution to the relationship. If that's the case, the *entire way* that reciprocal relationships work in general collapses. The actual relationship is then replaced with an internal-to-human facsimile to which more-than-human beings contribute their dualistically constrained, material qualities and capacities. They become receptacles, not relational partners. In response, I suppose some might suggest that *all* relationships are just a "self" acting and assessing meaning and value to an "other's" actions, but such a conceptualization is itself dualist in an ontologically individualist sense.

Saunders' dualistic portrayal of human-nature relationships is far from unique, as the rest of this chapter will show. When seen through the critical lens afforded by a human-nature relationship model based in a well-articulated, interhuman model, much of what are advanced as reciprocal, interdependent theories of human-nature relationships are not particularly descriptive of reciprocity, and in that sense, as a relationship as humans understand them between each other. Again, I don't attribute this to more-than-human beings lacking the qualities and capacities necessary to form this kind of human-nature relationships, but to dualistic conceptual underpinnings that *post hoc* collapse the structure and dynamics of such relational experiences by obviation of more-than-human contributions to them.

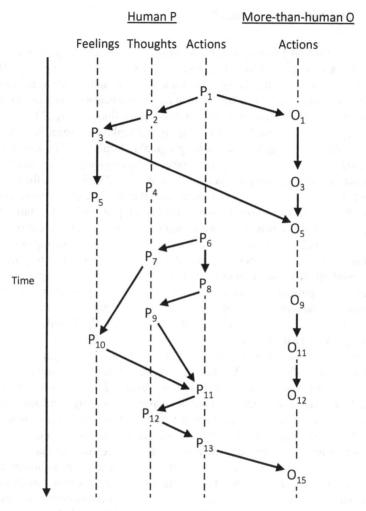

Fig. 5.2 Modified interdependence diagram from Fig. 4.1 with thoughts and feelings removed entirely

5.2   Place Theories

"Place," "place attachment" and "sense of place" form a family of theories (hereafter referred to collectively as "place theories") developed from the sociology, human geography and environmental psychology disciplines to explain the role of place in the human experience. Dualisms operate at many levels in these theories.

5.2.1 Definitions of "Place"

The Oxford English Dictionary (OED) defines "place" as: "A particular part or region of space; a physical locality, a locale; a spot, a location" ("Place", 2018). In place theories, however, the definition tends to be more anthropocentric. For example, Jorgensen and Stedman (2001) say that place "is a center of meaning or field of care that emphasizes human emotions and [human] relationships" (p. 233). Williams, Patterson, Roggenbuck, and Watson (1992) say that "[p]hysical space becomes place where we attach meaning to a particular geographic locale" (p. 31). Jorgensen and Stedman (2001) also say that "'a place…takes in the meanings which people assign to the landscape through the process of living in it'" (p. 233). So, while the OED definition of place and locale are equivalent, in place theories locales are purely physical and only become "places" when meaning or value are given to them by human beings through their activities there. In such a view, more-than-human beings are construed to have neither influence over what that meaning or value, nor standing as relational Selves in the contributions they do make. While a human attachment to place is certainly bound up with what activities a human or group of humans engages in there, that can only be positioned as the totality of the definition of place if one presupposes that more-than-human beings are not Selves. This is not something that can be uncritically averred, and as I'll explore in the Parts III and IV, is based almost entirely on assumption and a dualist hegemonic reinforcement thereof.

Posthumanism does counter such notions of place in that it would generally force an acknowledgement of the more-than-human contribution, and would prevent any privileging of humans as sole possessors of anything relational. Unfortunately, because of the largely material accounting they make of relational interchange, and their tendency, at least in those working in experimental fields, to frequently equate "nonhuman" things like manufactured "equipment" or more general "nonhuman matter," the elements that I refer to as more-than-material (and that are present and vital in close relational human-nature interchange) are still denied to more-than-human beings. More-than-human beings then only become the *sponsor* of those things as they cohere around humans. Thus, as Selves-in-relation, even if humanist boundaries are dissolved and the material accounting more egalitarianly distributed, we only become more confused as to whether a bird outside my window is responding to me relationally or if she becomes only a more well-accounted-for vehicle for my projections.

5.2.2 Instrumentalization

One effect of anthropocentrically removing Self-ness from more-than-human beings has is to make instrumental value for more-than-human beings the only kind of value obtainable. For example, place theorists Greider and Garkovich (1994) say,

> Our understanding of nature and of human relationships with the environment are really cultural expressions used to define who *we* were, who *we* are, and who *we* hope to be at this

place and in this space. Landscapes are the reflection of these cultural identities, which are about *us,* rather than the natural environment. (p. 2)

By "we" and "us," the authors mean human beings, thus the value that the authors see for more-than-human beings is as a vehicle for the formation of human cultural identity. Again, it's true that social and cultural elements are influenced by place, but place's sole value cannot uncritically be assumed to lie in serving this purpose. The only way it can is when one takes value to be a purely human construction. But this again cannot be assumed, it must be openly argued and supported, yet this does not occur in the place literature.

Another example of instrumentalization is in two subtheories of place attachment called *place dependence* and *place identity.* Place dependence is described as a "functional [place] attachment" (Williams & Vaske, 2003, p. 831) that indicates "how well...the place serve[s] instrumental values or [human] goal achievement" (Jorgensen & Stedman, 2001, p. 236). Place identity is described as "an emotional [place] attachment" (Williams & Vaske, 2003, p. 831) that is a "repository for emotions and [human] relationships that give meaning and purpose to *life*...As such... [it] has been described as a component of self-identity...that enhances self-esteem" (p. 831). It's clear that in these characterizations the central value of more-than-human beings is as instrumental to humans. Now, if these characterizations were presented as fitting into a larger web of relationships, where the more-than-human beings with whom humans are relating are simply left out of the discussion for the sake of focusing on how an individual becomes involved with place in general, such characterizations might be less problematic. That's because, implicit in such a portrayal would be the knowledge that there *are* others in the relationship, contributing what they do to it. For example, when discussing an individual human's connection to her human family, one might focus on what she gets out of those interactions while knowing that she puts something back into them, and others contribute and take from those connections as well. The issue in the context of these place relationships, however, is that it's assumed that the individual or group of humans get nothing from the place itself, they only create attachment, sense of place, or other place affiliation by the activities in which *they* engage while in the merely physical locality. Without an examination and discussion of the role of the more-than-human participants, the assumption that there simply isn't one is dualistic. And again, while posthumanism forces an accounting here for the contribution of the more-than-human participants, by making the contribution largely material while accepting humans as privileged relational beings, other than a more fulsome account of the material inputs, understanding the more-than-human contribution to the model of close relational structure and dynamics is not advanced. Further, while it's true that posthumanist accounts do allow for more-than-human relational contributions, such as Haraway (2008) suggesting that she and her dog share "the contagion of love" (p. 16), because that contribution is still materially rooted, as I discussed in Chap. 3, the dog's love is still positioned as lower down in the relational hierarchy compared to the human—that is, the human is still dualistically privileged with more relational qualities than the dog, and therefore the love they share is bounded in a way that may not reflect the existential relational arrangement. Working one's way down the hierarchy, when one gets to slime molds or my houseplant, there is little left as the basis of relationality that I suggest is possible, and that has largely been ruled out by posthumanists and place theorists based on an originating dualist assumption.

In general, intrinsic meaning or value for more-than-human beings has been largely, conceptually eradicated in place theories. For example, Williams et al. (1992) say that "...a place's value is assigned by individuals, groups, or society, without necessarily involving a strong correspondence between the physical attributes of the place and its meaning" (p. 32). Such an eradication is a two-step process. First, more-than-human beings are dualistically denied all but physicality and then, because mere physicality is generally taken by modern humans to carry little or no intrinsic meaning or value, by extension more-than-human beings can have none. While in itself this definition may not give theorists any conceptual trouble, when applied to the notions of place attachment that Williams et al. are addressing—a term carrying a distinctly relational connotation—it quickly becomes so. By rooting the notion of attachment in human conceptualization and its projection onto the material more-than-human world, the very notion of attachment loses meaning. Within such a conceptual framework, a well-engineered virtual reality experience would have to be thought of as equivalent.

Such a substitution for more-than-human beings by human artifice strikes this author as intuitively troublesome. Williams et al. (1992) see this as well when suggesting that instrumentalization "perpetuates the notion that recreation settings are theoretically interchangeable...[and] even reproducible" (p. 30). Of this, they say that it is "[o]f particular concern [given] that recreation settings are very often one-of-a-kind places that cannot be designed or engineered like so many makes of automobiles" (p. 30). These authors attribute this problem to the extreme instrumentalizing tendency of the natural resource and recreation management fields, who look at more-than-human nature as a "commodity" (p. 30) to be managed, and "as means rather than ends" (p. 30). Williams and Vaske (2003) rightly counter such purely instrumental perspectives by noting that "society increasingly values natural resources in ways not easily captured by the...production metaphor of 'use' and 'yield'" (p. 830). But, while these authors recognize the problem, their conception of an alternative fails to move their theory of attachment outside of an instrumentalizing, and thus dualist, framework. By contrasting a "use" and "yield" perspective with one where "society increasingly values natural resources" in more benign ways, the valuation is still wholly within the purview of the human valuer. This reveals an ontological commitment to the valuation of more-than-human beings that is wholly consistent with that which produces the "use" and "yield" approach. In their view, any benefits that accrue to more-than-human beings from less aggressive forms of "use" are still incidental, and carry no inherent ethical or ontological force of their own. Use is use. If a land manager doesn't want to clearcut a forest tract because he knows humans recreate in it and in so doing, valuing it as a unique physical locale, he has no standing to contradict the timber company that wants the trees for pulp production. Since all value is determined by one group of humans vs. another—the value of recreation vs. the value of timber production—the valuation

process never escapes the orbit of human instrumentalization that is its core orientation. By leaving intact a human-nature relationship that denies more-than-human
value outside of human utility, a natural place is still just Williams and Vaske's
"automobile," even if it has become one of a kind.

5.2.3   More Extreme Forms of More-Than-Human Reduction

Beyond anthropocentric definitions of place and the instrumentalization it leads to,
at times place theories reduce more-than-human beings even further. For example,
Jorgensen and Stedman (2001) refer to "place" as encompassing the "plethora of
concepts describing the relationship between people and *spatial settings* [emphasis
added]" (p. 233). Here, the more-than-human contribution to place is as "space"
which has the salient features of being devoid of any inhabitant at all. Just as
European settlers arriving in North America pretended to have found "virgin territory" which was actually populated by various Native American tribes, here a place
populated by more-than-human beings is emptied of them by the very *post hoc*
conceptualization process intended to portray their role in human interaction with
place. Tuan (1977) uses this kind of terminology when referring to place by saying,
"[W]hat begins as *undifferentiated space* [emphasis added] becomes place as we…
endow it with value" (p. 6). At its furthest extreme, the characterization of more-
than-human nature as empty space occurs in Jorgensen and Stedman's (2001) quotation of Ryden's statement that, "'a place…is much more than *a point in space*
[emphasis added]…but takes in the meanings which people assign to that landscape
through the process of living in it'" (p. 233). While I don't think Ryden meant that
more-than-human nature actually only exists geometrically in one dimension (literally what a "point in space" is), his description is characteristic of the degree to
which more-than-human beings are conceptually reduced in place theories.

Part of the cause of this reduction can be traced to the evolution of theories such
as place attachment from the environmental psychology and human geography disciplines (examples of which are Raymond, Brown & Weber, 2010; Williams et al.,
1992). These disciplines root their notions of external-to-human contributions to
"place" in built environments. Thus, it makes some sense that place attachment
theory might emerge from such disciplines with very little room for a definition of
"setting" beyond inert space. Leaving aside the implications of treating even built
environments in this way, I note that the reductionist assumptions made about built
environments cannot be transferred uncritically to more-than-human ones. That is,
no real argument has been presented by these theorists to establish that more-than-
human nature is as empty or inert as, say, a cinderblock-walled classroom. If place
theorists assume that, despite any differences between built and natural environments, the contributions of the more-than-human are still purely material, this itself
is predicated on an assumption about the qualities and capacities of more-than-
human beings—on ontological commitments as to the attributes of more-than-human
beings—that is rife with the dualisms I have noted thus far. Thus, they cannot be
uncritically accepted.

While posthumanists have inverted this paradigm by suggesting that *relata* are the product of relations, this still does not allow one to escape the materialist trap. For example, if one were to manage a "place" such as a marine reserve, I suggest that most modern humans would recoil at any kind of management decision to eradicate migrating Right Whales so that recreational activities such as powerboating could be made safer for the humans choosing to attach to place in this way. One might counter my suggestion by noting that a sports fan would feel a similar sense of loss of attachment if the stadium in which he grew up watching his favorite team play was demolished. But, I'd point out that while there is attachment to both the stadium and the whales, it's possible (and I will argue below, true) that the latter is more like the attachment one feels toward a spouse. What's more, the whales would have contributed to the attachment in ways that the stadium has not. One cannot uncritically suggest that a human's attachment to his spouse is like his attachment to a stadium primarily because the former attachment is reciprocated by a relational Self—and the reciprocation *contributes to the attachment.* I suggest that the only reason that more-than-human beings can be seen as not having the capacity to attach to, and be the subject of attachment with, humans in a "spousal" way is due to dualistic ontological commitments, not to any evidence or justification offered by place theorists. While my view echoes that of posthumanism, by leaving intact a materialist ontology, they still instantiate a relational hierarchy with humans at the top. While they might counter such a suggestion by saying that there is no top or bottom, just "asymmetry" or difference (see in-depth discussion of asymmetry in Chap. 3), when it comes to understanding close human-nature relationships, I see little difference in effect. Some more-than-human beings can contribute, and others are dualistically prevented from doing so by a material accounting that denies them, either as individuals or as Barad's (2007) intra-actions, the qualities and capacities to make such contributions. Again, this cannot be uncritically accepted.

Though by and large absent, defenses for dualist constructions of more-than-human beings as passive, material backdrops *have* been offered. One example is the chain of logic that says that, because the same physical setting can carry different place meanings for different humans, the physical settings themselves can't have meaning or significance of their own (intrinsic). An example of this is in Gustafson (2001), who defends his anthropocentric definition of place by referring to Massey's claim that "places do not necessarily mean the same thing to everybody" (p. 6). Greider and Garkovich (1994) expand upon this notion by saying that "The open field is the same physical thing, but it carries multiple symbolic meanings that emanate from the values by which people define themselves" (p. 1). Essentially, what's being argued is that because there is a physical locale and it means different things to different human beings, that it can't have any value, meaning, or Selfness of its own. But such a claim is poorly constructed. To my point, the very same dynamic of differential perception occurs in relationships between human beings. No two human beings will interpret the qualities, meaning, or value of a third human being in the same way even though that third human being is the same physical entity. But, to conclude from that because of this, the third human being has no inherent significance, value, meaning or Selfness beyond that which is projected onto her by the other two humans would be a mistake. Therefore, it does not follow that because a more-than-human being has material presence and is perceived differently by different perceivers that he or she has no value, meaning, Selfness or other relational capacities and qualities.

Ultimately, the only way that such a claim can be made about more-than-human beings is if there *already exists* a supposition that more-than-human beings have no relationally relevant qualities beyond the superficially material. That is, there is an *a priori* assumption that more-than-human beings *can't be* relational Selves. What is being offered, then, as support for the position that more-than-human beings aren't relational Selves is the assumption that they can't be. This is circular and thus invalid.

5.2.4   Is Unilateral Attachment Possible?

As I've explored above, place theories assume that sense of place or attachment to place result from the unilateral, human manufacture of feelings of connection projected outward onto a passive, material-only more-than-human world. Given that little in the way of explicit argument for such a structure and dynamics is offered in these theories, one might hope to find a clearer justification for such a mechanism in some of the theories upon which place theories are based. For example, place attachment theory is rooted in "attachment theory," which is the study of the attachment humans form with each other (Bretherton, 1985). But, instead of the unilateral mechanism of place theories, here one finds its opposite. Human attachment theorists Collins and Collins and Feeney (2000) say that "[human] attachment theory explicitly acknowledges that social support is a *dyadic process* that involves the interaction of two distinct behavioral systems: *the attachment system* and *the caregiving system* [emphasis added]" (p. 1053). Human attachment theory, then, is foundationally reciprocal, involving equal emphases on the contributions of the one attaching and the one with whom the attachment is made.

Drawing parallels between human and place attachment theories, in place attachment the "attachment system" would belong to the human and the "caregiving system" to the more-than-human. But, in the transformation of attachment theory into place attachment theory, the caregiving system is entirely lost. Since no argument is made for its abandonment, one must assume that either the caregiving system was inconsistent enough with pre-existing ontological commitments to the marginal role of more-than-human beings that it was simply ignored, or that this ontological commitment sponsored an assumption that no justification was necessary for the disposal that took place. Either way, such a radical departure from the structure and dynamics of interhuman attachment calls into question the utility of attachment as an explanatory framework for the feelings that theorists suppose are formed by humans toward more-than-human beings. Ultimately, either attachment as an explanatory framework, or the belief that it can form unilaterally, must be abandoned.

As I note in the Chap. 2, one of the hallmarks of dualist interpretations of the nature of more-than-human beings is a denial of all but passive, material roles in the human-nature relationship. Since no strong evidence or argument is offered to support these definitions of more-than-human beings in place theories, it is now highly likely that a *post hoc*, dualizing reduction is responsible.

5.2.5   Problems with Relational Structure and Dynamics

As stated in the beginning of this section, "place" in place theories is located almost entirely within the human. This presents real problems for the notions of attachment, bonds, or relationship upon which place theories depend for their coherence. Such relational elements are commonsensically understood to be *between* two entities, yet place theories position these elements as unilateral feelings that develop inside an individual human in response to what appears to be a relationally unimportant, material set of external elements. In interhuman relationships, no such facile reduction of such radically relational terms would be permitted, and so I suggest here that, a) without supporting argumentation and b) in light of what I've discussed so far about the obfuscating effect of dualisms, the same intolerance must apply to discussion of human-nature relationships.

As an example of the problems that arise from reducing what are foundationally relational terms to individual ones, I note Williams et al.'s (1992) statement that "emotional bonds [with places] are associated with long-term relationships to places" (p. 32). But, if a bond with a place correlates with a long-term relationship with that place, is it really sensible to think of the relationship as one that exists wholly inside the human and only "acted out," as it were, in an otherwise relationally inert natural setting? If so, what is the point of using terms like "bond" or "relationship" at all? Such a statement would be far more sensible within the place theory literature if it read something like this: *emotional attractions to the affective and cognitive elements a human internally associates with certain physical settings correlate with pre-existing, long-term internal connections that a human has made between his idea of himself and his idea of the importance of an external, physical setting.* When bonds and relationships are filtered through the main concepts of the place theory literature in this way, it shows just how labored and human-self-referential they can become. That, and just how out of keeping they are with the particular experiences of closeness many humans have with the more-than-human beings that populate and/or constitute "places."

Not only do these anthropocentric, instrumental, reduced definitions of relational terms become labored and awkward, at times they aren't even applied consistently by the theorists advancing them. For example, Beckley (2003) suggests that "sociologists have rarely turned their attention explicitly to the relation between humans and places themselves" (p. 106). Here, "place" seems to stand for external more-than-human beings. And while one might critique Beckley for not being clear on the use of term since he is not a sociologist by training, sociologists theorizing about place show this kind of confusion as well. Jorgensen and Stedman (2001) quote Stokols and Shumaker's definition of place dependence as an "occupant's perceived strength of association between him or herself and specific places" (p. 234). Jorgensen and Stedman themselves say that some "have suggested that [place] attachment 'involves an interplay of affect and emotions, knowledge and beliefs, and behaviours and actions in reference to a place'" (p. 234). In these examples, place seems again to mean more-than-human beings and the emotional response to them a separate, internal-to-human response. Are these verbal missteps the result of theorists meaning "locale," "location," or "space" and just using the term "place" accidentally? One would think that if a theorist was clear on her construction of the notion of place, that she would not make such a mistake. If it's not a confusion of terminology and the theorists mean place as they've defined it in these statements, then the statements themselves quickly become circular. The human's "ideas, thoughts and feelings" (Johnson, 1998, p. 5) about place become responses to...her ideas thoughts and feelings about....her ideas thoughts and feelings. It is self-referential, circular, and at some point insensible. I think the most likely explanation for this confusion is neither a mistake in terminology nor a collapse into circularity, but instead is a lack of clarity on the part of theorists as to what really does constitute a "place." Here is where I believe the lay, OED type of definition of place reasserts itself. In the very act of trying to define place as something other than what most people take it to be, at times place theorists betray that they, too, don't always take place to be the largely human creation they claim it to be.

A further problem introduced by this confusion lies in its confounding effect on attempts to measure human relationships with places. For example, one scale item from Jorgensen and Stedman's (2001) work asks respondents to "identify a position on the scale that best reflected their relationship to the *place* in question" (p. 235). If, in such scale items the authors intend for respondents to understand the authors' own construction of the term "place" as an external locale infused with the respondent's own meaning and value, then the authors aren't gathering support for their construction of the concept of place, only those elements of a respondents' construction that already align with it. If there are other elements that are not in alignment, they are missed. An alternative possibility is that Jorgensen and Stedman think that respondents see "place" *as* an external, relational Self, while the authors "know better" and correctly *post hoc* differentiate the physical space from the respondent's "ideas, thoughts and feelings" about it. But, since empirical work is meant to offer support for such *post hoc* interpretations instead of imposing that interpretation through a *pre-existing ontological position*, this explanation is flawed. In addition, if the conflation of the two notions of place (physical locale impregnated with value/meaning vs. OED definition of distinct locale) is intentionally included by researchers, then the authors are guaranteeing confusion over the term by participants and undercutting the validity of any results that support their constructions of place. If, on the other hand, the confusion of terms is unintentional, it reveals that theorists themselves are confused, which leaves the same undercutting effect intact.

5.2.6   Acknowledging the Absence of More-Than-Human Beings from Place Theories

In the disciplines that contribute to place theories, some few theorists have noticed the absence of substantive consideration of more-than-human beings. For example, Mueller (2002) says of the place attachment literature that there is a general "failure to mention the natural world" (p. 19). In addition, Murphy (1995) critiques sociology by that "sociology…has been suspicious of…any claims that the natural has influence over human relations" (p. 690). Gustafson (2001), who refers to Canter saying that "the influence of physical [setting] attributes on psychological and behavioural processes deserves more attention" (p. 6). Fox (1997) reinforces this when she says,

> It is rare to discover a [written] piece about a relationship with nature outside a role as backdrop…or resource that contributes to the human recreational experience…If leisure is a focus or connection with nature…, it is crucial to examine this relationship and how it may be supported or inhibited by the philosophy and frameworks inherent in current leisure philosophy and practice. (p. 164)

5.2.7   Conclusion

Place theorists suggest that attachment to, or sense of, place is something almost wholly constructed inside human beings. This contradicts some human experiences of external locales and their inhabitants as relational Selves, and threatens to make meaningless the very notion of interacting with place as most understand it. These theories are right to recognize that humans can feel attachment, closeness and other strong relational feelings with more-than-human beings, but by leaving intact dualist ontological commitments that force the substantive relational action to be confined wholly within the human, they fail to move the discussion of what constitutes the human-nature relationship beyond self-serving human definitions and the practices based upon them that proliferate today. If human relationship with place is not relational at root, the human-nature relationship doesn't really exist. Thus, it has no power to govern human actions that, whether conceptually acknowledged or otherwise, operate wholly within its bounds.

5.3   Connection to Nature Theories

I label the loose constellation of discourse considered in this section "connection to nature" theories. These theories mostly emerge from the psychology disciplines (and to a lesser extent, sociology) and involve exploration of both cognitive and affective dimensions of human connection with nature.

5.3.1   The "Nature" of Connection

Connection or *connectedness* with nature is defined in various ways in this body of literature. Schultz (2002) says that "[c]onnectedness refers to the extent to which an individual includes nature within his/her cognitive representation of self" (p. 67). Such a notion of connection is rooted in Aron, Aron, Tudor, and Nelson (1991) suggestion that the degree of closeness two humans feel in relationship with each other is measured by the degree to which each individual cognitively defines her "self" as including the other person. Here, then, connection is not an external, or necessarily externally produced, relational condition, but instead is an internal-to-human construction. Schultz (2002) affirms this when he says that connection "is a psychological [notion where]…the extent to which an individual believes that s/he is connected to nature has cognitive, affective and behavioral components" (p. 62). This indicates that Schultz's view of the human-nature relationship does not follow my close relational model's reciprocating interdependence along vectors of thought, feeling and action and instead is largely anthropocentric. Mayer and Frantz (2004) mirror Schultz's construction when, in wondering about the effects of connection to nature they say that they must "evaluate whether this sense of *feeling* [emphasis added] connected to nature leads to…ecological behavior" (p. 504). Here, instead of asking whether a human's *connection* leads to ecological behavior they ask if the *feeling* of being connected does. Thus, what would've been part of an external relational dynamic of actual connection is relocated inside the human, making any actual relational dynamic *a posteriori* to the unilateral human feeling that is its progenitor.

The work of Hinds and Sparks (2008) also carries an anthropocentric view of connection by reducing the more-than-human relational partner's influence. For example, in referring to connections with more-than-human beings, they say that "direct experiences of an object" (p. 110) show that "[r]epeated exposure to an attitude object may also be instrumental in the growth of positive affective connections with that object" (p. 110). Here, the more-than-human being is an object, and don't necessarily carry any particular qualities to provoke the positive affective connections humans feel toward them. The only dynamic that creates those connections is the repetition of exposure and the human response to it.

Nisbet et al. (2008) also subscribe to connection as a psychological notion, but do so at least in a *potentially* more relational (i.e., external to any particular individual) manner. They suggest that their construct of "nature relatedness" (p. 2) describes "individual levels of connectedness with the natural world" (p. 4) and also "encompasses one's appreciation for and understanding of our interconnectedness with all other living things on the earth" (p. 4). Here, it is a welcome change to find more "room" for the possibility that more-than-human beings could relationally stimulate the connection whose appreciation and understanding is the authors' focus. However, the focus is still on human appreciation and understanding. Thus, the focus still rests almost wholly on the human instead of on the structure and dynamics of any *interaction* between human and more-than-human that would produce closeness. Once more, the more-than-human being as relational Self is pushed to the periphery. While posthumanism does promise to force the dislocation of the human from the center of attention along with a more open accounting for more-than-human beings, because it is materially based and carries with it a dualistic hierarchicalizing effect, the qualities and capacities of those beings are still inaccurately positioned as less relational.

Though in connection to nature theories psychological treatments are the predominant ones, not all maintain such a view. For example, Dutcher, Finley, Luloff, and Johnson (2007) say that "[c]onnectivity attempts to describe the perception of a force or essence that holds the universe together—the same essence or force that runs through all creation" (p. 479). In this sense, they see the connection between human and nature as an energetic or spiritual (they use the term "creation") phenomenon, and so position it as an external, relational condition that humans perceive. Whether they do so in a way that escapes the dualisms that distort accurate interpretation of close human-nature relational experiences I will discuss further below.

5.3.2   External Connection

In noting that many theorists see connection as an internal-to-human construct, given that there must be a connection to someone or something outside the human, it's useful to explore just what they hold the actual, external connection to be. My suspicion is that, just as in place theories, the external interchange is purely physical or material. Of course there is a material element to such connections, but even when attending to the reality of that connection, as sustainability theory or posthumanism does, I don't believe it forms the basis of closeness. Therefore, we must press past the dualistic assumption of a material-only interchange to see what else might require attention and rehabilitation.

Schultz (2002) provides some clues when referring to Aldo Leopold, who explains that "human activity will be guided by the impact that it will have on the natural environment...[thus] we must know about nature: about ecology, about plants and animals, and about the effect that our behavior has on this ecology" (p. 65). This indicates that the actual external connection is thought by Leopold to be ecological, and thus largely material. To support this, Schultz suggests that the "implicit consideration of the similarities between humans and nature" is the result of such ecological knowledge and leads to human and nature being "connected symbiotically" (p. 65). Adding to that, he quotes a reference that Leopold makes to Darwin as support for the kind of "similarities" he means. Because Darwin's explanation of evolution is materially based, Schultz's invocation of Darwin's theory implies that he, too, takes the external connection to be material.

As a result, any feelings or experiences where, as Mayer and Frantz (2004) put it, "people experientially view themselves as egalitarian members of the broader natural community...[and] feel a sense of kinship with it," (p. 505) have to be explained through a materialist ontological lens. This is dualistically reductive, and seems to fly in the face of the definition of "kinship," which is a familial term, carrying with it a much broader notion of connection than the purely material. For example, if I were to suggest that my feeling of kinship with my two sisters was due to our material interconnection, this would seem to miss a core meaning. While many holding to a materialist ontology would ultimately attribute even that connection to material origins, I think it's safe to note that there is an important difference between my connection with my sisters and Mayer and Frantz's description of connection with more-than-human beings. If, in response, one suggests that sibling relationships are qualitatively different than human-nature relationships, I'd suggest that such an assertion is based on an *a priori* ontological assumption that more-than-human beings cannot contribute what humans do to relationships in general. But, one of the main purposes of this book is to take direct aim at this assumption and, as I will reiterate throughout this work, suggest that there is little or no *evidence* presented in these theories to support this assumption.

While the vast majority of the connection literature has this orientation, I noted above that Dutcher et al. (2007) are an exception. They take a community sociology approach that "emphasizes an intuitive rather than a material connection between people and their environments" (p. 478). I'll explore the ontological implications of this alternative stance with respect to the qualities and capacities of humans and more-than-human beings next.

5.3.3 Internalized Connections

Another way that more-than-human beings are dualistically reduced in connection theories of human-nature relationships is by portraying those relationships to be the sum of their individual participants. Combined with an unexamined assumption of the relational inferiority of more-than-human beings, human individuals come to dominate. I'll reiterate here that while posthumanism decries this very thing, *how* more-than-human beings come to be accounted for in posthumanist theories when dualistic materialism remains in force is yet to be convincingly articulated. As an example of humans dominating, Schultz (2002) says, "At the heart of the discourse on human-nature relations is the recurring theme about a *relationship* with nature… But at the core is the [human] individual, and his or her understanding of his place in nature (p. 66)." Here, the relevance of the human-nature relationship is shunted to the human. The irrelevance of the more-than-human being in such *post hoc* conceptual rearrangements can be seen elsewhere in Schultz's work, such as when he discusses possible "uses" of nature. He asks, "What value does nature have? An economic value, recreational value, aesthetic value, religious value?" (p. 64). We must note that regardless of which values he cites, all are anthropocentric, and instrumentalize more-than-human value. This is repeated in others' work, for example in a scale item that Hinds and Sparks (2008) develop to ask what the result of respondents engaging with the natural environment would be. The only possibilities offered are anthropocentric, such as "…allow me to experience beautiful scenery… help me escape the stresses of life…[or] give me a sense of connection with nature" (pp. 112–113). Even in the last response, the connection itself is not the subject, but the *sense* of it, which again one must assume is wholly inside the human.

I contrast this view with Dutcher et al.'s (2007) attempt to build "a spiritual community in which humans and nature are members" (p. 480), and where "[n]ature as a part of community can be understood intuitively" (p. 480). It seems here that not only are more-than-human beings and humans peripheral to the centrality of relationship, but that the nature of the connection is intuitive, or more-than-material in my ontological frame of reference. Their work does hold promise as a relational approach to connection, but since Plumwood (1993) rightly suggests that dualisms lurk everywhere, we must dig further to understand if their intuitive "identification and connection" (Dutcher et al.'s 2007, p. 480) yields a nondualistic portrayal. Given the promise of their views to add more-than-human beings as relational Selves based on a more-than-material element, it strikes me as particularly important that they robustly support such an assertion. Alas, they do not.

What support they do give for a more-than-material basis for connection comes in the form of quotations of others that hinder as much as aid their efforts. As just one example, they quote Finley as saying that what's needed is a "holistic viewpoint…, according to which natural resources can be seen, not only as attributes of the physical environment, but as attributes of the social and cultural order as well" (p. 481). Finley's statement carries a dualizing characterization of more-than-human beings as material "natural resources." In addition, these more-than-human beings serve human sociological and cultural ends, but never either their own or that of the relationship. Thus, though Dutcher et al. attempt to put humans and more-than-human beings on equal relational footing, by leaving unaddressed the dualistic ontological commitments to human superiority in the others that they use to support their work, their efforts ultimately don't carry us as far as I believe we need to go. I'll explore their views a bit more below.

5.3.4 Inclusion of More-than-Human Other in Human Self

If more-than-human beings are dualistically negated as all but material relational partners, theorists are much freer to move more-than-human beings wherever they'd like within the structure and dynamics of human-nature relationship conceptualizations. For example, as mentioned in the beginning of this section, Schultz (2002) develops a notion of connection by extending Aron and Fraley's (1999) work with interhuman close relationships. He quotes them saying, "The basic element of interpersonal closeness is cognitive, an overlap of knowledge structures of self and other, such that in a close relationship each individual includes aspects of the other as part of his or her notion of self" (p. 336). Schultz (2002) notes that in Aron et al.'s (1991) work, not only is there "overlap with many shared qualities…[but t]aken to the extreme, self and other become one" (p. 68). Schultz (2002) follows on this by saying that the overlap "is the central aspect of inclusion with nature. Individuals who define themselves as part of nature have cognitive representations of self that overlap extensively with their cognitive representations of nature" (p. 68). Mayer and Frantz (2004) echo this notion by suggesting that certain human actions with more-than-human beings "lead to a greater self–other overlap" (p. 504). One such act is "perspective taking" (p. 504), where the self takes the perspective of the other.

But, it is imperative that a differentiation be made here. On the one hand, Schultz (2002), Aron et al. (1991), and Mayer and Frantz (2004) suggest that in self-other overlap "self and other become one" or the "other" becomes "confused with the self" (Aron et al., 1991, p. 242). On the other hand, Dutcher et al. (2007) see connection as the "understanding [that] people and nature...[are] part of the same community" (p. 479). In the former construction, there is a conceptual collapse of self and other. In the latter, there is a similarity that is never confused with existential sameness. Further, if one digs more deeply into the basis of the notion of "perspective taking" proffered by the *self-other merging* group of theorists from which these human-nature relational concepts are derived, one does *not* find actual merging, but instead a specific warning against it. For example, Davis, Conklin, Smith, and Luce (1996) state, "It is important...to be explicit regarding what it means for the mental representations of two social objects to be merged" (p. 714), and go on to say,

> [M]erging in this sense refers simply to the fact that the two mental representations come to share an increased number of features. The self and the other are merged, therefore, in the sense that the features associated with each one are increasingly intertwined, rather than remaining as...[non-overlapping] sets of descriptors. (p. 714)

In Davis et al.'s account of merging, then, there is no possibility of the "other" being "*confused*" with the self." The merging isn't of identity, but of commonality of individual traits. Unfortunately, this distinction is largely lost on the nature connection theorists referenced above. For example, Mayer and Frantz (2004) say that when attempting to foster ecological behavior through expanding oneself, "'[I]f the self is expanded to include the natural world, behavior leading to destruction of this world will be experienced as self-destruction' (Roszak, 1995)" (p. 504). Nisbet et al. (2008) echo this usage by stating that "damage to the planet is seen as damage to the self" (p. 3). Even, Dutcher et al. (2007), who refer to sameness only in the context of membership in the same community, succumb to this:

> One of the more interesting implications of connectivity [is] the elimination of the debate about whether helping behavior is truly altruistic or merely a function of enlightened self-interest. Connectivity regards helping behavior as serving both self and others in the context of a diminished consciousness of the distinction between the two. (p. 479)

I note here that such descriptions of altruism-as-self-aid are both anthropocentric and instrumentalizing of the other participant (be it human or more-than-human). If helping the other is motivated either consciously or otherwise by the "fact" that it is really a helping of the self, then the altruism loses its meaning *as* altruism.

I believe that the danger in conceptualizing the human-nature connection as a merging of more-than-human with human self-identity is best characterized by what ecofeminist Whatmore (1997) calls the "imperialism of the self" (p. 45).

She describes this phenomenon as one where the human individual's wants, qualities and agency subsume all with whom the human interrelates. Such subsuming action is sponsored by dualisms that reduce or negate more-than-human beings to the extent that their identity, and their contribution to any connection perceived by humans, is either irrelevant or *post hoc* conceptually eradicated. Such dualistic activity does not necessarily operate consciously, however, since I see the conflation of the notions of self-other identity with self-other commonality as an outgrowth of *a priori* ontological assumption rather than any observations or conscious logical argument. To wit: when a more-than-human being has been dualistically reduced to the point of relational irrelevance, how much easier is it for the "other" in self-other merging to be subsumed in a human's own identity? The result is that, when connection theorists look at connections between human and more-than-human, they have difficulty differentiating between merging and commonality. As support for this, note that Mayer and Frantz (2004) see Leopold's views as evidence of merging of identity, yet their explanation of his views carry absolutely no such connotation:

> For Leopold...[humans being part of the broader world means] understanding the extent to which people experientially view themselves as egalitarian members of the broader natural community; feel a sense of kinship with it; view themselves as belonging to the natural world as much as it belongs to them; and view their welfare as related to the welfare of the natural world. (p. 505)

I believe that Mayer and Frantz have captured Leopold's views perfectly. But, if one considers oneself a member of a broader community, feels a sense of kinship, and belongs to the natural world, it seems discordant to then suggest that this broader world becomes a literal part of one's own sense of identity. If *belonging* is "really" an internal state, what meaning does it have as a relational term? Very little, in my estimation.

Another issue with the notion of self-other merging or overlap, especially in a human-nature relational context, is the determination of the basis of that overlap. If it is the "many shared qualities" (Schultz, 2002, p. 68) between human "self" and more-than-human "other", as Schultz suggests, one is left to wonder exactly what those qualities are. In Davis et al.'s (1996) research on inter*human* perspective taking, some of the "adjectives" (p. 715) or "traits" (p. 715) for which overlap was sought were immaturity, shyness, pleasantness, being carefree, being imaginative, etc. But, if more-than-human beings are only taken to contribute materially to the human connection with them, then which of those qualities would engender a feeling of connection?

Further, if the adjectives or traits that overlap are seen as having only a material basis, one must question whether they'd have the power to invoke the "extreme" self-other merging which connection theorists suggest can occur with more-than-human beings when humans feel connected with them. Place theorists have argued that strong feelings of relationship with physical elements of one's environment can occur, but I've already argued against the likelihood of unilateral formulation of strong relational feelings. What's more, one need not fall back to such a position when no case has been made against more-than-human beings being relational

Selves that contribute far more to the human-nature relationship. Thus, either connection with more-than-human beings is real (i.e., reciprocal along material and more-than-material lines) and the material-only characterization of more-than-human beings must be abandoned, or one must characterize as fantasy the feelings of connection that humans have, regardless of the object of one's connection. I suggest that the former is far more in keeping with the close relational experiences many humans have with more-than-human beings. Further, by eliminating dualisms and challenging the dualist thinking that allows one to assume the absence or irrelevance of more-than-human relational Selves, there is nothing ontological impeding the operation of a reciprocal human-nature relationship along more-than-material as well as material lines.

5.3.5   Collapse of Relational Structure and Dynamics

As I've noted more than once above in this volume, one outcome of allowing dualisms to operate in theories of human-nature relationships is that more-than-human beings are reduced to passive, material backdrops of instrumental-only value to the central human endeavor. This reduction, in turn, collapses any relationally reciprocal structure and dynamics in human-nature relationships. One excellent example of this effect in connection theories is from Schultz (2002). He says,

> One of the central aspects of a close relationship is a feeling of intimacy - the feelings of closeness and affection in a relationship. Intimacy involves a sharing of oneself with another, and a deep level of knowledge about the other. This knowledge about the other person produces a feeling of closeness…Although intimacy typically develops through a process of self disclosure, it seems an easy extension to suggest that people can have a sense of intimacy, or at least caring, for an animal or place. (p. 68)

Schultz's portrayal of intimacy here has at least three problematic features in the context of the point I just raised. First, I suggest that intimacy is innately reciprocal. While it's true that "Intimacy involves a sharing of oneself with another" by failing to point out that intimacy equally involves the sharing of the other with the self, any consideration of it is downplayed or obscured. Another problem with Schultz's description of intimacy occurs when he roots intimate feelings for another in the acquisition of "a deep level of knowledge" about the other. This causes me to wonder if it is really *knowledge* that inspires intimacy, or if instead it is a sharing of feeling—a reciprocating positivity that is as emotional as it is epistemological. Yet a third problem is what I see as a transformation of intimacy into a unidirectional human emotion. In the last sentence of Schultz's description, he says that "people can have a sense of intimacy…*for* [emphasis added] an animal or place" (p. 68). Throughout the human-nature relationship literature I have found this use of prepositions revealing as regards the ontological positions of those employing them. For example, Müller, Kals and Panza (2009), the authors talk about adolescent affinity *toward* nature, which makes the affinity a human-originated, relational element unilaterally projected toward the passive more-than-human. But, they seem to forget this in places, once using the terminology "affinity with" (p. 59) in the body of the article and then also mistakenly using it as part of the title when it appears in the running head of the article! In Schultz, he uses "for" to follow "intimacy." But, "for" is not the preposition most often associated with intimacy. That preposition is "with." Intimacy is a reciprocal relational element involving contributions from both relational partners. The only way it falls into this more awkward, unilateral construction is if more-than-human beings have been conceptually eliminated as contributing substantively to the intimacy. That this elimination appears fully formed in the foundational conceptualization process of what human-nature relationships are like (in Schultz's (2002) work along with many others) is evidence for a) its pre-existing establishment and b) its uncritical acceptance.

If more-than-human beings don't have the qualities and capacities to contribute relationally to intimacy, it's not surprising that Schultz can say that "it seems an *easy extension* [emphasis added] to suggest that people can have a sense of intimacy...for an animal or place" (p. 68). Contrast this with ecofeminist King (1991) when he notes that there are "real difficulties in...fostering the growth of concrete, multi-faceted, caring relations among individuals, societies, and...more-than-human beings" (p. 79). If it's solely within the human domain to create intimacy, then another human, an animal, or one's cobblestone driveway are equally viable candidates. I note that seeing human-nature relationships in this way not only does a disservice to the capacity for feeling and thought in more-than-human beings that I explore in later chapters, but makes humans look, frankly, emotionally stunted. To Schultz and many others, though, it's not the human who is stunted, it's that this brand of intimacy really is non-relational, and thus simple. It is a garden hose one turns on, the waters flowing over whatever ground they find and only in our minds does the fruit of such watering come to bear. I feel confident in saying that to characterize human intimacy this way is controversial. Therefore, either the use of relational terms such as intimacy must be abandoned in the context of human-nature relationship theories such as these, or the possibility of more-than-human beings being relational Selves must be revisited.

As with place theory, even theorists who espouse connection as a predominantly internal-to-human construction show deviations from such an interpretation. One might ascribe such deviations to the simple fallibility of human theorizing, but for one who takes the pragmatist and ecofeminist approach that what is reliably experienced is what is true, such deviations offer potential support for the position that such theories are just inaccurate. As an example, there are items on the Connection to Nature Scale (CNS) that Mayer and Frantz (2004) develop that appear to contradict the dualist conceptual underpinnings of the connection to nature theory that they espouse. One item from their scale is "I often feel like I am only a small part of the natural world around me, and that I am no more important than the grass on the ground or the birds in the trees." Another is "I recognize and appreciate the intelligence of other living organisms" (p. 513). Any interpretation of affirmative responses to such items as support for connection as an internal-to-human construction projected onto material more-than-human beings borders on the outlandish. Just as in my examples of place theory scale items above, here again one finds the language of reciprocal connection, with more-than-human beings as relational Selves, coupled with an interpretation that is its diametric opposite. I take such incongruity to be support for both the veracity of relationally reciprocal human-nature connection and, even if it is at some unconscious level, theorists' awareness of it.

5.4 Sustainability Theories

One does not have to look deeply into today's environmentally related endeavors before encountering the term "sustainability." The US Environmental Protection Agency (USEPA) says that sustainability is "a guiding influence for *all* of our work" (USEPA, 2015). US News & World Report (Gandel, 2013) says that Sustainability is one of eleven "hot majors" for college students. And while some critiques of the concept suggest that "[t]oo often the word becomes appropriated as a band-aid, cure-all additive that can be applied as environmental/ecological veneer" (Progressive Reactionary, 2009) its central notion *has* found purchase on a sweeping scale in modern society. Its central notion being the desire for humans to somehow live within their environmental "means" for an indefinite period of time.

In one sense, the notion that we ought not borrow endlessly against our environmental future is obvious. If we live beyond our "means," wasting "resources" that cannot be replenished as quickly as they're consumed, then sooner or later the "bill" is going to come due to some future generations of humans and more-than-human beings alike. At this level, then, to advocate sustainability is simply good accounting. But, if all that was required to begin to make better the worst of our global environmental problems—climate change, loss of biodiversity, etc.—was to correct an error in accounting or to fill in some knowledge that was heretofore lacking, as I suggested in Chap. 1, this would've been accomplished long before now. In part, sustainability's belief in this class of remedy overlooks the fact that modern societies depend for their very existence and functioning on unsustainable human-nature relationships. As Spehr (1999) notes in his Marxist critique of sustainability, "destruction of nature and non-sustainability are part of the social program of industrial capitalism" (under "Introduction").

So, then, it's possible that what sustainability is attempting to address is not that modern humans don't know what they're doing to the environment, it's that the very workings of modern societies—their social and economic structures—*require* unsustainable environmental relationships. Because of this, as Spehr notes, the industrial practices of modern societies cannot "just be willfully discarded" (under "Introduction") simply by putting industrialism on "a diet" (under "Capitalism on a Diet?"). Of course the potential incompatibility of modern economic/social structures and environmental well-being has been explored by others (Hawken, 1994, Vedeld, 1994), but what has garnered less attention is just how this incompatibility came to exist in the first place. Why is it, after all, that some humans feel enabled to *choose* to sacrifice more-than-human beings as part of good economic or social practices, and why is that the rest of society is constructed in such a way that humans can either shield themselves from this reality or see it as an acceptable cost?

As I've suggested thus far, ontology plays a part in more-than-human beings being treated this way. Bonnett (2002) notes this when he says,

> Western culture…is increasingly dominated by a set of motives which preclude the possibility of an approach to environmental issues which is genuinely open to nature. In such a culture, everyday values will need to be examined with a view to radical transformation. (p. 12)

As it relates to sustainability, Jickling (2002) says that "the only just and respectful way to ensure sustainability…lies in challenging societies' most basic assumptions about human-nature relationships" (p. 151).

But what would such a challenge look like? Is it simply to admit the need for a focus on the human-nature relationship? At some level, that focus is already part of sustainability's core orientation. For example, in UNESCO-UNEP's (1995) document *Re-orienting Environmental Education for Sustainable Development*, it is noted that "Sustainable development requires that…the harmonious relationship between man and nature should become a global issue and an everyday reality" (p. 34). In addition, when discussing sustainability, USEPA (2015) says that "To pursue sustainability is to create and maintain the conditions under which humans and nature can exist in productive harmony to support present and future generations." Imran , Alam and Beaumont (2014), in the context of discussing sustainable development, suggest that "the human–nature relationship should be redefined to establish a more well intentioned and harmonious one" (p. 138). And Jenkins (2009), in discussing sustainability theory, says that "part of the challenge of sustainability is to understand the mutual relations of humanity and nature" (p. 383).

In the context of those definitions (and to respond to Jickling's (2002) challenge), then, it seems necessary to explore exactly what sustainability theorists believe constitute the harmonious and mutual relations they advocate. "Harmony" and "mutuality" are fairly strong relational terms, and whether the creators of those definitions consciously intended it, their use of these terms reveal the deeply relational nature of the unsustainability problem. That, and the difficulties presented by holding a facile, or unexamined, definition of them. Harmony and mutualism have specific definitions that require certain capacities or conditions of its participants that are not generally examined in the sustainability literature. Mutualism, which can also be thought of as reciprocity, suggests that each participant must give something to the other to promote the well-being of both. As I'll show in this section, there is virtually no mention of in the literature what humans give to more-than-human beings. Instead, humans are portrayed as so self-interested that the only kind "giving" they are capable of is that of "taking less." Harmony, on the other hand, implies agreement or synchronized effort between participants. But, most of the sustainability discourse only considers human activity, and when it does, does not seek to align it with more-than-human activity but instead only unilaterally tries to fit it into the supposed capacities of more-than-human beings to withstand it. This strays fairly far from common understandings of harmony.

In order for sustainability to be coherent around the relational concepts it routinely employs, it is of foundational importance for theorists to examine the ontology in which they ground them. Without a thorough exploration of the definitions of humans, more-than-human beings and their relationships, it's impossible to know what harmony and mutuality mean in the human-nature relational context that is sustainability. Bonnett (2002) echoes this notion when he says that "an adequate response to our environmental predicament…requires a metaphysical [could be read as: ontological] transformation" (p. 12). As I've suggested above, I believe that dualist conceptual underpinnings are at work in most modern definitions of the human-nature relationship, and the analysis below shows this to be the case in the sustainability literature as well.

In this section I will not add to the already robust critique of sustainability along the lines of (a) the potential incompatibility of economic and environmental focus to which I refer above, (b) some of its deeply anthropocentric orientations, or (c) its instrumentalism. Instead, my critique will work along the dualist lines of the materialism and individualism that are prevalent in the sustainability literature to show that even in critiques that address some of the more straightforward dualisms I first list, subtler forms of it still threaten to confound achievement of sustainability's admirable goals of relational mutualism and harmony. In the process I will also consider posthumanist perspectives on sustainability in order to point out that their prescription for addressing its inherent anthropocentrism is insufficient.

5.4.1 Human "Nature"

Since it is my contention that the possibilities for human-nature relationships depend to a great extent upon what definitions are held regarding the nature of humans and more-than-human beings (i.e., ontology), I note that one of the main impediments to sustainability's mutually relational goals is its conception of human nature. Sustainability, as it's generally prescribed, depends upon two main types. The first is of human beings as universally self-interested and fundamentally consumptive. The second expands such notions to admit the possibility of altruistic or reciprocal inclinations, but usually does so either by characterizing these inclinations as a function of education or behavior modification (and thus still not of human *nature*), as ultimately sourced in anthropocentric ends (and thus not really altruism), or by failing to see them *as functions of* the human-nature relationship within which they are embedded.

The first type of human nature definition is exemplified by Rees (2010), who attributes the anthropocentrism of modern humans to genetics and seeks a cultural solution to activate sustainability. He says that for humans, "natural selection favors those individuals who are most adept at satisfying their short-term selfish needs… Humanity's well-known tendency to discount the future…has almost certainly evolved by natural selection" (p. 16). He also suggests that humans are consumptive by nature, as in his example of the "gold rush" mentality of nations trying to access petroleum reserves as arctic ice retreats due to climate change. Of this exploitive behavior, he notes that it is "the typical human response to any resource trove" (p. 16). The sourcing of selfish and consumptive behavior in unavoidable human genetics and the natural selection that has supposedly produced them falls apart, however, when one acknowledges that it is modern Western societies and their ancient Greek predecessor that are anomalous in their human-nature relationships. It is this particular strain of societies, what I refer to in this volume as simply modern societies, that are anthropocentric and consumptive in ways that other societies are not and have not been (Ingold, 2000; Washington, 2013). Given this, and the fact that evolution and its mechanism of natural selection operate on much longer time scales, one would be more accurate if one were to theorize that the selfish, consumptive orientation of modern societies is not a product of evolution, but instead is a break from it, and the genetics that should be predisposing us to enter into and maintain the mutually beneficial and reciprocal human-nature relationships that are the end-goals of sustainability.

Another example of the assumption of a selfish and consumptive human nature is seen in Goodland (1995), who says that "Protecting human life is the main reason *anthropocentric* [emphasis added] humans seek environmental sustainability" (p. 71). In the context of the author's discussion, he is not referring to *some* humans as anthropocentric, or as all humans as sometimes anthropocentric, but is instead suggesting that all humans are, by nature and fundamentally, anthropocentric. He also suggests that the fundamental role of more-than-human beings in relation to humans is as a source of consumption or a receptacle for waste. Through a relational lens, I believe this portrayal to be quite inaccurate. One contrary example can be found in Snyder (1990). When speaking about the old European system of a commons, he makes the point that it was at once a physical area and also a "community institution that…defines…rights and *obligations* [emphasis added]" (p. 30). In his account, there is a sense of the reciprocal where one not only uses the land, but also must do her part in its upkeep and maintenance.

Moving back to defining human nature as fundamentally consumptive, one finds another example in Johnson (1993). He says, "The notion of sustainability…seems to offer a middle ground between choosing extremes: ill-informed, inefficient, short-term, and unsustainable consumption on the one hand and no consumption at all (sustainable for the environment perhaps but not for humans) on the other" (p. 11). What he seems to be suggesting is that the only possibility for human relationship with more-than-human beings is one of consumption (without it, the author believes we would not be able to sustain ourselves). But the root of the word "consume" is the Latin, *consŭmere*, which means "to destroy, wear away, exhaust, devour, to use up, to swallow up" ("Consume", 2018). In that sense to *consume* generally means to use in a way that fails to then restore—that is, it is neither mutual nor reciprocal. One need only look at the practice of restoration ecology (Falk, Palmer & Zedler, 2006) to note that there are non-consumptive orientations for humans to adopt in their relationships with more-than-human beings.

If one were to practice low-impact farming, hunting and other food gathering for a group of people, and then continually invest effort into replenishing or otherwise promoting the well-being of the land and its more-than-human inhabitants, then it's possible to actually achieve a net gain of promoting overall more-than-human well-being. In other words, "taking" is unavoidable, whereas consuming—with its unidirectional, non-reciprocal connotation—is not. What's more, if "taking" is inextricably linked with subsequent restoration and/or promotion of more-than-human well-being, then it will become only one element of the healthy, whole functioning of place, and of humans within it.

Another example from sustainability where human self-interest is the norm is in Carpenter (1998), who says "The problem of sustainability is...human nature. We are all aware that humans willingly pay almost any magnitude of long term, uncertain cost for the most trivial immediate certain gain to their personal well-being" (p. 291). Jepson (2004) also notes this view of humans as having short-term, self-centered thinking when he notes in humans an "inherent disinclination to extend our sphere of concern [beyond ourselves]" (p. 4) and a tendency to "let our preferences be the guide to our decisions rather than the facts" (p. 4). All the traits detailed so far can be conceptually condensed, I believe, to a portrait of human tendency toward being self-centered and, if one reduces these descriptions to their unvarnished core, lazy both physically and intellectually. Granted, one cannot deny that these elements are strong in modern societies, but this is not the equivalent of offering robust evidence for the view that it is "human nature." The latter notion positions them as biological imperative, and as such, implies a much more limited ability to change them.

It cannot be underestimated what effect these self-centered definitions of human nature have on prescriptions for sustainability. If humans can only consume, then the mutuality suggested by sustainability theorists is unobtainable. Mutuality requires reciprocity. By definition, it is both taking *and* giving. But, if it's only within the human capacity to take, and the range of possibility of action is along the spectrum of too much taking and very little taking—then mutualism is simply not possible. That, or the "taking less" in which humans might choose to engage has to be defined, rather awkwardly, as our own version of "giving." Not quite so charitable, when viewed in this way. Based on this ontological conception of humans, the prescription for sustainability has to be targeted at limiting, by whatever external means necessary (laws, population restrictions, etc.), human influence. Goodland and Daly (1996) suggest exactly this when they say that "[r]educing impacts of human activities upon the environment can be achieved only by...(1) limiting population growth; or (2) limiting affluence; or (3) improving technology" (p. 1011). In their view, humans themselves can't actually change. They are taken as a constant— a constant that can be counted on to impact the environment in certain uniform, measurable and unsustainable ways when gathered at a certain scale. Further, by saying that these three ways are the "only" ways to reduce human impacts is to suggest that human individuals possess neither the innate capacity, nor the will, foresight or intelligence to undertake changes to their own behavior.

Such a view of human nature hampers efforts of environmental education to promote sustainability as well. If it's human nature to act selfishly, then it will always be an "uphill climb" to get humans to act altruistically or relationally.

This explains the tendency in at least U.S. environmental education circles to try to modify human *behavior* (Noel Gough, personal communication, May 11, 2009) instead of calling for a deeper exploration of the human-nature relationship as a means to recognizing different human *traits* available for those relationships. It's also why, in the process of relationally analyzing the sustainability literature, one tends to find anthropocentric calls for human sustainability juxtaposed, without any apparent recognition of the logical tension in doing so, with calls for mutual or harmonious human-nature relationships. If human nature is anthropocentric after all, then the outer limit of defining mutual relations *would* resemble a reduction in the consumptive, human half of the reciprocal human-nature relationship that Johnson (1993) or Goodland and Daly (1996) recommend.

Other authors do attempt to move beyond such a constricted view of human nature, but in doing so they still leave intact the implicit notion that humans are only self-centered by nature. Wheeler (2013), for example, suggests that "human potential is shaped by the surrounding social and cultural environment…[such that it can] counter pessimistic views of human nature as warlike and competitive" (p. 21). In acknowledging social and cultural influences, he says that a change in human potential is to be achieved by "evolv[ing] towards more conscious, compassionate and sustainable modes of existence" (p. 21). By offering evolution as the mechanism for that change, however, he is saying that it *is* human nature to be warlike and competitive (i.e., self-interested) currently, and it is only through the "nurture" of a better way that humans can change their behaviors and become more sustainable in the future.

Given that the economics disciplines play a substantial role in sustainability theories, one would expect their traditional definitions of human nature to have some influence here as well. Siebenhüner (2000) notes this when he says that in economics there has been a predominant "assumption of a rational, self-interested, and utility-maximizing individual" (p. 17). He disagrees with such characterizations, and explores the research into human nature that offers an alternative view. As it applies to the human-nature relationship, he notes that

> many animals arouse our special sympathy, that landscapes full of variety appear to us spontaneously as beautiful, and that flowers usually make us feel better…it seems to be a universal human trait to feel some kind of happiness in intact natural scenery. These feelings must also have proven to be helpful for the survival of human beings. (pp. 19-20)

So, here it is not just human potential, but basic human nature that is cast as less self-interested and more relational. But, while Siebenhüner's efforts make a positive contribution by trying to move away from purely self-interested conceptions of human nature, through a relational lens he still ultimately portrays these traits as emerging unilaterally from the human, instead of as a product of the human-nature relationship. For example, he says that "Sustainers inevitably need a certain emotional relationship *towards* nature which leads to caring and respectful action… [and] a kind of permanent and positive emotional affectation" (p. 19). As my examination of place and connection theories earlier in this chapter suggests, here again the human is relationally isolated from more-than-human beings, and the reciprocal dynamics of common relational terms are reduced to the unilateral "relationships *toward*" instead of "relationships *with*."

Even when the author suggests that these traits might be the *product* of the human-nature relationship instead of its determiner, as when he says that flowers make us feel better, his portrayal of the relationship is both anthropocentrically instrumentalized and materialist. He shows this by suggesting that the source of these human responses is that "all humans share needs for a natural environment which allows them to survive properly" (p. 20). Here then, what inspires the more-than-material emotional response of the human is an interchange whose ends are the material survival of humanity, not an emotional or relational response as an end in itself—as an ontological origin place. In addition, those emotions are, ultimately and anthropocentrically, buttonhooked back to human welfare. The flowers make *us* feel better.

Caldwell (1998) also does a nice job of acknowledging external effects on human behavior when he says that it "may be not so much basic human nature that must be changed…[but instead] those cultural circumstances that shape the way in which human behavior finds expression" (p. 7). As with Siebenhüner (2000), Caldwell (1998) doesn't extend the influence over human behavior to more-than-human beings since the external influence he cites is "cultural." Also as with Siebenhüner, he seems to suggest that basic human orientations are still self-centered, and that what needs to be changed is the societal constraint on that orientation. He says that for "development to be sustainable, ways must be found to direct perceived self-interest…away from unsustainable short-term interest" (p. 11). Ultimately, whether he means that this perceived self-interest is part of human nature, or that it is a perception that is potentially erroneous he does not specify. Regardless, his suggestion that society ought to manipulate this perception instead of replace its conception with a less self-centered one leaves open the idea that a self-interested orientation is still the "baseline" of human nature with which we must work.

In light of research that suggests that altruism is as innate a human tendency as self-interest (Fehr & Fischbacher, 2003; Singer, Seymour, O'Doherty, Stephan, Dolan, & Frith, 2006; Warneken & Tomasello, 2006), one is left to wonder whether the descriptions of an anthropocentric "altruism" like those to which I refer in Siebenhüner, are the only valid interpretations. Couldn't an alternative explanation position altruism as an evolutionary adaptation not for self-interested survival, but for maintaining others and through that, perhaps, the whole of nature? Kropotkin (1902) certainly lays the groundwork for such a notion when he suggests that, at least amongst the same species, there is not a "bitter struggle for the means of existence" (p. vii) but instead "Mutual Aid and Mutual Support carried on to an extent which made me suspect in it a feature of the greatest importance for the maintenance of life, the preservation of each species, and its further evolution" (p. ix). Therefore, the notion of altruism as mutual aid or cooperation for the furthering of a group, community, or potentially the wider world certainly seems plausible from this more ontologically relational perspective.

From such an orientation, the innate human capacity for self-interest could be conceived of as the means of perpetuating self and species long enough for it to make a contribution to the well-being of the whole. At a material level, this contribution could be to an ecosystem or the biosphere. At a material *and* more-than-material level together, it could be the contribution to an emotionally, intuitively *and* materially interconnected community of humans and more-than-human beings. For example, Jacoby (2001

), in describing the perspectives of the early European settlers that lived in what became the Adirondack Park in upstate New York, he notes the heartbreak that many felt when interloper hunters from the city shot the buck that they had all, as a community, refused to hunt "out of respect for his great size, endurance, and beauty" (p. 60). Of the death of this buck at the hands of the alien hunters, a local poet wrote his lament, We miss Brave Golden from his herd, we miss him from his home,

> We miss him from each grove and glen through which the king did roam;
>
> Our hounds will never strike his track to make the valley ring;
>
> The stranger's cruel, deadly shot laid low our noble king. (p. 60)

This kind of human-nature relationship is a far cry from that which predominates in the sustainability literature—that of humans evolutionarily crawling out of a brute, self-interested past toward a more relational future. And if sustainability's view is misplaced such that altruism really is an *earlier* evolutionary mechanism for maintenance of mutual or harmonious relations, then conceptions of self-interested humans "flip" to become part of a *later* development. Perhaps the view of human nature as self-interested is the product of the dualist worldviews that have only risen to prominence relatively recently—in the last three or four centuries and concomitant with the Scientific Revolution and the Enlightenment. In such a scenario, these worldviews move us ever further away from how we've conceived of ourselves in relationship with more-than-human beings in the more distant past: as fundamentally reciprocal and harmonious.

To portray humans as having had a precedent relational orientation is supported by Merchant (1980), who notes that pre-Enlightenment/Scientific Revolution views of nature were of nature as a living organism of which humanity was only a part. One example that reinforces Merchant's view of this progression is that of the "enclosure movements" that took place in England from the 15th through the 19th centuries. They were, in essence, the privatization of publicly held lands. Of the damage these enclosures did to those for whom the commons meant a way to live sustainably and in mutual relation with more-than-human beings, Boyle (2002) says that it was a "loss of a form of life...disrupting traditional social relationships, views of the self, and even the relationship of human beings to the environment" (p. 14).

Ultimately, a view of human nature as innately relational and altruistic would radically alter the foundation upon which sustainability theories have largely built their prescriptions. Instead of sustainability being aimed at promoting an evolutionary step requiring "only" a quantitative adjustment to our innately selfish human behavior, it becomes a reclamation project. The thing being reclaimed is a relational definition of human nature and with it, the more-than-material conditions of more-than-human beings with which relational humans might once more intuitively and reciprocally interact. Or, if one takes such a wholesale replacement of self-interest with altruism to be too extreme, at the very least the project becomes a clearing away of the misconception that humanity is *solely* self-interested. In the process, the innate human qualities and capacities that are foundationally altruistic and, thus, relational, can once more be accentuated as we work to improve human-nature relationships.

5.4.2 Loss of More-than-Human Identity and Individuality

In exploring the effect of dualisms on the structure and dynamics of the improved human-nature relationships for which sustainability theories advocate, another element which emerges is the dearth of individuation that occurs when describing the more-than-human world. While posthumanism rightly seeks to reject ontologically atomistic constructions of individuals, as I discussed in Chap. 3, one of the values of maintaining individuals as individuals-in-relation is that it makes room for what we come to know as a Self. We modern humans acknowledge these Selves in our fellow humans without really even thinking about it most of the time. We wake up day after day, expecting the spouse lying next to us to be as he or she was the day before. And it's true that who we take our spouse to be from one day to the next is pretty much what he or she ends up being. In the sustainability literature, however, the recognition of Selves in this way (or at least their importance to sustainability) is lost when discussing more-than-human beings.

I trace this loss to the failure to acknowledge the thoughts, feelings and subsequent actions of more-than-human beings in the human-nature relational contexts of sustainability, which itself is traceable to dualist modern ontological commitments that are pre-loaded with the assumption that such qualities and capacities do not exist. Thus, anthropocentrism doesn't cause this mischaracterization of more-than-human Selves, it is an expected effect of it. If there is no relational partner, then what more concern need one have about not treating more-than-human beings as living, individual Selves? Not even the lively matter of posthumanism will free us from this flattening. We need (and *have*) *living* Selves, not "lively" matter."

One way that this loss of individuation expresses itself in the sustainability literature is that more-than-human beings are conceived of not as living individuals or groups of individuals, but instead collectively (and often abstractly), as processes or systems. Terms such as "life-support systems" and "ecosystem services" predominate in the literature—their very coherence bound up in human-derived benefits and in a lack of the mutuality or harmony that sustainability attempts to encourage.

Wu (2013), for example, refers to more-than-human beings as "natural capital" when saying,

> Weak sustainability permits mutual substitutability between natural capital (e.g., ecosystems and mineral wealth) and human-made or manufactured capital (e.g., factories and urban infrastructure) to the extent that a system is considered sustainable as long as its total capital increases or remains the same. (p. 1003)

That more-than-human beings are seamlessly swapped with factories or other man-ufactured goods shows the extent of the anthropocentrism at work in such theories. This is accentuated by the fact that Wu, in referring to Daly (1995), notes that it should be considered "absurd" to take the position that "no species could ever go extinct, nor any nonrenewable resource should ever be taken from the ground, no matter how many people are starving" (p. 1003). The only reason it is considered absurd, however, is that the sponsoring, unexamined ontology permits elevation of human importance over the more-than-human world. This is far easier to accom-plish when the more-than-human world is stripped of its Selves and instead charac-terized in utterly instrumental and collectivized terms such as "capital."

Costanza and Daly (1992), being ecological economists, undertake a similar description when they characterize more-than-human beings and the results of their lives' activities as "a stock or population of trees or fish [that] provides a *flow* [emphasis added] or annual yield of new trees or fish, a flow that can be sustainable year after year" (p. 38). Here one encounters the instrumentalizing effects of eco-logical economic lenses. The identity of trees and fish are in their role as stocks and flows that contribute to human stores of natural income and capital. Any mutuality that accrues to trees or fish resides in promoting this contribution of theirs to human endeavors. This is borne out in the authors' suggestion that "[e]cosystems are renewable natural capital. They can be harvested to yield ecosystem goods (such as wood)…[or to] yield a flow of ecosystem services when left in place" (p. 38). Others provide similar, instrumentalizing characterizations, such as Ayres, van den Bergh and Gowdy's (1998) description of "market-priced extractive resources such as for-est products, fish or minerals" (p. 6), Cairns' (2003) characterization of "nature's resource reserves" (p. 46), and Kane's (1999) suggestion that it's even possible to explore how "new technologies can resolve problems of ecological sustainability by replacing certain ecosystem functions while permitting continuing growth of the gross national product" (p. 17). Especially in Kane's perspective, ecosystem func-tions are abstract ends of a supply chain with little or no connection to the plants, animals, insects, microbes, and other more-than-human beings that provide the "services."

The same problem can be seen with the widespread use of the term "life-support systems." Goodland (1995) talks of "emphasizing environmental life-support sys-tems" such as "atmosphere, water, and soil" (p. 69). Of these systems, he says they all "need to be healthy, meaning that their environmental service capacity must be maintained" (p. 69). Here Goodland equates the health of more-than-human beings with their ability to provide humans the "goods" and "services" humans want, and seems ignorant of the possibility that the health of a more-than-human being may require that humans no longer ask it to provide goods or services of any kind. Thus, here the more-than-human individuals that contribute services have had the very definitions of their well-being instrumentally and anthropocentrically co-opted.

Beyond these terms, Goodland refers to the "raw material inputs" (p. 71) or "assets (such as soil, atmosphere, forests, water, wetlands), which provide a flow of useful goods or services" (p. 73). Barrett and Grizzle (1999) refer to the more-than-human contribution as "forests, soils, water, [and] wildlife" (p. 30), with only the last referring to actual nonhuman animals, the rest being abstracted, collectivized entities or "abiotic" elements where the individual more-than-human contribution to them goes unexplored. Goodland and Daly (1996) refer to natural capital as the "stock of environmentally provided assets (such as soil and its microbes and fauna... atmosphere, forests, water, [and] wet- lands) that provides a flow of useful goods or services" (p. 1005). Here, while microbes and fauna are mentioned, they only garner mention by way of their "useful" contributions. Hinterberger, Luks and Schmidt-Bleek (1997) speak of "material flows" (p. 4) of "ores, coal, water, soil, timber etc." (p. 8), where again the more-than-human contribution is obscured, and the effects of the use of such material flows on the mutuality or harmony of human-nature relationships is absent. Gladwin, Kennelly and Krause (1995), while discussing human dependence on ecosystem service "transactions" (p. 875), speaks of the human disassociation from the "ultimate sources of life—the sun, photosynthesis, biodiversity, food chains, and biogeochemical and nutrient cycles" (p. 875). In this context, Gladwin, Kennelly and Krause take the ultimate sources of human life to not be the individual members of the food chain such as cows, deer or birds, or the detritivores that make the nutrient cycles possible, but instead the abstracted processes in which they are involved. In this sense, individual more-than-human contributions are obviated, or seen as secondary to the processes which we consider central, centrality being equivalent to the processes from which we would like to continue to benefit. In a systems thinking-based approach, this focus on process might be expected. But, even in those approaches, the contributors to those system processes are usually positioned as being of equal importance to the flows between them. In general, more-than-human beings in this context are represented not as individuals, but as invisible contributors to processes abstracted from the workings of their lives and reduced to equivalence with the role they play in human well-being or survival. This is thoroughgoingly anthropocentric.

To help underscore the loss of individuality to which I am referring, I offer an anecdote from my own experience reading about a wharf fire in Boston years ago, when a lobster company's entire warehouse full of live lobsters was lost in a fire. When discussing the damage, the story spoke of "the loss of 60,000 pounds of lobster" (Associated Press, 2008). It struck me that in the account, by referring to the weight of the animals and using the word "lobster" singularly, that their individual lives had been doubly lost: once to fire, and a second time in their utility to human ends. If, in contrast, the article had stated that what was lost was "60,000 pounds of lobsters" their individual lives would've begun to emerge, and could've been pushed even further if the loss was stated as "30,000 lobsters." In being referenced only with regard to their commodity value to humans, the lobsters had been denied their individuality, and in the process had been dualistically negated as relational Selves.

In contrast I offer a statement from Oakhurst Dairy president, Stan Bennett (Oakhurst Dairy News, 2008), in response to the company's rejection of the use of Bovine Growth Hormone with their milk cows. He says that

> [q]uestions and concerns from consumers over the use of artificial growth hormone, decisions in Europe and Canada to ban the substance *and our concerns for the health of cows that willingly give us their milk every day* [emphasis added] all contributed to our decision to keep artificial growth hormone out of the production of our milk. (para. 2)

Here, the ecosystem good of "milk" or service of "milk production" isn't abstracted from the animals that give milk. The cows here are not only acknowledged, but their well-being mutualistically considered in relation to their contribution to human welfare. Further, because their giving of milk is seen as a reason for subsequently treating them in a certain way, in this sense the relationship has become a guiding constraint for human interaction.

One can see what difference an acknowledged and well-defined more-than-human relational partner makes when trying to define and promote a mutual or harmonious human-nature relationship for sustainability purposes. When those partners are obscured through abstraction and instrumentalization, relationships lose meaning as relationships. Either that, or the only constraining factor in relating with the sources of those functions and services is their ability to contribute to anthropocentrically defined processes and, in the context of sustainability, to do so for an indefinite period of time. Because of that, then, human-nature relationships in such portraits of sustainability are wholly anthropocentric, and thus by definition, can be neither harmonious nor mutualistic except as unintended side-effect.

I attribute much of this highly anthropocentric orientation toward more-than-human beings to the sources of much of that literature, the environmental economics and environmental management disciplines. Environmental and ecological economics are right in the novel attempt to attribute value to what more-than-human beings provide, but by forcing those contributions into highly anthropocentric contexts, they limit the human's ability to qualitatively change the relationship to the point where humans can reciprocate with more-than-human beings and meet sustainability's broader, more egalitarian, goals.

For example, in the context of wild wolf management in Canada, Jickling (2002) characterizes the tension between environmental management and sustainability as the fact that "managers need to manage" (p. 148). What he means is that it is the perspective of environmental managers to see every human-nature relationship as a management problem. Therefore, not only the relationship, but in Jickling's example, the wolves themselves, are seen through the lens of needing to be manipulated for human ends—to be *managed*. As such, once again anthropocentric orientations reduce human-nature relationships to something less than mutual or harmonious exchanges. If more-than-human beings are seen as having only this reduced value and role, then it's possible to understand how a mutuality or harmony could, at times, emerge in the reduced state it has from these perspectives. Unfortunately, the ability of such reduced concepts to help sustainability meet its relational goals is severely limited.

5.4.3 What Ought to Be Sustained?

Another area where the dualisms inherent in the sustainability literature come to bear is in sustainability's myriad attempts to answer the question: "What ought to be sustained?" For many authors, this is the foundational question for the development of sustainability theory and practice, with Jenkins (2009) suggesting that it is sustainability's "new kind of moral question" (p. 380) while Wilkinson (1990) says that answering it should be "the first objective in approaching sustainability" (p. 18).

In the context of the role scientists play in advancing sustainability, Lélé and Norgaard (1996) attempt an answer to this question by saying that what ought to be sustained depends largely on "what kind of future is desired" (p. 355). For these authors, because this desire is human desire, it is rooted in a "quicksand of [human] worldviews and values" (p. 355). To ensure that this desire does not devolve into mere whim, they suggest formulating desires based on "social values" (p. 355) with consideration of "social process[es] and...tradeoffs against other social goals" (p. 355). Noss (1993) also focuses on human desire and values when suggesting that the answer to the question is "essentially an issue of [human] goal setting...[with] more attention [paid] to where we wish to head...and...what values are behind those objectives" (p. 19). Wilkinson (1990), too, echoes this emphasis on human values and desires by warning that if humans are too narrow-minded, some things that we might want to sustain won't be sustained. For example, he points out that a single focus on timber extraction, which itself can be a sustainable practice, could come at the expense of other things such as the "health of certain fish and wildlife populations. Soils on steep slopes. The recreation economy. Old-growth ecosystems. Views. Beauty. Majesty. Wonder" (p. 19). Finally, Gale and Cordray (1994) center the answer to this question on values when they note that whatever a "particular group has decided should be sustained...typically reflects some value (economic, biological, aesthetic, cultural, historical)" (p. 313).

While the authors in the previous paragraph focus on the values that underpin desire, others bypass those values entirely, or take them to be self-evident in the desires upon which they focus. For example, Hueting and Reijnders (2004) suggest that in order to determine what should be sustained, the "focus should be on the processes that underlie the persistence of life support systems" (p. 257). Above, I critiqued the tendency of sustainability theorists to dualistically conflate more-than-human beings with the ecological processes to which they contribute, and here one finds a strong expression of that. Thus here, the values are those of economics, resource stocks and life support systems from which humans want to continue to benefit indefinitely. In Solow's (1993) watershed ecological economics essay, he addresses what ought to be sustained more broadly when saying that sustainability is an "injunction not to satisfy ourselves by impoverishing our [human] successors" (p. 181). There is no mention of what values ought to underpin our self-satisfaction, though. Therefore one might assume that a somewhat hedonistic or superficial orientation underlies this prescription. In it, what is satisfying is then *equivalent* to what is valuable. Kane's (1999) sustainability definition echoes this orientation when he suggests that sustainability "hinges on underlying assumptions of what is important for the continued...happiness of the human species" (p. 17). Whether Kane's "happiness" is a "lower" one rooted in utilitarian John Stuart Mill's (1879/2002) notion of "a beast's pleasures [that] do not satisfy a human being's conceptions of happiness" (p. 240) or if he means the kind of "higher" happiness that produces Mill's "greatest good" (p. 158) is unclear. But, that Kane feels no need to qualify it leaves open the possibility of equating happiness with the "lower" form that is pleasure. Solow (1993) reinforces the notion of a lower, or simple human satisfaction over a higher, more enlightened desire when suggesting that the trouble we face in defining sustainability lies mainly in not knowing "the tastes...[and] preferences...of future generations [of humans]" (p. 181).

Whether superficial or value-driven, human desire as the arbiter of what ought to be sustained introduce several problems. First, there is an assumption in such perspectives that the only source for the answer to the question: "What ought to be sustained?" *is* human desire, and either individual human values or collective social ones. Second, this assumption is rooted in the belief that individuals or groups of humans *can* source their values outside of their human-nature relationships. Such a belief itself is sponsored by two dualistically wielded elements. There is a substantivist ontology that positions the human as antecedent to any relationships with more-than-human beings that she develops (see Chap. 6 for a more extended exploration of substantivism). Then there is the portrayal of the more-than-human being as making no substantive contribution to human desire or values largely because her contribution has been dualistically reduced to the material, if it is considered at all.

In addition to these problematic elements, one must also note that letting human desire and values operate in a vacuum to determine what ought to be sustained is the precise ontological framework within which *unsustainable* modern human-nature relationships have arisen in the past. Thus, what one finds in such approaches is precisely the "quicksand" from which Lélé and Norgaard (1996) hope to extricate us. But, if I want to clearcut the Tongass National Forest in Southeast Alaska, one hundred acres per sale, and you want the material biodiversity of that forest to be preserved, whose values decide? Lélé and Norgaard suggest the "force" of some socially arrived-at values. But if the society also values material wealth and, in deeper and perhaps less discussed ways, power over nature, then these social values will lead to unsustainability. In other words, at the end of this path the quicksand awaits regardless of the values. But this is only the case if one allows oneself to remain *within their ontological frame of reference*. If our last recourse to determining what ought to be sustained *can* only be human values because more-than-human beings and the human-nature relationship have been conceptually reduced to the point of failing to have the power or standing to influence determinations of what ought to be sustained, then what avenue does one have left in determining what ought to be sustained other than what humans want to sustain by whim or deeply held values?

If, on the other hand, one's ontology is reoriented and more-than-human beings become relational Selves making material and more-than-material contributions to the human-nature relationship, then not only does another avenue present itself, that avenue is unavoidable. No longer can the mutualism and harmony of human-nature relationship be reduced to human desire and values. Granted, some might suggest that mutualism or reciprocity are not themselves included in all areas of sustainability theory and so should not by necessity be included in the process of determining what ought to be sustained. The group of sustainability theories fitted under the rubric of "sustainable development" (Hopwood, Mellor & O'Brien, 2005; Lélé, 1991) is one set of theories that come to mind. Theirs is a guiding goal of development in the modern human sense of economics, poverty extinguishment, and other humanistic orientations (Hopwood et al., 2005). But if what is meant by sustainability *in general* is sustainable development's attempt to keep humans doing whatever they want in perpetuity, then I'd suggest that sustainability as a concept differs in no substantive way from the more environmentally rapacious societal paradigms (e.g., capitalism) it is meant, in part, to supplant. It's just a supposedly more benign form of it. If sustainability is just human desire exercised in perpetuity, then sustainability loses most of its meaning as a distinguishable concept, and as the distinctly *environmental* concept it has ineradicably become.

If it *is* environmental, on the other hand, then the mutualism and reciprocity of human-nature relationship inevitably take center stage, where what ought to be sustained is not determined by human values and desire, but by the values and desires of both humans and more-than-human beings, themselves constrained and determined by a primary and precedent, mutual and harmonious human-nature relationship. When expunged of dualistic reductions, this means that the well-being of human and more-than-human making contributions at material and more-than-material levels must be the end goal.

This is all a circuitous way of saying that if by sustainability we seek to constrain human activity so that it is *environmentally* sustainable, first and foremost we must submit the activity of choosing what ought to be sustained to just this constraint. Freeing human desire and values from such limitations, and only afterward attempting to "retrofit" them so that they are sustainable, is a paradigmatic case of putting the cart before the horse. Lélé and Norgaard (1996) remain stuck in this place when they say that social values "*determine* [emphasis added] whether the objective to be kept undiminished [i.e., what is be sustained] should be human material wealth, human spiritual well-being, or the well-being of all living beings" (p. 355). Here, social values beget potentially more harmonious and mutual human-nature relations instead of the inverse: human-nature relationships determining humans' social values.

Granted, Lélé and Norgaard's suggested reliance on social values may do some good if those values are in keeping with the relational focus of sustainability, but without explicitly imposing an external-to-human relational constraint on those values, it affords us no guarantee. To see just how directionless sustainability can become if human considerations are employed outside a human-nature relational context, I offer the words of ecological economists Hinterberger et al. (1997). When answering the question of what ought to be sustained, they say that it is "human society that should be sustained" (p. 11). While ensuring the survival of humanity certainly ought to be one of the hoped-for outcomes, I note here that such a statement is almost completely empty. First, "human societies" are far from uniform, especially in their relationships with more-than-human beings, thus the degree to which their practices are, and can be, sustainable, varies quite widely. Second, certain human societies—namely modern ones—are centrally implicated in unsustainable societal mechanisms, structures, and practices. Therefore, the authors cannot mean that *all* human societies (and thus, all their practices) ought to be sustained without qualification. In such a statement, one quickly arrives back at having no guiding context by which to select which societal values, desires or activities ought to be sustained.

Wilkinson (1990) and Noss (1993) both articulate a set of guiding social values that are somewhat closer to what would result if the human-nature relational context I suggest above were employed. Noss, when considering forestry practices, rightly suggests that sustainability thus far has been "utterly anthropocentric…[and] in need of significant revision" (p. 17). But, he goes on to say that even "Ecological arguments can suggest many different things, depending on the goal" (p. 20). For example, of our ability to "design…a perfectly functioning tree farm [that is]… quite diverse in native species and genotypes" he says that this is "not good enough [because]…it is too much like a machine" (p. 20). He also says that such a design may ignore other forest qualities that "are seldom considered in forestry or ecology but yet are important to us in immeasurable ways" (p. 20). These values can "include wildness and naturalness" (p. 20). Here, Noss is recognizing what Lélé and Norgaard (1996) have, that human values alone can result in vastly different sustainability prescriptions. And while wildness and naturalness are values that I believe are essential components of harmonious and mutual human-nature relationships, he still culls them from an anthropocentric root. That a tree farm is "too much like a machine" is a reflection of his values and nothing more. Wildness and naturalness are, without an explicit sourcing *in* the human-nature relationship, also something valued by humans. He argues for these sorts of things because they are "important *to us* in immeasurable ways." This is still anthropocentric and instrumentalist. Thus, while I see his values as likely *products* of harmonious and mutual human-nature relationships were those relationships to be defined first and offered as a constraining context, by pinpointing their source in human values alone, Noss' (1993) prescriptions never escape categorization as part of Lélé and Norgaard's (1996) "quicksand" of worldviews and values.

Wilkinson's (1990) discussion falls into a similar anthropocentric trap. As I mentioned above, he believes that some of the values to consider in sustainable forestry are "health of certain fish and wildlife populations. Soils on steep slopes. The recreation economy. Old-growth ecosystems. Views. Beauty. Majesty. Wonder" (p. 19). Like Noss (1993), the scope of what ought to be sustained is broad indeed, with values that seem likely to emerge from the context of harmonious and mutual human-nature relationships. Yet he, too, still positions these elements as human-only values, or at the least doesn't further qualify them as anything else. This leaves them vulnerable to interpretation as means for human ends. For example, Jenkins (2009) says that it is "human dignity [that] requires access to natural beauty" (p. 382), implying that Wilkinson's (1990) beauty could be taken as a means to human dignity. Thus, the interpretation of Wilkinson's values can still be anthropocentric and instrumental.

Ultimately, Wilkinson comes closer than most to a relational approach when he speaks of the need for humanist input into sustainability. Of writers in his home state of Montana and their contributions, he says they "have written of the places they know" (p. 20) and have produced a literature "rooted in place and in the *eternal truths that rise up from discrete places* [emphasis added]" (p. 20). Whether he knows that he is speaking of the fact that truth can rise up *from places* or human relationship with them is unclear, but his descriptions do carry the seeds of them doing just that in our conceptions of them. Unfortunately, he never explicitly reorients the broader set of human values so that they are definitively *produced by* the human-nature relationships for which such places are the context.

The importance of submitting human values and desires to constraint by something external to humans alone is perhaps best articulated by Houck (1998), though he does so in a different context. Regardless, from his approach a directly relevant parallel can be drawn. In discussing ecosystems-based land management policy and the possibility that humans should be considered part of the ecosystem in formulating initial management plans, he recommends against this. He says that "once humans and their impacts are put into the definition of an ecosystem, the term loses all objective meaning. We simply manage for whatever we want" (p. 2). His rationale is that if "Chicago…[or] Hiroshima, circa 1946" (p. 2) is considered a "natural" part of the ecosystem, then there is no way to discern a healthy ecosystem from an unhealthy one, and thus to manage toward one ecosystem state over another. To resolve this problem, he suggests splitting the notion of ecosystem management into "ecosystems" on one hand, and "management" on the other. The former, at the beginning, does not contain humans so that one can see just what a healthy ecosystem looks like (based on the assumption that without human interference, ecosystems are largely healthy). Once that's established, and there is a standard or goal, then human activities can be introduced into its constraining context.

I think a similar technique can and should be applied to the process of defining what ought to be sustained. Here, we must split the definition of mutual or harmonious human-nature relationships from isolated human values and desires that could very well corrupt those definitions into "whatever we want" as human desire for skyscrapers as part of their well-being butts up against a bird's desire for good nesting locations.

What is a mutual or harmonious human-nature relationship without unilateral human desire then? At its root, it must be a relationship where the well-being of both human and more-than-human in relationship are held as the central focus and irreducible goal. What differentiates this portrait of mutual relations from those that sustainability currently offers is that here, human action is constrained not by an imperative of human survival and what humans individualistically desire, but instead by the good functioning of human-nature relationships where both human and more-than-human well-being are promoted and maintained. Given my contention that well-being is not only material, and that more-than-material elements can be contributed by human and more-than-human alike, suddenly mutual and harmonious human-nature relationships take on characteristics of the close human-nature relationship for which this book makes the case. A good example of this appears in the "Materialism of Biocentric/Ecocentric Relational Approaches" subsection below in this section, where the Menominee tribe maintains both a material and more-than-material relationship with the more-than-human forest they "manage" and in doing so, promote the well-being of both their people and the more-than-human inhabitants of the forest. Thus, what ought to be sustained is the well-being of those within the confines of the well-being of the relationship.

Once these kinds of mutual and harmonious relationships are defined, as Houck suggests, only then should we bring human desire back for consideration of its contribution or detraction from determining what ought to be sustained. Of course engineering the process this way should appear quite constraining to human societies accustomed to thinking that their activities ought to be governed by their own values and desires outside of any human-nature relational context, or with only minimal recognition of the constraint of material "natural resources." In such a constraining context even if I *want* to build a shed in my backyard, I must consider many things. I must consider where and from whom the materials come, and whether this disrupts the material and more-than-material well-being of birds and other more-than-human beings in that location—or worse, destroys the mutual care that has developed between humans and more-than-human beings in that place. I must also consider whether my installation of the shed disrupts the well-being of those in my own yard. And so on. As one might suspect, such deliberations would substantially slow unsustainable human activity in modern societies, as is only proper. Steel mills, for example, become eminently problematic in such a context. But, I suggest it is just such a context that is necessary if human-nature relationships are to be mutual and reciprocal, and if sustainability is to be meaningfully achieved.

At this juncture, some might suggest that the task of defining a mutual and thus sustainable human-nature relationship is itself subject to the vagaries of human values. In response I suggest that this can only be the case if one believes such definitions to themselves not be the products of actual relations with more-than-human beings as relational Selves. At the very least, approaching the problem of sustainability as something that must be solved *in conjunction with* more-than-human beings forces the definition that emerges to actively consider and include more-than-human beings in ways that sustainability most often doesn't attempt to achieve. It also forces the definition of sustainability to emerge from something outside human imagination alone. Lastly, it moves context-less human desire out of its position as determiner, and into a position of being determined by the relationships that are the hoped-for products of those theorizing about sustainability.

When looked at in this way, the question of what ought to be sustained is ultimately moot, since in its very asking is carried an all-too-easy assumption that it is for humans alone to answer. In the end, I believe that the only basis for discovering what ought to be sustained that does not devolve into a "quicksand of worldviews and values" or "whatever we want" is the acknowledgement of the ontologically material and more-than-material relations with more-than-human beings that we have, with the work becoming the definition and establishment of what is required from humans in the harmonious and mutual relationships for which sustainability rightly calls. Only after we do this, and as a result see what *is* sustained, can we know that this is what ought to be sustained. In the end, our good relations will, as a natural byproduct, determine sustainability entirely for us.

5.4.4 Human Activity as Environmental Activity

Another area where the sustainability discussion suffers from its rooting in a self-interested concept of human nature and a dualistic reduction of more-than-human beings is in its frequent focus on what ought to constitute "human activity" in a sustainability context. To illustrate, I turn to environmental educator David Orr (1992), who in his book *Ecological Literacy: Education and the Transition to a Postmodern World* says that "all education is environmental education" (p. 90). What he means by this is that all that is learned in schools is actually learned in the environment and subsequently put into practice in the environment. Because of this, all education is unavoidably environmental—and what's more, this is ontologically so. I mention Orr's words in this context because I believe the same sort of perspective must be applied to the individualist notion of "human activity" in the sustainability literature. Adapted to my purposes, Orr's words become: "all human activity is environmental activity."

I recommend this as an ontological starting point for understanding human activity because, at present, this activity is often viewed as conceptually separable from its environmental context. For example, the three pillars of sustainability are economic, social and environmental (Blackburn, 2007), as depicted in Fig. 5.3. This equates roughly to what Wu (2013) refers to as the "triple bottom line" (p. 1002) view of sustainability. Ontologically speaking, this construction means that economic and social activities are sometimes taken to be intelligible in isolation from environmental considerations. At other times in the sustainability literature, however, economic and social activities are portrayed as being contained within an environmental context (see Fig. 5.4)—for example, in discussions of how to conduct economic activities without eroding natural resources and the ecosystem services those resources provide (Goodland, 1995, p. 70). This is what Wu (2013) refers to as "strong sustainability" (p. 2013) where "natural capital" is not seen as substitutable by "human-made capital" in such a system (p. 1003).

While these two portrayals represent two distinct ontological concepts, and thus are not compatible, since human activity is ontologically environmental (at the very

Fig. 5.3 "Three pillars"
view of sustainability

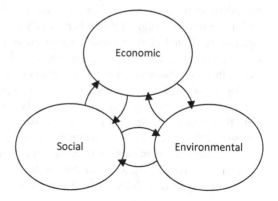

Fig. 5.4 Environment as
ontological context for
economic and social pillars

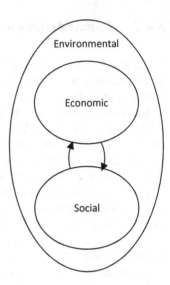

least in material terms) then the conceptual position represented in Fig. 5.3 is the
less accurate of the two. Furthermore, it is precisely the human ability to conceive
of the interplay of these three pillars as individualizable elements that contributes to
unsustainable practices in the first place. After all, if it's ontologically possible to
separate the social and economic from the environmental, then it's possible to
undertake activity that can be considered "human activity" while ignoring its impact
on the environment—without seeing it necessarily as environmental activity. Again,
I believe dualisms operating in the sustainability literature are largely responsible
for sponsoring the possibility of such obfuscation.

Pressing further, through a human-nature relational lens, I'd suggest that even
Fig. 5.4 is inaccurate. By positioning the environment—that is, more-than-human
beings—as *context* instead of as relational Selves with whom real material and
more-than-material interchange is taking place—those Selves are still dualistically
reduced. This is thoroughly anthropocentric, with human economic and social
activity cast as foreground against the backdrop of more-than-human beings.
Regardless of the focus of scholarship *on* the social or the economic (with its inherent human focus), I suggest that it is precisely the anthropocentric casting of ontologically relational elements into foreground and background that sponsors, and
serves as an ongoing reinforcement of, the very dualisms that work against sustainability. In a close human-nature relational ontological context, I offer Fig. 5.5 as one
ontological arrangement better suited to the goals of sustainability.

Examples of how human activity is currently portrayed as being in isolation from
more-than-human beings abound in the sustainability literature. For instance,
Johnson (1993) says, "To be sustainable is the goal of a growing list of human
activities that place demands on the environment" (p. 11). Of course the recognition
that human activities do place demands on the environment is laudable, but in making such a suggestion, Johnson is also implying its complement, that there is an

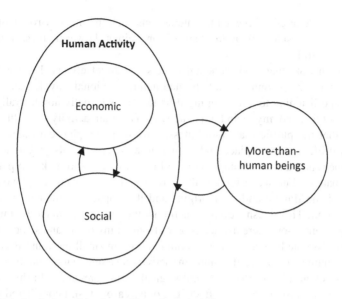

Fig. 5.5 Relational view of the three pillars of sustainability

albeit diminishing list of human activities that *do not* place demands on the environment. To suggest such a thing is to engage, at a material level at least, in self-deception. From breathing to clearcutting forests, human activities place demands on the environment. A relational ontology insists that the actions of one must, by necessity, be recognized as an action toward the other. In human-nature relationships, then, all human activity is also inescapably environmental activity.

In another example, Goodland (1995) says, "The environment has now become a major constraint on human progress" (p. 69). Here again, the recognition that the environment ought to factor into the measure of the wisdom of human activity is well placed, but the fact that the environment only becomes relevant or visible when its limits are reached, and otherwise doesn't garner consideration, is dualist. The presence and contribution of more-than-human beings to "human progress" is and has always been ontologically unavoidable. Further, it is the very capacity of humans to have conceptually eliminated more-than-human beings from their conceptions of "human progress" that seems central to the problem of unsustainability.

When more-than-human beings are considered as inseparable from human activity in the literature, they are often positioned as secondary or marginal. For example, Jepson (2004) says that "the level and character of human activity must be tempered by an appreciation of the effects of that activity on natural resources and characteristics" (p. 4). But by suggesting that the activity itself must only be "tempered" by "appreciation" means that the activity is not seen as a *centrally* environmental activity. It is only to be adjusted to be more aware of one of its "side" effects. But, at the very least materially, environmental effects can only *be* conceived of as side effects because humans have the benefit of knowing their intentions, and of arranging the "clutter" of the actual interactions such that the environment is conceptually to the "side." To more-than-human beings, however, these effects are direct and central.

To demonstrate that environmental effects are direct effects, I'll reinterpret a human activity in this ontologically human-nature relational way. If I am sitting at my desk and listening to music on my iPhone, that is actually not centrally what I am doing—this is not my main human activity. The main activity is me first largely ignoring that the plastic that contributes to my iPhone's physical reality is from crude oil drilled in some place and refined in another. It is also me ignoring the South American lakebeds from which the lithium is taken to make up part of my iPhone's battery. Further, it ignores the copper and other metals employed for the circuitry of the iPhone and their origins and the impacts of their extraction and transformation. This doesn't even broach how the music was made by musicians, their instruments—or where my desk comes from, or my room, and so on. The point is that this "simple human activity" is human only in small part, and is actually a relational (human and more-than-human) activity on a quite large and diverse set of scales. Its actual, direct effects are ontologically environmental. In this view, my listening to the music is the side effect. The only reason it isn't considered that way in a modern societal frame is because of anthropocentric tendencies coupled with a technological ingenuity that allows us to hide from ourselves the environmental activities that constitute the bulk of the activity. Sustainability has arisen in response to the reality that our hiding our environmental activities from ourselves is increasingly difficult and costly on many levels.

In another example of the literature treating more-than-human effects as side effects, Jenkins (2009) asks, "can human activity successfully maintain itself and its goals without exhausting the resources on which it depends" (p. 380)? This example underscores another feature of anthropocentric interpretations of human activity. Here, the effect of the activity on more-than-human beings is seen as being of *post hoc* consideration to the formulation of the activity itself. As I suggested above when discussing "What Ought to Be Sustained?", if the human-nature relationship is not held to be the constraining context and determiner of the formulation of human activity, then trying to retroactively fit that human activity within some relational consideration is far more difficult. As my reference to Houck (1998) in that discussion suggests, independent conceptions of human activity are not the best guides for sustainable environmental considerations. Instead, those considerations must be formulated within the bounds of what we consider to be good and mutual human-nature relationships. In this view, Jenkins' question might be better phrased as follows: *Can human actions be recast relationally so that both their origins and goals are understood to be emerging from, in relationship with, and directed toward, the lives and activities of more-than-human beings?*

Finally, while in this discussion I've largely explored the material interconnection that ought to constrain "human activity" in the sustainability literature, next I will discuss the weaknesses of such approaches in the segments of the sustainability literature that do try to acknowledge more-than-human beings as more substantive relational partners.

5.4.5 Materialism of Biocentric/Ecocentric Relational Approaches

Much has been made of a split between anthropocentric and ecocentric approaches to cultivating sustainability (Barrett & Grizzle, 1999; Bonnett, 2002; Corral-Verdugo, Carrus, Bonnes, Moser, & Sinha, 2008; Gladwin, Kennelly & Krause, 1995; Kopnina, 2013). When authors critique anthropocentrism's role in promoting a mutual human-nature relationship, most of the time the alternative offered is bio- or ecocentrism, since both of these orientations hold that value accrues to humans and more-than-human beings alike, or at least that neither takes precedence in the value accorded the systems to which both are taken to contribute essentially. Those supporting ecocentrism as an alternative to anthropocentrism hold that "moral decisions [must] take into account the good of ecological integrity for its own sake" (Jenkins, 2009, p. 382) and that the solution to "environmental problems [lies] in [developing] a working relationship with nature to resolve conflicts between society and nature" (Barr, 2003, p. 229). Since ecocentrism is still at its core ecological, and as such ontologically materialist, then it's possible that dualisms still operate in theories proposed by those offering it as a solution to anthropocentrism. In this discussion, I'll explore the ways in which ecocentrism in particular is relied upon as a guide for sustainability, yet because of its materialist roots, it still fails to alter the dualistically reduced view of more-than-human beings.

Shrivastava (1995) speaks of ecocentrism when discussing the development of an ecocentric management strategy. He recommends such a strategy "because nature is fundamental to all life, and certainly human welfare depends on it" (p. 127). While he is correct in his suggestion, his account begins by immediately instrumentalizing more-than-human nature. He is also correct in noting that a "natural ecosystem...is a network of connected interdependent organisms and their environments that give and take resources from each other to survive" (p. 127). But, when he undertakes to construct a model of human sustainable activity to mimic it, he fails to see that the interface of human and more-than-human must also contain this reciprocity. As an example he discusses industrial ecosystems, saying that they are a "network of organizations that jointly seek to minimize environmental degradation by using each other's waste and by-products and by sharing and minimizing the use of natural resources" (p. 128). So, while within the human industrial system he suggests there is ecosystem-like reciprocity, in fact he's bifurcated the human-nature ecosystem by isolating the human as an ecosystem unto itself and set the more-than-human world somewhere outside of it. There is a give and take amongst humans in the industrial ecology, but the relationship between humans and more-than-human beings is still only a unidirectional "use" of more-than-human beings to feed into this artificially isolated sub-ecology. In a broader view, what Shrivastava is representing is ultimately anthropocentric, and fails to be reciprocal in the context of the sustainability discussion. Thus, while in his model ecocentrism may "give" some standing to more-than-human beings and reduce the destructiveness of human activity, it does little to move the role that more-than-human beings are accorded in such definitions from the reduced, dualist positions they currently occupy in most modernist thinking.

Imran et al. (2014) recommend a more ecocentric approach to sustainable development. They attempt to promote an egalitarianism rooted first and foremost in ethics. For example, they say that an "ecocentric approach...gives intrinsic and moral values to nature in which all living beings, including humans, have needs for survival and well-being" (p. 137). In the context of my discussion, however, how they come to define "well-being" is of critical importance. If it attends to the happiness and closeness both within more-than-human relations and with human-nature relationships, then such an approach might have real promise. But, after making their statement about well-being, the authors go on to say that such an ecocentric approach "adopts a systems-based approach that views the universe as an interconnected and interdependent web of sub-systems" (p. 137). In the latter portion of the statement, one finds the more mechanistic, and less relational, language found in both material scientific understandings of life on Earth and the systems thinking approach well known in environmental conservation (e.g., Salafsky, Margoluis, Redford & Robinson, 2002), amongst other fields and disciplines. Throughout the article, the author seem to vacillate between two poles of thinking on human-nature relationships. On the one hand, some statements indicate that their prescription for sustainable development could accommodate the close human-nature relationships whose possibility I am arguing for in this book. For example, they say, "Sustainable development has to be reviewed as a holistic approach in the light of the ethical codes that attach moral values to both humans and non-humans" (Imran et al., 2014, p. 137). On the other hand, they seem to allow the materialism inherent in modern ecology to reduce more-than-human beings. For example, they say, "This new perspective with a strong ecological ethic could make a significant positive contribution to life and natural processes that guide life on Earth" (p. 139). Note the focus on "natural processes," which points to a dualistic reduction of more-than-humans that I discuss above in this section. In a way, the focus on material systems is betrayed by the ways in which sustainability is discussed even in the most general terms. Well-being is mentioned, but it either goes undefined, or is glossed over for more material foci. Is love a natural process? In one sense, of course it is, but is that one of the processes to which Imran et al. refer? It's unclear, but the focus on ecology suggests it's not likely. Finally, a more lengthy quote reveals what I believe to be a dualist problem. They say,

> The reinterpretation of sustainable development...must incorporate nature and the environment as subjects and not view these as objects, which it does in its present interpretation. This all-inclusive definition will be one in which the concept of present and future generations does not only point to fulfilling the needs of humans but the needs of all species without compromising not only the ability of future human generations but the future ability *of all the ecological and environmental processes* [emphasis added] to fulfill their own needs. (p. 140)

This quotation is essential in illustrating just how subtly, yet pervasively, some forms of dualism work in the sustainability literature. Imran et al.'s article speaks directly to equating human and more-than-human interests and moral standing.

Such an effort is to be applauded. But, even though elsewhere in the paper the authors speak to expanding the original Brundtland (1990) definition of sustainable development to include future generations of more-than-human beings along with those of humans, in the quote above this egalitarianism is missing, and is one of the two dissonances that indicate dualism at work. In this first dissonance, "human generations" are juxtaposed with abstracted "ecological and environmental processes." The latter is absent any recognizable individuals, at least with respect to any individual being distinguished from the processes to which he or she contributes. The second dissonance lies between these "ecological and environmental processes" and the notion that such processes have "needs." *Beings* are in possession of needs. Processes are the result of beings in possession of needs—they do not have their own needs except in the most abstract sense, like an engine "needing" oil. When authors such as these either overtly or unwittingly collapse more-than-human beings with the processes to which they contribute, they are either discarding such beings as relational Selves, or are downgrading the importance of their individuality relative to the processes to which they contribute. Instead of more-than-human generations, you have a biotic engine with some set of "needs." One could argue that this is what ecocentrism is, and would apply equally to human and more-than-human. But, I'd note that nowhere in the article do the authors conflate *humans* with any processes to which they contribute. If they did, I feel comfortable concluding that it would be discomfiting to most readers.

Why then, is referring to more-than-human beings in this manner acceptable? I contend that it is because of a presupposition that there is nothing substantially more that more-than-human beings can contribute to human-nature relationships within the context of sustainability. Thus, even though ecocentrism is held in a seemingly strong ethical manner by these authors, one cannot ultimately conclude that the more-than-human "well-being" to which they refer equates to anything more than the proper functioning of a "cog" in the material "machine" of ecosystems and their functioning. Such a view is ultimately insufficient for two reasons. First, it is undergirded by a dualistic assumption that this is all there is to the more-than-human contribution to human-nature relationships. Second, as I argued above, sustainability aspires to a reciprocity and harmony that I suggest requires more than mechanistic interconnection. I'll explore the latter point in more detail in the context of my discussion of the Menominee tribe's forest management practices just below.

Noss (1993) is another author who recommends a "biocentric or holistic concept of sustainability [that] focuses on sustaining natural ecosystems and all their components for their own sake, with human uses included only when they are entirely compatible with conservation of the native biota and natural processes" (p. 26). From a material perspective, this is a wonderfully strong portrayal of sustainability, leaving no room for humans to materially interact with more-than-human beings unsustainably. However, from my close relational perspective, it still has weaknesses that threaten true sustainability. First, it constricts human relationships with more-than-human beings to ones of "use," a unidirectional relational condition. Second, it implies that the interactions humans have with more-than-human beings are interactions with "biota and natural processes" and as such are predominantly material and impersonal. As I mention about the loss of individuality when describing the more-than-human world in the previous discussion about Human Activity as Environmental Activity, here again it occurs in Noss' "natural processes." While such a material view of more-than-human beings would explain why "use" is considered the only relationship possible with more-than-human beings, as I'll argue in later chapters, I believe that more is both possible and necessary for true sustainability.

Callicott and Mumford (1997) also offer a biocentric/ecocentric prescription. They add to Noss' argument by saying that since most conservation and preservation efforts are anthropocentric, a notion of "ecological sustainability" should be employed instead. As an example, they recommend that "a proposed economic venture…should be deemed unworthy of undertaking…if it will compromise the health of the…ecosystems on which it is imposed" (p. 36). For them, the solution is the measurement and maintenance of ecosystem integrity and health understood biologically/materially. As examples of the sort of ecological sustainability they espouse, they offer the forestry practiced on the Menominee Indian reservation in Wisconsin and *jhum* agricultural practices in northeastern India.

As with Noss (1993), theirs is a reciprocal and rigorous view of material sustainability, but from a relational perspective, it is also lacking in ways that prove troublesome to sustainability's goals for harmony and mutualism. Perhaps the best way to underscore these weaknesses is to note that in both the examples they cite, the material sustainable practices are *deeply* rooted in more-than-material human-nature relationships. For example, the Menominee's ecocentric forestry practices are sourced in their "belief that each life form is a '*person*' to be respected for its *knowledge* and *power* [and that this] pervades the pursuit of the *material conditions* [emphasis added] of life" (Nesper & Pecore, 1993, under "Values and Practice"). Callicott and Mumford (1997) neither acknowledge nor mention this element of the Menominee's relationship with the more-than-human forest. They are far from alone in this practice, and indeed it might seem unfair to ask them, as ecologists, to consider relational elements that fall outside their disciplinary focus. But, I believe that ecology and other materially-dominated disciplines must begin to do exactly this in order to see that more-than-material elements are inextricably tied to material processes—that one can neither succeed nor even be sensible without the other. In that sense, these more-than-material relational elements are as much a part of the ecology as material ones. Ultimately, the very ability of modern humans to conceptually isolate the material from more-than-material elements both obscures their deep interdependence and tends to reduce or negate the contributions of the latter. Callicott and Mumford seem to inadvertently undertake this kind of reduction by not mentioning the more-than-material sources of the ecocentric practices they hold up as examples. In so doing, they are either making the assumption that the source of these practices is "really" also only material/physical—that the more-than-material elements don't really exist—or that whatever more-than-material elements do exist are secondary in importance to ecological considerations, and therefore not worthy of mention as long as the ecological practices can be put into effect.

These two possibilities are reinforced by the fact that the only place that more-than-material influences are mentioned by the authors is when they're dismissed as anthropocentric relics of a bygone preservationism. In describing their own approach to sustainability as moving beyond this, Callicott and Mumford take a contemporary preservationist tack where "biota is valued for its own sake…and [biological] reserves are selected, delimited, connected, and managed in accordance with the best available science, irrespective of their conventional recreational, esthetic, or spiritual appeal" (p. 35). Their dismissal of all but materialist perspectives here is thorough. While their intention appears on the surface to be reasonable—the elimination of shallow pursuits such as recreation in the face of ecological imperatives—the breadth of their dismissal is overreaching. For example, spirituality is being equated in its superficiality with recreation, and characterized as a human *use* of more-than-human beings. But, such a notion is itself rooted in the idea that spirituality is a wholly human creation projected onto an externally material reality. This, instead of it being a human response to more-than-material/spiritual elements in external reality, including the more-than-human world. While I won't pursue the argument further here for why I take spiritual elements to be externally real (and thus any assumption that they are human creations is erroneous), the remainder of this book will attempt to lend credence to the possibilities of both material and more-than-material ontological elements, and expose others' rejection of the more-than-material as based purely in *a priori* materialist ontological commitments.

Thus, I take Callicott and Mumford's construction of spirituality as a superficial, and therefore less important, human-nature relational element to be in error. If their ontology is dualist, as it appears, it explains why they turn to ecocentrism's intrinsic valuation of "biota" as the means to grant standing to more-than-human beings in sustainability. Ultimately, however, characterizing all more-than-human beings *as* "biota," and as such to be known best through the material lens of science, neglects the central role of more-than-material elements in the Menominee's and others' relational approaches, and the possible reality of those elements.

This is not to say that everyone who wants to practice sustainable forestry need adopt the *particular beliefs* of the Menominee or come to hold *their particular kind of more-than-material knowledge*, but it is to say that more-than-material sources of material actions cannot be ignored or downplayed through a dualist elevation of material needs or outcomes. By ignoring those sources, it reduces or negates them *as* sources. I feel safe in suggesting that without those specific kinds of more-than-material elements, the Menominee's material practices would never have been developed, and without their continued influence, those practices would not be sustained. Therefore, what ultimately seems missing from Callicott and Mumford's account is the understanding from Menominee knowledge that more-than-human beings have more-than-material qualities and capacities of equal or greater importance to those of humans, and the Menominee's knowledge of these elements is absolutely essential to a human-nature relationship that is sustainable on a material level as well. Again, this does not mean that the more-than-material elements are a human concept that evolved as means to help the Menominee materially survive—as materialist evolutionary perspectives might suggest. Instead, both material and more-than-material elements must be acknowledged in their own rights as ontological bases, and must ultimately constrain the actions of humans in their relationships with more-than-human beings.

Jickling (2002), for his part, recognizes this, though he doesn't necessarily put it in these terms. When talking of working with students regarding the value of wolves and the issue of Canadian wolf-kill programs, he says,

> Students are encouraged to remain open to possibilities beyond sustainability—to accept sustainability for what it is, but to also explore what it is not. For example, the language of sustainability may be used when trying to optimize "harvest," but it cannot describe the magic of wolves howling on a winter's night... (p. 146)

I believe Jickling sees the same limitation of sustainability that I do—that it fails to account for the more-than-material elements of human-nature relationships that make the *desire* and willingness to sustain both more-than-human beings and humans possible. Again it cannot be overstated that the more-than-material elements should *not* be promoted instrumentally *so that* we materially survive, but instead as ends in themselves. Where I differ from Jickling, however, is his locating these more-than-material elements outside the sustainability context. He says above that we ought to "accept sustainability for what it is, but to...explore what it is not." I suggest instead, and in no uncertain terms, that sustainability is an empty, anthropocentric exercise with little possibility for success if it excludes the magic of wolves howling from its formulations of "what ought to be sustained." That doesn't mean that some like wolf howls and others like roosters crowing...and how do we decide which to sustain? That would be dualistically conceiving of these fundamental more-than-material elements as passive, material utterances whose meaning and power resides only in unilateral human response to, or valuing of, them. I suggest instead that the magic of wolf howls is a more-than-material quality *of wolves*, and as such is something to which many humans have and could have close relational responses. Therefore, they must form an indispensable part of mutualistic and harmonious human-nature relationships—of sustainability.

This work is dedicated to making the argument that without the close relational understandings of the human-nature relationship that emerge from admitting expanded material and more-than-material elements to the consideration of human-nature relationships, that sustainability, along with many other prescriptions for improvement of the human-nature relationship, will fail. The magic of howling wolves belongs in the very heart of sustainability—at its ontological roots. Without it, dualisms threaten to marginalize or negate the qualities and capacities of more-than-human beings (the magic that wolves and their howls possess and that humans perceive, instead of the magic that humans conceive of in response to the material wolf and its material howl) that make human-nature relationships the interpenetrating more-than-material and material structures and processes that many people experience them to be. Without it, sustainability simply cannot succeed.

5.4.6 *Posthumanist Considerations*

An excellent example of the application of posthumanist perspectives to sustainability is provided by Alaimo (2012). In her essay, she rightly calls out the anthropocentrism inherent in mainstream sustainability dialogue when noting that the "human centered discourses of sustainability"(p. 562) have a "tendency to render the lively world a storehouse of supplies for the [human] elite" (p. 558) and reduce that world to "the material for meeting [human] material 'needs'" (p. 562).

Alaimo claims that a cure of sorts lies in the "recognition that one's very self is substantially interconnected with the world [and that] the material self cannot be disentangled" (p. 561). Because of this, she suggests that sustainability "would... benefit from the new-materialist insistence on material agencies set forth by Donna Haraway, Nancy Tuana, Karen Barad, Bruno Latour, Andrew Pickering, and others" (p. 563). To further that claim, she notes that "[w]hile the epistemological stance of sustainability offers a comforting sense of scientific distancing and objectivity" the "transcorporeal subjects" that such new materialism affords "are often forced to recognize that their own material selves are the very stuff of the agential world that they seek to understand" (p. 561). Alaimo suggests that embedding humans in what Barad might call the intra-action of sustainable activity will help "formulate more complex epistemological, ontological, ethical, and political perspectives in which the human can no longer retreat into separation and denial" (p. 563).

While all this is true at a material level, as I note in my critique of posthumanism in Chap. 3, if the basis of interaction with the more-than-human world is still only material. Regardless of how broadly subsuming of relational elements that materiality is positioned to be, it still dualistically hierarchicalizes and limits the array of beings with which meaningful relational interchange is possible. As an illustration of this limitation, I analyze Alaimo's suggestion that in sustainability we should "embrace Braidotti's sense of becoming, which boldly calls for a 'non-rapacious ethics' 'for the hell of it and for the love of the world'" (p. 563). There are two key dualist issues in such a statement that are of note as it pertains to my discussion.

First, if one looks closely at the suggestion of founding an ethics, in part, "for the love of the world," the origin of the embrace is not the more-than-human world, nor the human's relationship with it. When doing something "for the love of the world" we are not necessarily doing so *within* the context of any relationship because one might exist—that is, the relationship might not be the source of the love. As a result, our ethics may not be an *outgrowth* of the relational love we share with that world. In *interhuman* relationships, one might reasonably equate these two in the sense that it would seem odd if one were to do something for someone out of love without that love being rooted in the reciprocity of the relationships itself. But, as I've noted throughout this volume, for example in my discussion of Schultz's (2002) notion of an "intimacy *toward*" the more-than-human world in the Connection to Nature Theories section of this chapter, there is a sense in certain conceptions of human-nature relationships that the human is unilaterally projecting such a thing as love outward onto the rest of nature from some internally isolated, human-only point of origin. I concede that posthumanists argue that we are materially embedded in the world, so one could argue that Alaimo's suggestion *must* come from one's relational entanglements with that world. But, nowhere in Alaimo's discussion is this possibility explored, and so it is unclear what influence such a relationship has on this love.

In the posthumanist literature more generally, at times such products of relationships are treated in an oddly facile manner, as if noting the existence of such a thing as love is sufficient proof of its origins in the material intra-action subsuming human and more-than-human alike. Haraway's (2008) love with her dog fits this characterization nicely. She speaks of material interconnection and the capacity for, and presence of, the "contagion called love." But that's where it ends. She's more interested in the implications of it and less so the origins. The reason for this is, I believe, an assumption that the origins are well settled in the material. But, they can only be taken to be settled in this way if one assumes a material ontology, one that is usually accompanied by some murky phenomenon such as emergence or contagion as the delivery mechanism. As I'll explore in later chapters, emergence is an empty concept that grows directly out of an assumption of material reality and therefore ought not to be taken as sufficient to explain a material origin of love, nor as the foundation for any more fulsome explanation that follows.

That the *relational* underpinning of such love for the world is a weak one is further reinforced by the other reason Alaimo quotes from Braidotti: that this ethics ought to be created "for the hell of it." I recognize the joy with which such an ethics is advanced—the sense of freedom, of *joi de vivre*, and of rebelliousness against a humanist paradigm that comes through not only in Braidotti's account, but in Barad's, Coole and Frost's and so many other posthumanists as they lay out the basis for their materialist end-run of humanist domination and oppression. But, while one can forgive posthumanists this sense of relief, one cannot reasonably embrace it as a legitimate basis for an ethics of human-nature relationships. It has little to do with the relationship which it is supposed to enable, and again, treats the origin of the human contribution as emerging from within the human, rather than from the context of the human-nature relationship itself. While new materialist and posthumanist views suggest this isn't possible, when the connection is seen as mostly material, the *relational* contribution of the more-than-human Self fades from the structure and dynamics so that the interdependence collapses. It is no wonder that in such scenarios, more-than-human beings fail to even garner mention as a relational, motivating force. Instead, in this "for the hell of it" context, the human is free. She deploys a human-nature relational ethic because she wants to and for no other reason. This is a relationally unbounded reason. It is an anthropocentric one. It borders on the whimsical. If the origins of "the hell of it" is not relationally reciprocal in an ecofeminist's sense of relationalism (see Warren, 1990) or in my sense of interdependent relations, in the end such inducements to ethical behavior don't look particularly different from humanist ones.

While Alaimo may be right to conclude that "we must hold ourselves accountable to a materiality that is never merely an external, blank, or inert space but the active, emergent substance of ourselves and others" I argue that when such exhortations don't actually examine the ways in which this will occur—it *may* be a result of leaving it to some future discussion—it could just as easily be the result of lurking dualisms that turn the "asymmetry" of heterogenous *relata* into a mechanism delivering the same humanist hierarchy that posthumanists are working to subvert. We need more than this to compel us to act sustainably.

5.5 Common Worlds Pedagogy

Common Worlds Pedagogy is an approach to early childhood education that employs a "conceptual framework developed to reconceptualise inclusion in early childhood communities…[where] common worlds take account of children's relations with all the others in their worlds—including the more-than-human others" (Taylor & Giugni, 2012, p. 108). Its orientation is anti-anthropocentric, with roots in postcolonial and posthumanist theory. In Taylor and Giugni's essay, statements expressing this orientation are numerous. For example, they state that "relations between ourselves and others – both human and more-than-human others – are of paramount importance" (p. 109). They say that, "the notion of common worlds encourages us to move towards an active understanding of and curiosity about the unfolding and entangled worlds we share with a host of human and more-than-human others" (p. 111). Lastly, they seek to "take up the ethical and political challenge of learning how to live well together and to flourish with difference" (p. 109).

With these authors, as well as many other posthumanist theorists, positivity accretes around recognizing the difference, heterogeneity, or "asymmetry" in human-nature relationships. I see the reason for this positivity resting upon two goals the acceptance of such differences accomplishes. First, there is the posthumanist suggestion that in order to have substantive human-nature relationships, everything no longer needs to be symmetrical between human and more-than-human in order for the relationship to be worthy of consideration. Put differently, we need not anthropomorphically project human or human-like qualities and capacities onto the more-than-human world in order to wedge for them standing in the relationships we have with them. The second accomplishment follows on the first. By recognizing these differences and labelling them as legitimate, humans have added something to the relational equation on the side of the more-than-human world that was, in more positivist or humanist theorizing, either overlooked or absent altogether.

These are both important factors to add to the calculus of human-nature relationships, but to inoculate against them carrying the vestiges of dualism, and thus ultimately working against posthumanism's goals of decentering the human, one must press further to see just what theorists take these differences to be. If difference is simple heterogeneity, then as Warren (1990) notes, difference *qua* difference is not dualist. It only becomes dualist when that difference is viewed through what Warren calls the "logic of domination" (p. 129). In that logic, the difference is taken to mean that the human is superior, and *because* he or she is superior, subordination of the more-than-human other is justified. Posthumanists, being intimately familiar with the workings of dualisms, would clearly recognize the distinction between difference as such and difference as lever to dominate. Yet, because of the deep influence of dualism on monist materialist ontological commitments, as I discussed in Chaps. 2 and 3, materialism itself works to conceptually diminish or negate the relational capacities of more-than-human beings.

In the common worlds pedagogy explained by Taylor and Giugni (2012), these "differences" are also rooted in a materialist ontology, and therefore, are still dualist. For example, their essay takes up discussion of a children's book's described encounter between a boy and a whale. The authors say of this encounter:

> Instead of simply substituting the whale for a human other, and translating this moment of recognition into an ethics of (human) care for other species (which was probably [the children's book author's] intention for her child readership), we propose a reading that suggests a new common worlds relational ethics. In this relational ethics, a significant cross-species encounter in common worlds, such as this one, begs the question: How might we live together in heterogeneous common worlds in a way that allows difference to flourish? (p. 112)

The authors ask an excellent question. But the way in which they determine the answer rests on a knife's edge. It is foundationally right to say that the whale and the human are importantly different and in ways that don't dualistically reduce the whale's capacities as they relate to the human's. And yet, if the balance shifts slightly in a dualist direction, it becomes all too easy to suggest that we shouldn't apply a human ethics of care to a whale because a whale is not the kind of being to which such a robust, interhuman kind of care ethics ought to be applied. According to Noddings' (1984) care ethics, for care to *be* care it must be reciprocal—not only must the one-caring behave in a caring manner, but the recipient must receive it *as care*. One wonders if Taylor and Giugni (2012) have this reciprocity in mind when they warn us off simply swapping a whale in for another human being. If the whale, in their minds, cannot receive the care as care, then this swapping will fail. If the authors take this to be the case, then to them whales must be different enough from human beings that they lack the qualities and capacities to reciprocate and otherwise be close and cared for by a human being in a way that humans feel close with one another—with the love, care, willingness to defend, etc., that go along with such care. It's interesting to note that Turner (1989) laments a lack of care and other emotional commitments when it comes to human attitudes toward more-than-human beings, but does not attribute this to the nature of those more-than-human others. Instead, he sources it in the fact that modern humans have become so detached from wild nature—the wild having become so abstracted from the daily lives of modern humans—that they fail to respond to a threat to the more-than-human world with the rage, anger or other strong emotions that he takes as an indicator of the deep care one ought to have. He admonishes us on this front when saying, "We must become so intimate with wild animals, with plants and places, that we answer to their destruction from the gut. Like when we discover the landlady strangling our cat" (p. 87). For him, at least, care *in its strongest form* is possible, appropriate and *missing* from modern human-nature relationships.

In Taylor and Giugni (2012), however, the lean toward dualist reduction of whales does, eventually, present itself. For example, when the authors' suggest that a whale shouldn't be substituted for another human being, they follow this suggestion by aligning their common worlds pedagogic theory with Haraway's notion of "Queer kin [that] encapsulates the possibility of sustaining relations with unlikely and very different but nevertheless significant others in common worlds, in ways that resist the temptation to minimise, negate, sentimentalise, anthropomorphise, or assimilate these relations" (pp. 112–113). As I've pointed out about Haraway (2008) already in Chap. 3, how she conceives of the difference between human and more-than-human still carries dualistic influences. By adopting Haraway's constructs, Taylor and Giugni's theories carry the same dualist signature. For example, we see in their discussion the explicit warning not to "anthropomorphise." But to attribute human qualities to a more-than-human other is only anthropomorphization when the more-than-human other does not actually possess the qualities in question. As Rollin (2017) notes, those who cry foul over supposed anthropomorphism may themselves be guilty of anthropocentrism when assuming, based in part on dualist thinking, that when the qualities in question are within the domain of human or human-like animals, then they must exclusively be so. Closeness in relations is a good example of where such anthropocentric assumptions may lurk.

Further, celebrating difference and elevating superiorizing difference over similarity because one fears, in the latter, the dualist charge of anthropomorphism, is unwarranted. It requires investigation, not supposition, to determine just what differences *and* similarities the whale and human have in Taylor and Giugni's (2012) exploration. If investigation is preemptively cleansed of its dualist assumption-making, then experiences like those of Nollman (1987), who carries out musical dialogues with whales in the wild, can accommodate tremendous similarity. What strikes him in his encounters with whales is just how clearly complex and intelligent whale thinking is, and how discernible it is as such in this context of his musical exchanges with them. For Nollman, intelligence is something that whale and human share. Thus, it's not a *human* quality at all. Inverting the arrangement, one might imagine whales judging the granting of intelligence to humans as a case of cetamorphism. When humans claim exclusive rights to certain traits or qualities, and then only credit other creatures for having them when other measures of similarity are applied (e.g., neural type and "complexity"), this is the height of anthropocentrism.

Neither can posthumanists save us here by chasing us off "traits" as being falsely positioned as static or unchanging things. Even though posthumanist theorists such as Barad (2007) argue that traits are dynamically generated and constantly changing in the flow of spacetime that powers intra-action, as I argued in Chap. 3, traits are still persistent and cohere around individuals- or Selves-in-relation. As such, they are invaluable indicators of the what is possible in relations between humans and the more-than-human world. Intelligence is a trait, regardless of how dynamic are its features. If intelligence is something shared by whale and human, then we can reasonably ask what else is shared as a trait. Is the ability to care and be cared for as well? With dualisms and the dualistically undergirded claim of anthropomorphism swept aside, it's still possible. Whether anthropomorphization is occurring in the

example that Taylor and Giugni (2012) use is difficult to discern without first identifying whether dualisms cause the authors to imply that not only are whales or dogs
or ants different from humans, they are either *less* capable of developing the kinds
of closeness that humans have with one another or have no capacity for it at all.

Dualism is, however, lurking in the authors' words when they reject the "recent
movement of nature kindergartens and forest schools" because it carries a "seductive appeal of a romanticised, immutable and originary nature, inhabited by its cor-
ollary, the innocent natural child" (p. 115). While children aren't innocent, and
while nature is surely not immutable, seeing the attempt to cultivate a certain kind
of human-nature relationship in such schools as a "harmonious Disney worlds"
(p. 113) fantasy because the authors' notion of the "reality" of the modern world
diverges from notions of the capacity of humans *to* fall in love with more-than-
human beings and therefore the world we all inhabit, is dualist to its core. It certainly cannot be offered up without justification, at least not without first being
exposed as dualist assumption about the possibilities for such relationships.

In fact, the equation of close human-nature relational experiences with romanticism is a common critique leveled those arguing that close relational encounters can
and do happen in human-nature relationships. This critique is usually characterized
as the act of some humans projecting their own ideas onto what is assumed to be a
material environmental *tabula rasa*. Cronon's (1996) "The Trouble with Wilderness"
is a perfect example. In it, he suggests that the last few centuries of human interaction with the natural world is littered with these kinds of projections. He says that
"the language we use to talk about wilderness is often permeated with spiritual and
religious values that reflect human ideals far more than the material world of physical nature" (p. 16). What he means by this is that within nature there are no such
qualities, that this world is material. This is reinforced when he says of Thoreau's
words about Mt. Katahdin that they "took the physical mountain on which he stood
and transmuted it into an icon of the sublime" (p. 12). His materialist worldview is
exposed in such suggestions, though he doesn't openly discuss it. That he doesn't
feel the need to do so is testament to the dualist power that monist materialism has
over modern thinking—so powerful that the materialism needs no justification.
Merchant's (1980) explanation of the conversion of views of the more-than-human
world as organic and holistic to that of biotic machine shows that monist materialism is no less a worldview rooted in assumptions and beliefs than any other view.
What view of human-nature relationships is most accurate will depend entirely on
the accuracy of how we assess the qualities and capacities of humans and more-
than-human beings, and thus, what the relationships between them will produce in
terms of closeness and other attributes. Part III of this book explores these foundations extensively.

One more quote from Taylor and Giugni (2012) is worthy of some examination.
At one point they say, "The process of continual questioning is central to any ethics,
but in a common worlds relational ethics, queer kin relations, with all their unlikely
intimacies, predictable asymmetries and radical differences pose significant challenges for living together" (p. 113). Dualism subtly saturates the entirety of this
statement. The "intimacies" can only be seen as "unlikely" if one *a priori* takes the

possibility of intimacy to *be* unlikely. The unlikelihood itself is sponsored by a dualism that takes its default inception point to be the "likely" impossibility of intimacy from which we must then move away toward a slightly more likely, yet still unlikely, intimacy.

The "predictable asymmetries" are also potentially dualist in that they are only predictable if the context within which they come to be defined as asymmetrical is accepted by those engaged in the discussion. What is predictable about them? Which ones are predictable? If, for the purposes of the point I'm making, I leverage the Haraway text to which the authors refer, I am on firm ground asking if the authors accept that animal experimentation is a predictable asymmetry, as Haraway suggests and as Weisberg (2009) rightly condemns as the epitome of humanism. Of course there are differences between humans and more-than-human others, but understanding the nature of those differences—the foundation of characterizing any interchange as asymmetrical—turns on whether dualisms have crept into the ontological commitments that underpin explanations of the qualities and capacities of the *relata.* If anything lasting is to be made of the human decentering exercises that common worlds pedagogic and other posthumanists have rightly undertaken, then dualisms that falsely position humans as more relationally adroit than their more-than-human counterparts must be called out as neither "predictable" nor, necessarily, as truly productive of asymmetry. If asymmetries are based in dualism, and not fact, then within the common worlds pedagogy the authors espouse, difference as the antidote to dualist eradication of more-than-human relational partners is ineffective. Instead, it becomes a trojan horse for the very dualisms that lead to a superior/inferior dyad in the first place, and thus eventually to the very eradication of the more-than-human from the structure and dynamics of human-nature relationships against which posthumanists are rightly attempting to fight.

Lastly, the "radical differences" the authors note are present indeed, but an honest accounting of what those differences might be, and from whence their designation as "radical," or as "difference" at all, comes, needs to be examined for dualist underpinnings. If the capacity for closeness is a radical difference, this book stands as counterargument, and seeks to pin such suppositions to the dualist origins I take them to have instead of to any quality or capacity of whales, dogs, ants, plants, or any other being.

Next, it's worth examining another common worlds theorist to see if the same dualisms appear. Pacini-Ketchabaw (2013) certainly begins her treatment of common worlds theory in a way that would put her perspective in line with my own. She opens by discussing

> the possibility that forests like Chickadee Woods know they are entangled with humans and nonhuman others…[and] the certainty that forests themselves know many stories: they know about practices of deforestation; they know about settlers' insistence on marking their boundaries; they know about progress; they know about violence, conquest, and more. (p. 357)

But, no sooner is she finished stating this, going so far as to refer to this Chickadee Woods as an "active collaborator," than she qualifies her deployment of the language of relationship. She says, "By saying that the forest knows stories, I am not attributing forests and their inhabitants human agency" (p. 357). Here again is the dualistically rooted end-run that persists in posthumanist theories. That is, posthumanists seek to decenter human beings in relational considerations, and see that one of the critical ways to do that is by suggesting that more-than-human others are active relational partners. But, because of the dualisms inherent in their scientific and materialist ontological commitments, they don't take them to be actually be active relational participants *in the way that other humans are*—and, incidentally, in the ways humans value the most and which help form the basis of their closest, strongest relational bonds. Because of this, posthumanists must employ an alternative means of admitting the more-than-human world, and do so by radically broadening the notion of terms like the ones Pacini-Ketchabaw uses in "collaboration" and "knowledge." These definitions end up so broadened, though, that they are no longer in keeping with the way in which they are used in interhuman relationships—that is, as being wielded by *one* who collaborates and *one* who knows. Again, Barad (2007) is right that such a "one" may not be metaphysically distinct, but as I discussed in Chap. 3, the identifiable individual- or Self-in-relation is ontologically irreducible, and essential to understanding the structure and dynamics of any relationship, including those between humans and beings in the more-than-human world.

The tethering of a knower to knowing and a collaborator to collaboration is so readily accepted in interhuman relationships that one might argue it's intuitive. Pacini-Ketchabaw's (2013) words certainly align with such a suggestion for humans when she differentiates forest agency from the human. "I am not attributing forests and their inhabitants *human* [emphasis added] agency" (p. 357), she says. Clearly for this author, human agency is a thing readily understood as in possession of an agent in her account, and thus in need of differentiation from more-than-human agency, which by implication, is not. In making this differentiation, the author reveals the deep dualism lurking in her posthumanist approach. For all of Barad's work to divorce agency from agents *cum* individuals, posthumanists still cling, consciously or otherwise, to the notion that agency in interhuman relationships is begotten by an agent, and in the more-than-human world it emerges through some opaque process from what I call in Chap. 3 the *microstructurally material*. By now, what flows from such ontological premises is fairly predictable. Possibilities for close relational interchange are conceptually forestalled, and any interdependence of relationship collapses except at the most superficial, material level. Actions, in Kelley et al.'s (1983) diagrams remain intact, but the thoughts and feelings that form interdependence in close relationships disappears from the more-than-human side of the interdependence dynamic.

This is precisely what happens in Pacini-Ketchabaw's (2013) work. She begins her exploration of the relationships that are possible and desirable between her, her students and Chickadee Woods by offering a quoted example of material reciprocity: "fungi are nourished by carbohydrates exuded from the plant roots, and they aid the plants by helping the roots absorb water and nutrients...Plant growth is greatly enhanced by mycorrhizal fungi, and some plants cannot grow without them" (p. 358). While such lessons in material interchange are valuable, they don't seem in any substantive way to shift the human anywhere from the center of the frame in human-nature relationships. That's because, in the interhuman relationships that we hold most dear, material considerations are secondary to love, care and other bonds of closeness. If it is only material interchange that counts as "collaboration" when the collaborator is more-than-human, I'm not sure that common worlds pedagogy, at least as Pacini-Ketchabaw has conceived it, is anything more than a glorified ecology lesson. Such things have their place and can, indeed, expose the wonders of the natural world to human learners, but there's nothing ontologically radical about such ways of viewing the world. After all, modern humans already have copious ecological knowledge and it doesn't do nearly enough to impede the ecological destruction wrought by the modern human project. That's why in this book I take such environmental problems not to be material, but to be problems of relationship at a personal or close-relational level. No matter how much material explanation there is recommending forest integrity, there will always be a human being who thinks that, even at times ecologically, it makes more sense to cut the forest down. That nothing in the author's common worlds pedagogy offers a baseline refusal to accept such possibilities speaks to the weakness of the relationship being cultivated, and its alignment with more typically modern, humanist approaches.

The author even suggests that the act of cutting down the forest might be reasonable in some circumstances. It's subtle, but makes an appearance when she and her class see a zoning proposal to cut down the trees in Chickadee Woods for the construction of an underground sewage overflow tank. Though she makes note of the shock and disbelief felt by both her and her class when seeing this proposal, at the same time she accepts as legitimate the desire to engage in such a project by saying, "Yet, given the area's urban development, it is clear that a sewage solution is urgently needed" (p. 358).

I suggest that a lack of an unalloyed horror on the part of the author shows the lack of real depth this "listening" to the forest knowers actually contains—and the real persistence of an anthropocentrism that weighs the urban human need to store waste against the lives of so many more-than-human others. I feel even more confident making the charge of anthropocentrism here by noting that nowhere does the author suggest that the Chickadee Woods forest knowers might be telling the humans surrounding it, including her class, to "buzz off" entirely. It's not even entertained that human pressure ought to be relieved by humans leaving the forest alone. Even in the reduced sense of more-than-human "Selves" as ecological cogs that the author is using, the forest may absolutely be telling them that. That this is not even considered shows the depth of the dualist, anthropocentric privileging of humans at work.

I contend that such privileging cannot be displaced until something other than material scientific wordplay elevating what is thought of as an inferior relational more-than-human partner is brought to bear. A dualistically privileged human is a dualistically privileged human no matter where they are placed in the frame of relational reference. Nothing about this will change until this superiority is punctured, and the qualities and capacities of the more-than-human world earnestly re-evaluated with dualisms cast aside.

For another example of this dualism at work, the author actually does use the language I did above to differentiate "learning about" from "learning with." She says, "What if forest pedagogies are not so much about learning about forests, but thinking with forests?" (p. 358). Given the traces of dualism in her work, the real question becomes with *whom* does she recommend we engage in this thinking? If they are dualistically reduced more-than-human others, we are right back where we started, for all the good intentions of the author. If they are primarily material, then "thinking with" really only means interpreting the material elements of their ecological role and synchronizing with them. She seems to be suggesting this latter interpretation when recommending that we enter into "the forest's ecology of selves" (p. 358) because in speaking to how these more-than-human forest selves shape us reciprocally, she says that "we might pay attention to the forest's liveliness, to its agentic inhabitants" (p. 361). This appears to be a reference to new materialist ontologies like Barad's (2007), ones that I've already argued do little to overcome the dualist reduction of more-than-human Selves.

Pacini-Ketchabaw (2013) goes on to describe, as an example of her "messy co-shapings" (p. 361), a blackberry bush "shaping us" when she says that "their fruit provides important nutrients to our bodies and to other species, such as deer and birds, who are plentiful in Chickadee Woods. The bushes tolerate poor soils, and so they thrive at the edges of the paths that have been created for our use" (p. 361). Here again we have a purely material description. I now quote her at length as she goes on to describe the interchange between the bush and one of her students:

> During one of our visits, a child who was walking in front of me picked up a long blackberry vine that lay on the path. The sharp prickles quickly attached to the child's hand, causing her pain and inscribing her body. The blackberry bush has changed this child's relation to the forest, and this newly established connection, in turn, reshapes the forest. This is no one-way relationship; the encounter has consequences for the bush, for the child, and for other more-than-humans who inhabit the forest. The futures of the blackberry bushes, with all their histories, and the futures of the child, with her histories, are entangled and in question: What are the (perhaps unintended) consequences of the frictions that occur when different kinds of bodies rub up against each other in the forest? (p. 361)

I must point out that this is not a strong description of a human-nature "relationship"—if it even rises to fit the definition of the word as the author attempts to leverage it into place as a building block for common worlds pedagogy. Getting pricked by a blackberry bush is an encounter, certainly, and one that will shape certain elements in both the child's and the bush's lives. But, to suggest that this is some foundation for a radical ontological move away from anthropocentrism is overstating things, unless one finds scientific forms of ecology to be radical, which I do not. In her description, she resorts to new materialist focus on the body and effects upon it having some influence over relationships. Not only is this insufficient to substantively change that relationship as I discussed in Chap. 3, but assuming that material interchange is the sum total of possibilities for human-nature relationships is based in dualist assumption, not in evidence.

In truth, Pacini-Ketchabaw's assertion that "what a child does and says is not more important than what the blackberry bush does or says" is just right. But if all the blackberry bush can really "say" (but not really *say*) is, "I've pricked you," then I'm not really sure anything substantive has been achieved in moving discussion away from anthropocentrism. Exhortation to align with scientific ecology's assessment that all parts of a material biological system play a role in the functioning of the whole is not the stuff of radical ontological transformation. How much more confusing, frustrating and disheartening, then, to hear the author use the language of interhuman interaction only to fail to document a single instance of it occurring. Instead, she moves quickly away from such language to either qualify it as metaphoric, or to tie it to such an expanded use of those terms that it renders them meaningless when one is attempting to ascertain the possibilities that exist in relations between humans and the more-than-human world.

Given that even uncontroversial mainstream materialist ontologies note that *some* animals have cognition, consciousness, feelings, intent and other attributes that enable them to participate in some way in meaningful human-nature relational interchange, the author's omission of them from her pedagogy of human-nature relationships is troubling. It also reinforces the main contention of this book, that belief in the improbability of close human-nature relationships is rooted in unexamined dualist assumption. Ultimately, with "relationships" as tenuous and materially superficial as those explored by Pacini-Ketchabaw, when the bulldozer comes to put in that sewage overflow tank, will the author or any of her students really have the relational foundation to stand in the way?

5.6   Critical Animal Studies

One other posthumanist field worth consideration as it relates to the structure and dynamics of human-nature relationships is Critical Animal Studies (CAS). This field also takes aim at ameliorating human-nature relationships by focusing on the relationship between humans and nonhuman animals. CAS theorists Twine (2010) situates animal studies as having

> re-evaluated the role and presence of nonhuman animals in a wide range of disciplines where they have been for the most part ignored. It is also unprecedented for the way in which it has politicized human–animal relations, a development which is even more apparent, as we shall see, in the variant *critical* animal studies. (p. 1)

The field is critical in particular of human/animal Cartesian dualisms, and because of that

> should also be read as an interrogation of the *human*, as a critical openness to rethinking what we understand as 'human' and to reconceptualizing its historical and cultural consolidation as constituted by being 'not animal'. Increasingly there are affinities and overlaps between animal studies and ideas of *post*humanism. (p. 2)

CAS theorist Best (2009) notes that the field is a "critique of human supremacism, Western dualism, and the human exploitation of nonhuman animals" (p. 10). He says that it is inherently politically active, and "openly avows its explicit ethical and practical commitment to the freedom of well-being of all animals and to a flourishing planet. It opposes all forms of discrimination, hierarchy, and oppression as a complex of problems to be extirpated from the root, not sliced off at the branch" (p. 12).

These are lofty and worthy goals, and as such, are a positive step in the improvement of human-nature relationships. For the purposes of the discussion here, I will explore the CAS literature in an attempt to discern whether the underlying ontology, both expressed and implied, contains any dualistic thinking. If it does, my goal will be to both demonstrate the ways in which it works against an ontology of human-nature relationships that will better aid CAS theorists in reaching their goals, and also to show the ways in which any assumptions about the nature of various more-than-human beings—some animal, some vegetal—operate in the interdependence model articulated in Chap. 4.

As is evidenced by the name of the field and the examples I've just given, CAS is focused on animals. To underscore this, one may note the absence of any field called critical *plant* studies, for example. Right away, this means that we are operating in a reduced sphere of human/more-than-human relations. This is not problematic in itself, it only has the potential to become so if the ontology that undergirds CAS's largely ethical focus itself is problematic—that is, if it contains dualisms which distort modern human understandings of the qualities and capacities of certain groups of more-than-human beings. To determine whether dualisms are in operation, one area to investigate is how CAS theorists justify their focus on animals. For example, if they place their focus there because they feel that there is a greater similarity between animals and humans such that animals are deserving of ethical consideration and other more-than-human beings are not, this is likely to be sponsored by some underlying dualism. Another way that dualism may be revealed is by examining the ways in which CAS theorists situate their field in relation to other disciplines and fields working to improve human-nature relationships. If they view their field as one part of a broader effort, choosing to focus their energies on animal exploitation because it is one of many worthy areas, this is unproblematic. If, instead, they see their field as the most important or central area for human-nature relational study, this could be underpinned by dualisms.

5.6.1 Sentience

One of the key justifications for taking up the cause of animals specifically is rooted in many CAS theorists' belief that because animals are sentient, they ought to be freed from exploitation and other instrumentalization (Best, 2009; Twine, 2012, Weitzenfeld & Joy, 2014). The OED defines a sentient being as one "[t]hat feels or is capable of feeling; having the power or function of sensation or of perception by the senses" ("Sentient", 2018). This is only one of many kinds of definitions one finds of sentience, however, with others including elements like the capacity for consciousness or the ability to feel pleasure or pain. But, at least according to the OED definition above, if sentience is the reason to include animals in efforts to eradicate exploitation, I note that it cannot be definitively concluded that plants are not sentient. To support this point and using the OED definition, one should ask which senses are meant, since plants perceive with their senses, they just don't happen to be human senses. For example, they are able to detect and respond to attacks by predators, releasing volatile organic compounds into the air that other plants can detect (Baldwin, Halitschke, Paschold, Von Dahl & Preston, 2006). They certainly couldn't detect an attack and have that unique response if they didn't have the ability to sense one. If one were to counter such a possibility by saying that this is not what is intended by "senses" because they aren't mind-based, this would be adding a qualification to sentience that the OED definition doesn't carry. It would also be restricting the definition so that it more closely aligned with *human* senses, which is anthropocentric and dualist. Further, and as I'll explore in depth in Chap. 10, it is an unsettled matter whether plants have minds.

As just one example of the use of sentience, when discussing actor-network theory (ANT) Twine (2012) says "…it is not new to recognize that ANT is of potential use to the study of human/animal relations (Whatmore & Thorne, 2000) in large part due to its insistence upon the agency of nonhuman actors" (p. 25). But, if one were to think that this capacity for agency meant including non*animal* actors with agency, Twine qualifies his statement in an end note by saying, "Of course there is a danger in ANT that in the granting of agency to animals that they become…conflated with non-sentient objects" (p. 36). It's interesting that he constructs a dyad with "animal" in one half and "non-sentient object" in the other. Such a dyad could be read in one of two ways. The first is that the dyad has, as one half, sentient animals and on the other, all remaining more-than-human beings that not only lack sentience, but are "objects." This reading would certainly reinforce my suggestion in earlier chapters that dualisms can act to reduce more-than-human beings to largely passive, material objects of instrumental value. The other way that the dyad can be read is that there are sentient animals at one end of a spectrum, non-sentient objects at the other end, and in the middle, unaddressed by Twine directly other than by implication, non-sentient beings that are not so dualistically reduced that they descend to the level of "object." But, even if this second reading is true and there is some space in the middle for non-sentient *subjects*, that they garner neither attention nor mention by the author indicates their reduced status in his theorizing.

In this section I did not address the fact that sentience as a boundary marker for ethical consideration is often critiqued for its anthropocentrism. I'd be remiss, however, if I didn't mention what I take to be the ontological inference one can make from the use of sentience or other human-like traits as a means of extending ethical consideration to the more-than-human world. In some sense, I take the focus on human-like traits such as sentience to be less about the qualities and capacities of more-than-human beings or their lack, and more about modern humans. What I mean is, we use sentience as a marker because it's *easier*. We can identify similarities— and thus avenues for relation—in animals in ways that we don't think about with non-animals. We do this, at some level, because it alleviates the need for us to do any "heavy ontological or relational lifting" to either interact with or extend substantial consideration to non-animals. To learn the language of relationships with plants, for instance, is challenging. Even if the choice of sentience as a cutoff point isn't completely about convenience, the body of literature speaking to the incredible capacities of plants for all manner of things like intelligence, memory, etc. is growing every day. Again, Part IV contains a more in-depth look at these specific capacities as they relate to the possibility of thoughts or feelings in plants. The point here is that if there are these similarities between plants and humans, it's then not as difficult to find ways to relate with them or, with dualisms taken out of the equation, to even become close with them. That we still don't undertake such relational efforts is testament not to the availability of information about how we might begin, but more to the hegemonic effectiveness of dualisms that persist in our ontologies, and our willing acceptance of them as they trigger in us a reflexive rejection of the possibility for closeness with these beings. I repeat here that evidence presented suggesting that plants cannot be relational partners always carries with it a pre-existing assumption of their incapacity in this regard. Thus and again, the relational structure and dynamics of interdependent relationships collapses under the weight of *post hoc* dualisms filtering out experiences that contradict these assumptions.

5.6.2 Veganism

From an ontological perspective, one particularly telling element in many CAS theories is the support given to veganism as an ethical stance. Weitzenfeld and Joy (2014), for example, say that "Vegan praxis is one means of…challenging the hegemony of speciesist institutions and anthropocentric ideology" (p. 4). About this praxis they say,

> Given the profound impact of carnism on the structure of human-nonhuman relations, vegan praxis—a counternarrative and practice in which more-than-human beings are not viewed or treated as appropriate for human consumption—is perhaps the most effective and direct way to subvert the speciesist complex. (p. 21)

Some interesting features emerge from this statement that highlight the authors' ontological perspective. First, it appears that they may not be removing plants and other non-animals entirely from consideration, since their aim is to eradicate speciesism overall, and may view veganism as a first step with animals that may lead to later steps with plants and other non-animals. Such a possibility is also indicated by the fact that the authors position veganism as a response to carnism and the unique pressures that carnism puts on human-nature relationships in general. They are right to point out that "Naming and deconstructing carnism…may enable us to make invisible naturalized and normalized practices and affects of speciesism visible objects to challenge and transform" (p. 21). In sum, their suggestion that veganism is a lever with which to open examination of problematic conceptions of human-nature relationships overall is reasonable.

Where their views become problematic however, is when they make clarifying statements about their advocacy for veganism. For example, they say that "carnism enables the oppression of the majority of more-than-human beings who are exploited for human ends" (p. 21). This is clearly a distortion, since one need only count the trees on a Southern Pine plantation in the state of Georgia or the plants in the endless miles of genetically modified corn in the midwestern United States to understand that there are multitudinous non-animal beings suffering from exploitation yet missing from the authors' "majority." Such an oversight is some indication that there is a hierarchy at work that is dualist in origin, one that positions animals as more worthy of anti-speciesist efforts than plants and other more-than-human beings. This is reinforced by the fact that so much of that abused corn is put to use feeding animals for the meat industry, and as such underwrites the carnism that the authors decry. It is undeniable that the oppression of plants and animals are inextricably linked and, in my estimation, come from a common origin. Another example of this dualist hierarchy appears when the authors say that "the perpetual, intimate, and deeply symbolic act of eating animals in large part defines the human-nonhuman relationship" (p. 21). This is another overstatement. That some cultures eat very little meat is one indication that such a broad statement is not accurate. I suppose the authors are right in the sense that human history has always been bound up with the eating of animals, but the statement carries more than a hint that the authors are taking the act of eating animals as more central and primary than eating other more-than-human beings. By leaving plants and other non-animal beings out of their considerations, the structure and dynamics of my interdependent model collapses for those beings. Thus, to hold such views of plants and other non-animals cuts off consideration of their relational qualities and capacities by assumption, not through any evidence.

Jenkins (2012) provides another example where the call for veganism is positioned as all that's necessary to vastly improve human-nature relationships while the relationship with plants and other non-animals is largely ignored. She says of veganism that it is a "responsive, affective ethics of nonviolence" (p. 505). But, the OED defines violence as "The deliberate exercise of physical force against a person, property, etc." ("Violence", 2018). Jenkins' suggestion that veganism is nonviolent contradicts this definition because, in order to be vegan, one must commit violence against all manner of non-animal more-than-human beings. The only way this can be considered nonviolent is if one *redefines violence* so that plants and other non-animals cannot be its victims. It appears that this is precisely what Jenkins is doing. More specifically, she appears to be assuming that if a being does not have sentience, then the effect of the violence need not be tallied, and in fact renders the act itself one that should not be considered violent. But, smashing a glass against a wall in anger is a violent act against the glass and the wall regardless of the capacity of the glass or the wall to respond to it in some way. If the wall was another human being or the family dog with the capacity to be hurt in ways that are not purely in the realm of physical damage, this would compounds *the effects* of the violence, but it would not change the fact that the act was violent. Ultimately, she appears to be moving the "violence" portion of the physically forceful act out of the act itself or the intent of the one committing it and into the response, or more accurately, the capacity *to* respond, of the one against whom the act is committed.

That Jenkins takes plant responses to be more akin to the glass and the wall than to the family pet seems to be the case. Further, even if one were inclined to allow such a radical alteration of the definition of violence, that the plant doesn't have the capacity to respond to violence *as* violence is far from clear. As mentioned above, plants can detect an attack by a predator and not only respond (Baldwin et al., 2006), but may also be communicating the situation to other plants so that they can be proactive in defending against the attack's most severe effects. I will explore in later chapters just what such a capacity in plants means for the possibility of a plant having a mind, and even feelings, but here only suggest that such a response to attack by plants certainly calls into question any conclusion that avoiding violent acts to *animals* amounts to all that is necessary to make human-nature relationships nonviolent. Again, by putting plants and other non-animals outside the boundary of consideration for violence allows deeply engrained and difficult to detect modernist, dualist assumptions about the incapacity of plants to engage with us relationally to operate unchecked in our modern perspectives on human-nature relationships. This has the effect of collapsing the structure and dynamics of interdependence that at this point in my discussion is still possible between humans and all beings.

Best (2009) engages in a line of argumentation regarding veganism similar to that of Jenkins (2012) when pointing out the "profound importance of veganism… for human liberation, peace and justice, and ecological healing and balance" (Best, 2009, p. 19). Again, such statements are a testament to the limits inherent in some CAS theorists' ontologies whereby taking up the killing of plants for food is thought sufficient to achieve overall "ecological healing and balance" (p. 19). It is a clear case of dualism in operation when CAS theorists advance overly facile ideas of the eradication of violence and the achievement of ecological healing and balance based only on modifications to the human/animal relationship. This is especially the case when doing so flies in the face of an ever-growing body of literature about plant intelligence and the capacity of plants for memory, trial-and-error learning, self/ other recognition and host of other capacities that I'll explore in Part IV.

Such blithe and assumption-based dismissals of non-animal more-than-human beings is also demonstrated when Best equates veganism with valuing animals as "living beings." Again, one must wonder why plants, also indisputably living beings, don't inspire a similar valuation. Calling for the elimination of the mistreatment of any category of more-than-human being is always right and welcome, but in drawing the boundaries around their theories in the ways that they have—by dualist assumption about the qualities and capacities of non-animals and thus the empowering of their irrelevance—some CAS theorists are in danger of allowing their theories to be rooted in unseen dualisms that work against their well-articulated goals of nonviolence and freedom from exploitation.

I would only add this: given that the subject of the CAS field is *animals*, I don't expect CAS theorists to openly advocate for non-animal more-than-human beings. That being said, I also think that some awareness, given the rise of posthumanist and other adventures into absolute, egalitarian ontologies, would at least give CAS theorists pause when speaking of stopping the mistreatment of animals as if this, and this alone, will lead to the achievement of an unblemished "peace and justice, and ecological healing and balance." Without this kind of expanded acknowledgement and/or consideration, in some sense CAS is really just a late-stage, dualistically sponsored, ultimately anthropocentric, animal rights movement. The sentiment is correct, but until CAS is positioned as one essential *piece* of a larger ecological healing and balance, I'm unclear on what their perspective adds to the overall push for radical change to modern concepts of human-nature relationships. While the structure and processes of interdependent human-nature relationships remain intact for some animals and humans under CAS, without the rest of more-than-human world provisionally included, the structure for them inevitably collapses.

5.6.3 Equating Animal with Nonhuman

There are numerous examples in CAS theory where acknowledgement of the more-than-human beings beyond the category of animal is given. For example, Best (2009) notes the "need for profound changes in the way humans define themselves and relate to other sentient species *and to the natural world as a whole* [emphasis added]" (p. 17). Even within the animal category, some theorists acknowledge that dividing lines have been drawn, and that some beings might be outside of consideration. For instance, Freeman (2012) points out the "paradox that mainstream animal rights philosophy itself could be perceived as excluding some animals (such as some invertebrates) who don't appear to qualify as fully sentient or conscious subjects" (p. 117). In responding to this, she says,

> All identity-based movements rely on boundaries and exclusions, even though they work on extending current boundaries to incorporate new groups, extending opportunities for equality. It's possible that a feminist ethic of care (Kheel, 2008) or phenomenological ethics (Oliver, 2010), both of which base ethical concern on inter-species relationships and emotional and empathetic experiences, can overcome some of the limitations which come with identity-based approaches. (p. 117)

Here she appears to recognize that whenever any boundary is drawn, some beings are excluded. As I mentioned at the outset of my discussion of CAS theories, if this recognition allows one to situate CAS within a larger context, then it's likely that such formulations of CAS are less susceptible to operating on dualist assumption. Freeman does just this by both pointing out the instrumentality of CAS theory in expanding boundaries for the good of all, and also in acknowledging that other approaches might address some of CAS's shortcomings.

Having noted this, there are still many examples within the CAS literature where such boundary drawing has a reductive effect on consideration of non-animals similar to that of taking veganism to be ameliorative of the human-nature relationship in its entirety. Here I'll discuss the problem of equating the category of animal to the more general category of nonhuman or more-than-human. Once again, Jenkins (2012) discussion is illustrative. To begin, she is precisely right in saying that even though some of the

> purported posthuman theories [of animal studies scholars] reject the ontological dualism between human and animal, hierarchical dichotomies reside within these theories' normative presuppositions…Until we recognize the lives of all animate beings as worth protecting, the hierarchical dualisms of human/animal, mind/body, and nature/culture will remain intact. (p. 505)

As one can see, she calls here for recognition for all animate beings (I'll leave discussion of the animate/inanimate dyad to Part IV of the book) while also, as discussed above, suggesting that "veganism is a necessary component" in the "dissolution of the human/animal binary" (p. 505). One can see here a false equivalence. On the one hand Jenkins creates a dyad containing humans and all animate beings and on the other is a dyad with only the human and the animal. Further, even though she calls for ameliorative action for all animate beings, her prescriptions are targeted only at the animal. Thus, in her work, animal and animate become one, where plants and other animate non-animals garner no consideration even though plants are, without question, animate.

Another example of such a false equivalence, one that effectively removes non-animals from ontological consideration, is from Best (2009). He recommends "a discursive approach [to CAS that] would analyze, for instance, how the Western world fractures the evolutionary continuity of human/nonhuman existence by reducing animals to (irrational, unthinking)—Others who stand apart from (rational, thinking) human Subjects." (p. 16). Here again, on one hand the dyad is presented as human/nonhuman, and all that is used to fill the nonhuman half of the dyad are animals. One last example from Jackson (2013) shows a similar pattern. She says, "Inanimate matter and nonhuman animals have affective power, shape human subjectivity, and alter human perception." As in my discussion of Twine (2012) above, here the two categories of more-than-human being she constructs are "inanimate matter" and "nonhuman animals." It is these two alone that the author sees as shaping human subjectivity, etc. The inanimate matter aligns with posthumanist views, and nonhuman animals to the subject of CAS. That non-animal more-than-human beings are left entirely out of such effects shows an assumption that they have none. This again collapses the structure and dynamics of possible closeness in human-nature relationships for non-animal more-than-human beings even though, with dualisms pushed aside, my discussion in Parts III and IV will show such interchange to be possible. Therefore, the only way that Jackson or these other CAS theorists can hold to such a notion, consciously or otherwise, is through dualist assumption that plants and other non-animals do not have the qualities and capacities to be relationally considered in modern ontologies.

5.6.4 Conclusion

To conclude my discussion of CAS, I'll note that the passion with which CAS theorists attempt to resolve issues of human-animal relations is laudable. When CAS theory situates itself within the larger set of fields attempting to improve human-nature relationships from various perspectives and addresses the mistreatment of all manner of more-than-human beings, their efforts are a welcome complement to myriad others. If, however, dualisms are allowed to operate in an unexamined fashion in CAS theories, evidence for which I have discussed above, then CAS is in danger of falling well short of its goals.

References

Alaimo, S. (2012). Sustainable this, sustainable that: New materialisms, posthumanism, and unknown futures. *PMLA, 127*(3), 558–564.

Aron, A., Aron, E. N., Tudor, M., & Nelson, G. (1991). Close relationships as including other in the self. *Journal of Personality and Social Psychology, 60*(2), 241–253.

Aron, A., & Fraley, B. (1999). Relationship closeness as including other in the self: Cognitive underpinnings and measures. *Social Cognition, 17*(2), 140–160.

Associated Press. (2008, May 30). 30 tons of lobster lost in Boston fire. *CBS News*. Retrieved from https://www.cbsnews.com/news/30-tons-of-lobster-lost-in-boston-fire/.

Ayres, R. U., van den Bergh, J. C., & Gowdy, J. M. (1998). *Viewpoint: Weak versus strong sustainability* (No. 98-103/3). Tinbergen Institute Discussion Paper.

Baldwin, I. T., Halitschke, R., Paschold, A., Von Dahl, C. C., & Preston, C. A. (2006). Volatile signaling in plant-plant interactions: "talking trees" in the genomics era. *Science, 311*(5762), 812–815.

Barad, K. (2007). *Meeting the universe halfway: Quantum physics and the entanglement of matter and meaning*. Durham, NC: Duke University Press.

Barr, S. (2003). Strategies for sustainability: Citizens and responsible environmental behaviour. *Area, 35*(3), 227–240.

Barrett, C. B., & Grizzle, R. (1999). A holistic approach to sustainability based on pluralism stewardship. *Environmental Ethics, 21*(1), 23–42.

Beckley, T. M. (2003). The relative importance of sociocultural and ecological factors in attachment to place. In L. E. Kruger (Ed.), *Understanding community-forest relations, General Technical Report PNW-GTR-566* (pp. 105–126). Portland, OR: U.S. Department of Agriculture, Forest Service Pacific Northwest Research Station.

Best, S. (2009). The rise of critical animal studies: Putting theory into action and animal liberation into higher education. *Journal for Critical Animal Studies, 7*(1), 9–52.

Blackburn, W. R. (2007). *The sustainability handbook: The complete management guide to achieving social, economic, and environmental responsibility*. Washington, DC: Environmental Law Institute.

Bonnett, M. (2002). Education for sustainability as a frame of mind. *Environmental Education Research, 8*(1), 9–20.

Boyle, J. (2002). Fencing off ideas: Enclosure & the disappearance of the public domain. *Daedalus, 131*(2), 13–25.

Bretherton, I. (1985). Attachment theory: Retrospect and prospect. *Monographs of the Society for Research in Child Development, 50*(1/2), 3–35.

Brundtland, G. H., & World Commission on Environment and Development, Australia, & Commission for the Future. (1990). *Our common future*. Melbourne, VI: Oxford University Press.

Caldwell, L. K. (1998). The concept of sustainability: A critical approach. In J. Lemons, L. Westra, & R. Goodland (Eds.), *Ecological sustainability and integrity: Concepts and approaches* (pp. 1–15). Dordrecht, Netherland: Springer Science and Business Media.

Callicott, J. B., & Mumford, K. (1997). Ecological sustainability as a conservation concept. *Conservation Biology, 11*(1), 32–40.

Carpenter, R. A. (1998). Coping with 2050. In J. Lemons, L. Westra, & R. Goodland (Eds.), *Ecological sustainability and integrity: Concepts and approaches* (pp. 290–308). Dordrecht, Netherland: Springer Science and Business Media.

Collins, N. L., & Feeney, B. C. (2000). A safe haven: An attachment theory perspective on support seeking and caregiving in intimate relationships. *Journal of Personality and Social Psychology, 78*(6), 1053–1073.

Consume. (2018, June). *Oxford English dictionary online*. Retrieved from http://www.oed.com

Corral-Verdugo, V., Carrus, G., Bonnes, M., Moser, G., & Sinha, J. B. (2008). Environmental beliefs and endorsement of sustainable development principles in water conservation: Toward a new human interdependence paradigm scale. *Environment and Behavior, 40*(5), 703–725.

Costanza, R., & Daly, H. E. (1992). Natural capital and sustainable development. *Conservation Biology, 6*(1), 37–46.

Cronon, W. (1996). The trouble with wilderness: Or, getting back to the wrong nature. *Environmental History, 1*(1), 7–28.

Daly, H. E. (1995). On Wilfred Beckerman's critique of sustainable development. *Environmental Values, 4*(1), 49–55.

Davis, M. H., Conklin, L., Smith, A., & Luce, C. (1996). Effect of perspective taking on the cognitive representation of persons: A merging of self and other. *Journal of Personality and Social Psychology, 70*(4), 713–726.

Dutcher, D. D., Finley, J. C., Luloff, A. E., & Johnson, J. B. (2007). Connectivity with nature as a measure of environmental values. *Environment and Behavior, 39*(4), 474–493.

Falk, D. A., Palmer, M. A., & Zedler, J. B. (2006). *Foundations of restoration ecology*. Washington, DC: Island Press.

Fehr, E., & Fischbacher, U. (2003). The nature of human altruism. *Nature, 425*(6960), 785–791.

Fox, K. M. (1997). Leisure: Celebration and resistance in the ecofeminist quilt. In K. Warren (Ed.), *Ecofeminism: Women, culture, nature* (pp. 155–175). Bloomington, IN: Indiana University Press.

Freeman, C. P. (2012). Fishing for animal rights in the cove: A holistic approach to animal advocacy documentaries. *Journal for Critical Animal Studies, 10*(1), 104–118.

Gale, R. P., & Cordray, S. M. (1994). Making sense of sustainability: Nine answers to 'What should be sustained?'. *Rural Sociology, 59*(2), 311–332.

Gandel, C. (2013, September 10). Discover 11 hot college majors that lead to jobs. U.S. News and World Report. Retrieved from http://www.usnews.com/education/best-colleges/articles/2013/09/10/discover-11-hot-college-majors-that-lead-to-jobs?page=2

Gladwin, T. N., Kennelly, J. J., & Krause, T. (1995). Shifting paradigms for sustainable development: Implications for management theory and research. *Academy of Management Review, 20*(4), 874–907.

Goodland, R. (1995). The concept of environmental sustainability. *Annual Review of Ecology and Systematics, 26*, 1–24.

Goodland, R., & Daly, H. (1996). Environmental sustainability: Universal and non-negotiable. *Ecological Applications, 6*(4), 1002–1017.

Greider, T., & Garkovich, L. (1994). Landscapes: The social construction of nature and the environment. *Rural Sociology, 59*(1), 1–24.

Gustafson, P. (2001). Meanings of place: Everyday experience and theoretical conceptualizations. *Journal of Environmental Psychology, 21*(1), 5–16.

Hawken, P. (1994). *The ecology of commerce: A declaration of sustainability*. New York: Harper.

Hinds, J., & Sparks, P. (2008). Engaging with the natural environment: The role of affective connection and identity. *Journal of Environmental Psychology, 28*(2), 109–120.

Hinterberger, F., Luks, F., & Schmidt-Bleek, F. (1997). Material flows vs. 'natural capital': What makes an economy sustainable? *Ecological Economics, 23*(1), 1–14.

Hopwood, B., Mellor, M., & O'Brien, G. (2005). Sustainable development: Mapping different approaches. *Sustainable Development, 13*(1), 38–52.

Houck, O. A. (1998). Are humans part of ecosystems. *Environmental Law, 28*, 1.

Hueting, R., & Reijnders, L. (2004). Broad sustainability contra sustainability: The proper construction of sustainability indicators. *Ecological Economics, 50*(3–4), 249–260.

Imran, S., Alam, K., & Beaumont, N. (2014). Reinterpreting the definition of sustainable development for a more ecocentric reorientation. *Sustainable Development, 22*(2), 134–144.

Ingold, T. (2000). *The perception of the environment: Essays on livelihood, dwelling and skill*. New York: Psychology Press.

Jacoby, K. (2001). *Crimes against nature: Squatters, poachers, thieves, and the hidden history of American conservation*. Berkeley, CA: University of California Press.

Jenkins, S. (2012). Returning the ethical and political to animal studies. *Hypatia, 27*(3), 504–510.

Jenkins, W. (2009). Sustainability theory. In W. Jenkins & W. Bauman (Eds.), *Berkshire encyclopedia of sustainability: The spirit of sustainability* (pp. 380–384). Great Barrington, MA: Berkshire Publishing Group.

Jepson, E. J. (2004). Human nature and sustainable development: A strategic challenge for planners. *Journal of Planning Literature, 19*(1), 3–15.

Jickling, B. (2002). Wolves, environmental thought and the language of sustainability. In D. Tilbury, R. B. Stevenson, J. Fien, & D. Schreuder (Eds.), *Education and sustainability: Responding to the global challenge* (pp. 145–154). Gland, Switzerland: IUCN.

Johnson, C. Y. (1998). A consideration of collective memory in African American attachment to wildland recreation places. *Human Ecology Review, 5*(1), 5–15.

Johnson, N. (1993). Introduction. In G. H. Aplet, N. Johnson, J. T. Olson, & V. A. Sample (Eds.), *Defining sustainable forestry* (pp. 11–16). Washington, DC: Island Press.

Jorgensen, B. S., & Stedman, R. C. (2001). Sense of place as an attitude: Lakeshore owners attitudes toward their properties. *Journal of Environmental Psychology, 21*(3), 233–248.

Kane, M. (1999). Sustainability concepts: From theory to practice. In J. Köhn (Ed.), *Sustainability in question: The search for a conceptual framework* (p. 32). Northampton, MA: Edward Elgar Publishing.

Kelley, H. H., Berscheid, E., Christensen, A., Harvey, J. H., Huston, T. L., Levinger, G., et al. (1983). *Close relationships*. San Francisco: W. H. Freeman and Company.

King, R. J. H. (1991). Caring about nature: Feminist ethics and the environment. *Hypatia, 6*(1), 75–89.

Kopnina, H. (2013). Evaluating education for sustainable development (ESD): Using ecocentric and anthropocentric attitudes toward the sustainable development (EAATSD) scale. *Environment, Development and Sustainability, 15*(3), 607–623.

Kropotkin, P. A. (1902). *Mutual aid, a factor of evolution*. New York: McClure Phillips & Co..

Lélé, S., & Norgaard, R. B. (1996). Sustainability and the scientist's burden. *Conservation Biology, 10*(2), 354–365.

Lélé, S. M. (1991). Sustainable development: A critical review. *World Development, 19*(6), 607–621.

Mayer, F. S., & Frantz, C. M. (2004). The connectedness to nature scale: A measure of individuals' feeling in community with nature. *Journal of Environmental Psychology, 24*(4), 503–515.

Merchant, C. (1980). *The death of nature: Women, ecology, and the scientific revolution*. San Francisco: Harper & Row.

Mill, J. S. (2002). *The basic writings of John Stuart Mill*. New York: Modern Library (Original work published in 1879).

Mueller, A. M. (2002). *Sense of place among New England organic farmers and commercial fishermen: How social context shapes identity and environmentally responsible behavior* (Ph.D.). (3057251).

Müller, M. M., Kals, E., & Pansa, R. (2009). Adolescents' emotional affinity toward nature: A cross-societal study. *Journal of Developmental Processes, 4*(1), 59–69.

Murphy, R. (1995). Sociology as if nature did not matter: An ecological critique. *The British Journal of Sociology, 46*(4), 688–707.

Nesper, L., & Pecore, M. (1993). The trees will last forever: The integrity of their forest signifies the health of the Menominee people. *Cultural Survival Quarterly, 17*(Spring), 28–31.

Nisbet, E. K., Zelenski, J. M., & Murphy, S. A. (2008). The nature relatedness scale: Linking individuals' connection with nature to environmental concern and behavior. *Environment and Behavior, 41*(5), 715–740.

Noddings, N. (1984). *Caring: A feminine approach to ethics & moral education*. Berkeley, CA: University of California Press.

Nollman, J. (1987). *Animal dreaming: (the art and science of interspecies communication)*. New York: Bantam.

Noss, R. F. (1993). Sustainable forestry or sustainable forests? In G. H. Aplet, N. Johnson, J. T. Olson, & V. A. Sample (Eds.), *Defining sustainable forestry* (pp. 17–43). Washington, DC: Island Press.

Oakhurst Dairy News. (2008, August 13). Five years after Monsanto ruling, Oakhurst dairy feels vindicated. Retrieved from http://www.oakhurstdairy.com/about/release.php?nID=1133

Orr, D. W. (1992). *Ecological literacy: Education and the transition to a postmodern world.* Albany, GA: State University of New York Press.

Pacini-Ketchabaw, V. (2013). Frictions in forest pedagogies: Common worlds in settler colonial spaces. *Global Studies of Childhood, 3*(4), 355–365.

Place. (2018, June). *Oxford English dictionary online.* Retrieved from http://www.oed.com

Plumwood, V. (1993). *Feminism and the mastery of nature.* London: Routledge.

Progressive Reactionary. (2009, January 3). Towards a critique of sustainability [web log comment]. Retrieved from http://progressivereactionary.blogspot.com/2009/01/towards-critique-of-sustainability.html

Raymond, C. M., Brown, G., & Weber, D. (2010). The measurement of place attachment: Personal, community, and environmental connections. *Journal of Environmental Psychology, 30*(4), 422–434.

Rees, W. (2010). What's blocking sustainability? Human nature, cognition, and denial. *Sustainability: Science, Practice and Policy, 6*(2), 13–25.

Rollin, B. E. (2017). *A new basis for animal ethics: Telos and common sense.* Columbia, MO: University of Missouri Press.

Salafsky, N., Margoluis, R., Redford, K. H., & Robinson, J. G. (2002). Improving the practice of conservation: A conceptual framework and research agenda for conservation science. *Conservation Biology, 16*(6), 1469–1479.

Saunders, C. E. (2003). The emerging field of conservation psychology. *Human Ecology Review, 10*(2), 137–149.

Schultz, P. W. (2002). Inclusion with nature: The psychology of human-nature relations. In P. Schmuck & P. W. Schultz (Eds.), *Psychology of sustainable development* (pp. 61–78). Norwell, MA: Kluwer Academic Publishers.

Shrivastava, P. (1995). Ecocentric management for a risk society. *Academy of Management Review, 20*(1), 118–137.

Siebenhüner, B. (2000). Homo sustinens—Towards a new conception of humans for the science of sustainability. *Ecological Economics, 32*(1), 15–25.

Singer, T., Seymour, B., O'Doherty, J. P., Stephan, K. E., Dolan, R. J., & Frith, C. D. (2006). Empathic neural responses are modulated by the perceived fairness of others. *Nature, 439*(7075), 466–469.

Snyder, G. (1990). *The practice of the wild: Essays.* San Francisco: North Point Press.

Solow, R. M. (1993). Sustainability: An economist's perspective. *Economics of the Environment: Selected Readings, 3*, 179–187.

Spehr, C. (1999). A more effective industrialism: A critique of the ideology of sustainability. Swordfish: Spirit of the Times with Fishbones - Political Perspectives between Ecology and Autonomy, 1(1). Retrieved from http://www.cddc.vt.edu/digitalfordism/fordism_materials/spehr.htm.

Taylor, A., & Giugni, M. (2012). Common worlds: Reconceptualising inclusion in early childhood communities. *Contemporary Issues in Early Childhood, 13*(2), 108–119.

Tuan, Y. (1977). *Space and place: The perspective of experience.* Minneapolis, MN: University of Minnesota Press.

Turner, J. (1989). The abstract wild. *Witness, 3*(4), 81–95.

Twine, R. (2010). *Animals as biotechnology: Ethics, sustainability and critical animal studies.* London: Earthscan.

Twine, R. (2012). Revealing the animal-industrial complex: A concept and method for critical animal studies. *Journal for Critical Animal Studies, 10*(1), 12–39.

UNESCO-UNEP. (1995). *Re-orienting environmental education for sustainable development: Final report for the UNESCO-UNEP inter-regional workshop on environmental education.* Paris: UNESCO.

USEPA. (2015, September 21). *Sustainability.* Retrieved from http://www2.epa.gov/sustainability

Vedeld, P. O. (1994). The environment and interdisciplinarity ecological and neoclassical economical approaches to the use of natural resources. *Ecological Economics, 10*(1), 1–13.

Violence. (2018, June). *Oxford English dictionary online.* Retrieved from http://www.oed.com

Warneken, F., & Tomasello, M. (2006). Altruistic helping in human infants and young chimpanzees. *Science (New York, N.Y.), 311*(5765), 1301–1303.

Warren, K. J. (1990). The power and the promise of ecological feminism. *Environmental Ethics, 12*(2), 125–146.

Washington, H. (2013). *Human dependence on nature: How to help solve the environmental crisis.* London: Routledge.

Weisberg, Z. (2009). The broken promises of monsters: Haraway, animals and the humanist legacy. *Journal for Critical Animal Studies, 7*(2), 22–62.

Weitzenfeld, A., & Joy, M. (2014). An overview of anthropocentrism, humanism, and speciesism in critical animal theory. *Counterpoints, 448*, 3–27.

Whatmore, S. (1997). Dissecting the autonomous self: Hybrid cartographies for a relational ethics. *Environment and Planning D: Society and Space, 15*(1), 37–53.

Whatmore, S., & Thorne, L. (2000). Elephants on the move: Spatial formations of wildlife exchange. *Environment and Planning D: Society and Space, 18*(2), 185–203.

Wheeler, S. M. (2013). *Planning for sustainability: Creating livable, equitable and ecological communities.* London: Routledge.

Wilkinson, C. F. (1990). Crossing the next meridian: Sustaining the lands, waters, and human spirit in the west. *Environment: Science and Policy for Sustainable Development, 32*(10), 14–34.

Williams, D. R., Patterson, M. E., Roggenbuck, J. W., & Watson, A. E. (1992). Beyond the commodity metaphor: Examining emotional and symbolic attachment to place. *Leisure Sciences, 14*(1), 29–46.

Williams, D. R., & Vaske, J. J. (2003). The measurement of place attachment: Validity and generalizability of a psychometric approach. *Forest Science, 49*(6), 830–840.

Wu, J. (2013). Landscape sustainability science: ecosystem services and human well-being in changing landscapes. *Landscape Ecology, 28*(6), 999–1023.

Part III
Human-Nature Relational Ontology

Thus far this book has focused on centering possibilities for human-nature relationships in ontology, articulating a sensible structure and dynamics of human-nature relationships if closeness and intimacy are thought possible, and examining the ways in which conceptual dualisms operate to affect our understandings of the limitations of those possibilities. The chapters that remain contain a conceptual exploration of close human-nature relationships using the interdependence model I articulated in Chap. 4. I also articulate and explore a human-nature relational ontological orientation that includes material and more-than-material elements, and in so doing, open up the possibility not only for more-than-human beings to be relational Selves, but for humans to develop close relationships with them.

Chapter 6
Foundations of Human-Nature Relational Ontology

This chapter explores some of the key ontological elements of the relational reality that I believe underpin the close human-nature relationships for which I advocate in this book. Because of my pragmatist approach, I will not conceptually synthesize these elements as part of an *a priori* grouping of theory, but instead will distill them *a posteriori* from various experiences that I believe to illustrate the foundationality of their features. Toward that end, I begin by offering a story of an encounter I had with a horse that will serve as a touchstone of experience from which I draw these features throughout the remainder of the book.

6.1 A Horse in Colorado

Once, when strolling about my friends' farm in Paonia, Colorado, I chanced upon a neighbor's horse pasture. It was separated from my friends' fields by a low, non-electrified fence, and as I climbed the berm that made me visible to the pasture's occupant—an iron-black stallion—I saw him at once set off at a gallop from the far end of the field. He moved quickly as I watched, first along the fenceline at the other end and then, arcing toward me, down the length of the field's northern border until, after approximately 30 seconds, he pulled up and was standing, sparkling like the night sky, just 20 feet in front of me. At that moment, as he snorted and pawed the earth on his side of the fence with more than a little ferocity, his message was crystal clear: This was not only his pasture and his trifling fence, but it was his whole goddamn universe. A universe in which I, at that moment, was most unwelcome. I could simply *feel it* coming off of him. His intent was clear, and so was mine (I hoped) as I calmly apologized and, turning my back with the hope he wouldn't jump the low fence and pursue me, slowly descended and disappeared into the thicket of corn through which I'd come.

© Springer Nature Switzerland AG 2019
N. H. Kessler, *Ontology and Closeness in Human-Nature Relationships*,
AESS Interdisciplinary Environmental Studies and Sciences Series,
https://doi.org/10.1007/978-3-319-99274-7_6

6.2 Relations as Ontologically Basic

I take as the most basic feature of my experience with the horse that relations between the horse and me exist. The horse influences me, I influence the horse, and it is impossible for this not to be so. The nature of those influences, and the impact that the qualities and capacities of each of us has upon them, is a matter of no small debate, as it forms the main line of inquiry in this book. But, I begin by suggesting that, ontologically speaking, whatever the nature of our relations, at root they irreducibly exist. In this sense, I take relations to be ontologically basic. Posthumanism would have no quarrel with this, though it might warn me off of any essentialist take on what qualities and capacities "I," as the human, possess. Likewise for "the horse." But, as I discussed in Chap. 3, while one should acknowledge the posthumanist reality of the indeterminacy of the boundaries of our being-ness in a dynamically unfolding relational reality, this is no cause for denying the simple, charged reality of the horse and me, inextricably embedded in our relational matrix, but there as Selves under the hot Colorado sun. While I understand Barad's (2007) anticipated counterpoint that I as a Self and the horse as a Self are manifestations of precedent and causally influential relations, as I also pointed out in Chap. 3, so too can be found any multitude of Selves causally influencing the formation of the relations in which the two of us now find our Selves. If, behind that, Barad wants to find yet more relations, then one can see where this ends. We're back to our paradigmatic chicken and egg scenario which I'll discuss in a bit more detail in the next section. To sum it up here, without Selves relations aren't relations. Instead, reality becomes a homogenous landscape stretching in all directions that Barad also rejects when she speaks of it in the context of whether separateness or difference are real. To this point she says that her ontology "does not take separateness to be an inherent feature of how the world is. But neither does it denigrate separateness as mere illusion… Difference cannot be taken for granted; it matters—indeed, it is what matters: (p. 136). Chap. 3 contains more discussion of the "relationship" between relations and *relata*.

One other point worth reinforcing here is that when I speak of relations in this chapter, I do so using my definition of the term from Chap. 1, and that I also discuss in depth in my consideration of posthumanist materialism in Chap. 3. To also reiterate here, relations are not the mere juxtaposition and/or dynamic interchange of the micromaterial. Instead, they are relations in the substantive, personal and ecofeminist sense, where both Selves and the relational matrix within which they are inextricably embedded are of coequal value and ontologically irreducible to one another or to any constituent parts.

The notion that reality overall is foundationally relational is reflected in the root of words like "existence," the etymology of which shows that it is formed from *ex-*, "forth" and *sistere*, "to stand out, be perceptible" ("Exist", 2018). Inherent in the meaning of the word, then, is the notion of standing forth, or being differentiable *from something else*, and thus inherently in some type of relation with that other thing. I pause to note that my use of the terms *relation* or *relational* in this context is intended to convey the reality of *interchange*, or *connection*, with no specific further qualities to these relations as of yet; they can currently, equally represent the relational sparseness of purely physical proximity or the complex interconnection of the closest of friends.

Many philosophical perspectives support a view of reality as ontologically relational. There is, of course, the aforementioned posthumanism. There is also American pragmatism, which I noted in Chap. 1 as one of the ontological lenses through which I would view the possibilities for closeness in human-nature relationships. William James (1912), an influential early pragmatist, cleaves to a relational ontology in his notion of "radical empiricism." He says that "pure experience" is the "primal…material in the world" and that "relation itself is a part of [that] pure experience" (p. 4). As to what types of relations are possible, James sees them as having "different degrees of intimacy" (p. 44) progressing from most distant or co-incidental to most intimately entwined. To "designate types of conjunctive relation arranged in a roughly ascending order of intimacy and inclusiveness" (p. 45) he suggests this list of terms: "With, near, next, like, from, towards, against, because, for, through, my" (p. 45).

Some have compared the relational orientation of pragmatism with that of existential philosopher Martin Buber (Lothstein, 1996; Pfuetze, 1967). Buber (1923/1970) believes that humans carry an innate "longing for relationship" (p. 77) that comports with a relational ontology. Buber says that this longing shows itself in human babies where, "Before any particulars can be perceived, dull glances push into the unclear space toward the indefinite…and…soft projections of the hands reach, aimlessly to all appearances, into the empty air toward the indefinite" (p. 77). To bolster his contention that reality is relational, Buber asserts that "[i]t is not as if a child first saw an object and then entered into some relationship with that. Rather, the longing for relation is primary" (p. 78).

Speaking specifically to ontology, Buber further suggests that "[i]n the beginning is the relation—as the category of being, as readiness, as a form that reaches out to be filled, as a model of the soul; the *a priori* of relation" (p. 78). To get to this ontological notion of relation, he first suggests that whenever a human being says the word "I" that she actually also instantiates the "other" that stands across from her either as a separate object (an "It") or as a connected Self. That Self is a "You" or a "Thou", depending on the translation of Buber. Therefore, when a person says "I," she is really saying one of two basic word pairs "I-It" or "I-Thou." While some may suggest that the youngest of children do not have a sense of relation *qua* relation—they have no way of distinguishing self from other and thus have any capacity to long for relations as such—Buber suggests otherwise. He says that even at the "earliest and most confined stage" of development, a child has the "primal nature of the effort to establish relation" (Smith translation, 1923/1958, p. 25). He explains by saying,

> Before anything isolated can be perceived, timid glances move out into indistinct space, toward something indefinite…You may…call this animal action, but it is not thereby comprehended…these acts [are not experiences] of an object, but [are] the correspondence of the child—to be sure 'fanciful'—with what is alive and effective over against him. (This 'fancy' does not in the least involve, however, a 'giving of life to the universe': it is the

instinct to make everything into *Thou*, to give relation to the universe, the instinct which completes out of its own richness the living effective action when a mere copy or symbol of it is given in what is over against him. (Smith translation, 1923/1958, p. 25)

To push beyond Buber's suggestion that the infant, while not beginning life in relation, begins life with the predisposition for it, care ethicist Noddings (1984) says that "the [human] infant, even the near-natal fetus, is capable of relation—of the sweetest and most unselfconscious reciprocity" (p. 89). She follows this with something essential as it pertains to my claim that closeness, and its companion caring, are possible in human-nature relationships. In the case of care with the human child, not only is the "child's capability to respond" essential, but so is the adult human's encounter with the child that produces an obligation on the part of the adult to the child. As it pertains to my point here, if the child were instead a more-than-human being in whom also resides the ability to reciprocate, then our encounter with him would also invoke in us an obligation to care—to not "ignore the child's [or more-than-human's] cry to live" (p. 89). If this ability is true, then we would at least have a foundation for care, and thus closeness. While the inverse case, that of a more-than-human being caring for a human may seem a bit more unusual to consider, examples of actions on the part of more-than-human beings to help humans in distress are not all that unusual (see "Animal Feelings" section in Chap. 9). Were such particular interactions to be sustained over a long period, there is nothing, at least ontologically speaking, precluding care and closeness from developing. In a yet more extreme case—that of a plant's ability to reciprocate or care—before dismissing such a notion, I'll point out that there has been no investigation whatever into whether such a thing has occurred, nor into the plant's ability to achieve her part in it. I'll also point out that thinking such an investigation frivolous is almost wholly sponsored by the dualisms that pervade modern discourse on the nature and capacities of plants and not in any evidence supporting such a view. I'll discuss this at length in the Feelings and Thoughts chapters below.

To return to Noddings, she states that "relations [should be] taken as ontologically basic" (p. 3), and roots this position in seeing "human encounter and affective response as a basic fact of human existence" (p. 4). While human/more-than-human encounter is also an unavoidable fact, whether that encounter can form the basis of closeness as Noddings' frames the dynamic, is subject to two constraints. The first is whether humans can have an affective response to more-than-human beings. Noddings thinks that some affective response occurs on the part of humans in response to more-than-human beings. She suggests that this response depends mostly on what kind of more-than-human being it is, and on the human's socialization around more-than-human beings. But, to her this is not enough to consider it universal, which is what she believes is needed to make such an arrangement ontological...ontological at least in the sense that it provides a universal mechanism for the development of care as it does between human beings (p. 149). But, even if human socialization works against human affective response with more-than-human beings, this is no basis to ontologically dismiss more-than-human beings as potential close relational partners. In addition, I suggest that a human not having a strong affective response to a particular more-than-human being or set of more-than-human beings does not imply that the human *cannot* have an affective response to *any* more-than-human being. To wit: it's possible for a human being to have *no* affective response to a certain other human being. This certainly does not preclude her from developing close relationships, or from caring about, humans other than that particular one for whom she does not have a caring response. The second constraint is whether there is an affective response on the part of more-than-human beings to humans. I suggest that this is necessary to forming closeness in my close human-nature relationship model in the Chap. 4, and explore the possibility that more-than-human beings do have feelings that can be shared reciprocally with humans in the Feelings chapter below.

To return now to Buber (1923/1970), one weakness in using the relational elements of his ontology to support my contention that relations are ontologically basic is that he does not position the relations themselves as ontologically basic, only the readiness for them. To respond to this, however, I'd suggest that his thinking is influenced in no small part by a *substantivism* or *individualism* that incorrectly positions the individual as existing prior to the relations, and that posthumanists like Barad (2007) have rightly critiqued. Even though the critique of substantivism has been robust in the posthumanist literature, I will undertake an exploration of substantivism here, as well, as it pertains to my treatment of relations in the ecofeminist sense that I differentiate from posthuman relations above. In this ecofeminist relational context, and as I've already noted, Noddings (1984) thinks that even the youngest of human beings are reciprocally relational beings. Yet substantivism positions relations as subordinate to, and wholly contained within, fundamentally independent entities or "substances." Again, while posthumanist arguments against substantivist positions are robust, because I do not adopt the position posthumanists like Barad (2007) contrapose—that of subordinating *relata* to their relations—I will explore here the possibilities of relevant *relata and* relations, each as contemporaneous with, and mutually generative of, the other.

With regard to substantivism and referring to my example of the horse encounter above, in one sense it may seem as if the existence of the horse and I precede any relations between us. Articulated robustly by Aristotle, substantivism positions the horse and I first as "unique in being independent things" (Cohen, 2012, "The Categories" section, para. 2) who only subsequently enter into our relations. Cohen says that Aristotle actually divided beings into a total of ten categories that include, along with the primary "substance," "quality, quantity, and relation, among others" ("The Categories" section, para. 1). Of Aristotle's view on primary substances, Cohen explains this by saying,

> [T]he items in the other categories all depend somehow on substances. That is, qualities are the qualities of substances; quantities are the amounts and sizes that substances come in; relations are the way substances stand to one another. These various non-substances all owe their existence to substances—each of them, as Aristotle puts it, exists only 'in' a subject. That is, each non-substance "is in something, not as a part, and cannot exist separately from what it is in" (*Cat.* 1a25). Indeed, it becomes clear that substances are the subjects that these ontologically dependent non-substances are 'in.' ("The Categories" section, para. 2)

Figure 6.1 depicts Aristotle's arrangement, specifically between substance and relations. In my estimation, however, there are two flaws in this substantivist view. First, just because a thing cannot exist separately from another thing, it does not follow that one is contained within the other or that the dependence of one upon the other is not reciprocated—that is, that the other is not dependent in equal measure upon the one originally dependent. The idea of complementarity presupposes *mutual* dependence, for instance in the organism lichen, which is the product of two beings—fungus and algae—that complementarily exist. Neither would exist without the other yet neither is a product of, or contained within, the other without the inverse also being true. The second flaw is that the substance Aristotle positions as containing relations did not come to exist, (and does not have the opportunity to continue to exist) without a series of past and currently occurring relations. In a view that accounts for these relations as existential and ontologically basic, the horse and I, while being differentiable, are never truly "real" outside the context of our past and present relations. Again, posthumanist arguments also suggest this, but in my treatment I do not subordinate the individual to the relation, the latter being construed by Barad (2007) and other posthumanists as the only "ontological primitive" (p. 429). Instead, I acknowledge the individual's causal influence as an individual-in-relation over the ongoing and future relations. In this view, the spatial and/or temporal scale of Fig. 6.1 needs to be expanded.

Figure 6.2 portrays this expansion, with substances 1 and 2 as part of a larger relationship, such as a human-nature relationship. Granted, one could always expand the scale again to create yet another substance inside of which the relationship that subsumes substances 1 and 2 is located (such as an ecosystem). But, at some point such expansions of substantive and relational scales becomes the example of the chicken and the egg, as I discussed above when countering Barad's claim that relations are ontologically more basic than individuals.

Further, I take relations to not only be material in origin. This is another divergence from posthumanist relationality. Putting aside posthumanism's active matter for the moment and only dealing with possibilities that either distinct substances or potentially more-than-material relations are primary and precedent, I suggest that the only way the origin of relations is in substance—perhaps back at the Big Bang— is if a materialist ontology is assumed. This is also true of Barad's agential matter in the sense that my more-than-material relational structures and processes stand as

Fig. 6.1 Aristotle's Substance as ontologically independent, with relations that depend upon it for existence

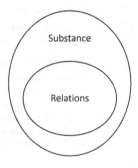

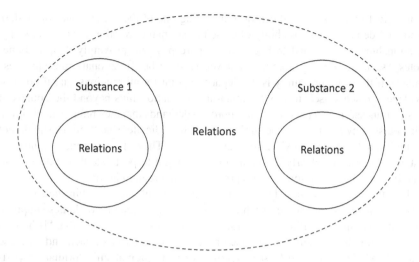

Fig. 6.2 Relations with superseding ontology, and substances that depend upon them for their existence

counterpoint to Barad's lively matter. The only way that it is matter that is agential is if one assumes a materialist ontology. What if the Big Bang wasn't a purely *material* event? What if love as an existential, more-than-material thing in the cosmos instantiated two things and, between them, sparks flew? Ultimately, the debate over whether reality is passive substance, lively matter, or more-than-materially relational in some way has been going on for millennia. My exploration here is not intended to resolve it. My purpose is only to show that, when materialist dualisms are removed from the ontological frame, it is equally plausible that more-than-material relational forces create substances, or matter. Even Barad's lively matter. Substantivism, or materialism, is only more plausible or acceptable as an origination point if a dualist ontology empowering the material is already in place.

6.3   Characteristics of Ontologically Basic Relations

In taking my kinds of relations to be ontologically basic, the characteristics of those relations must be discussed. If James (1912) is correct, those characteristics cover a wide range. A cursory reading of James would indicate that he takes the most distant relations, that of his "with" to be the most ontologically basic. This is evidenced by his referral to the "with" end of the spectrum as a "bare relation of *withness*" (p. 47), or when he suggests that "*A priori*, we can imagine a universe of withness but no nextness; or one of nextness but no likeness…" (p. 45) and so on. The latter quotation shows how intimacy appears to be constructed from the more basic "building blocks" of relations, also implied when he says that "there appear to be actual forces at work which tend, as time goes on, to make the unity greater" (p. 47). I note that this ontological view has figured centrally in setting one of the cornerstones of modern theorists' denial of the possibility of close human-nature relationships. Specifically, relationships are assumed to begin in the pure physical proximity that, as James notes, "seems to involve nothing whatever as to farther [relational] consequences" (p. 44). If this pure proximity is the departure point for all relations, and if no clear evidence presents itself to move human-nature relationships beyond the bounds of this "withness," then a parsimonious approach demands that they remain at this level. This parsimony is a form of Occam's Razor, which dictates that "entities should not be multiplied unnecessarily" (Smith, 2010, p. 12). Thus, if no closeness in human-nature relationships clearly exists, no theories ought to postulate that it does exist, and the more basic relations of pure proximity must remain in force.

But, there are unexamined ontological conceptual elements within such a stance that call for refutation, some of which a more nuanced reading of James supports. First, in response to calling out his range of relations, James notes that "[t]he universe of human experience is, by one or another of its parts, of each and all these grades" (p. 45). So here at least he recognizes that pragmatism's "primary experience," our source for knowledge and truth, can itself show all manner of connection from least to most intimate. Also, in contrast to his "bare relation of witness" being overemphasized by "ordinary empiricism" and "unduly ignored" by rationalism, he says that his "radical empiricism…is fair to both the unity and the disconnection. It finds no reason for treating either as illusory" (p. 47). This point speaks directly to what I take to be the misapplication of Occam's razor. That is, if *in experience*, intimacy or unity is perceived, then it is not a multiplication of entities to identify it as such.

But, we still have James' suggestion that "withness" is a building block of "nextness," etc., that at least implies that there is no intimacy without the origin of pure proximity. This means that we still have to contend with the notion that within intimacy there is supposed, always, the foundation of pure physical encounter such that whatever develops relationally between the horse and me, our existences begin apart. But, such a view has two dualistic hallmarks of note.

First, such a view strongly resembles substantivism given that the overarching theme is that two things begin apart and come together, growing in their togetherness as time passes. Thus, I suggest that within James' and others' notion that relations begin in pure proximity is an unseen "relational" element even less connected than "with"—that of "without." Even if James suggests that connection is ontologically basic, by starting at pure physical proximity, the assumption is that "without" either precedes it, or stands as the *conceptual*, asymptotic baseline against which all more complex or intimate connections are measured. In a substantivist or individualistic ontology, if the encounter itself begins in pure proximity, then this must have been preceded by a lack of it. The horse and I encounter each other, but before that, we were not together. I suppose one could play at "six degrees of separation" or the "butterfly effect" and relationally see that *some* connection always exists between us through our material intermediaries, but "withoutness" (or as asymptotically close as one can get in a materially interconnected world) is still the starting place. Posthumanists would also have us inhabit a foundationally connected reality, but because of their materialist ontological tendencies, pure, material interconnection is still the assumed foundation of those relations. I doubt that such posthumanist theorists would see this inception point as the logical, if not actual, extension of the atomism they rightly critique, yet that's precisely what it is.

Pure material proximity and its logical predecessor of separateness also puts some of James' statements at odds with each other. While he might interpret close relationships in primary experience as composed of withness, then nextness, and so on, there is nothing in the experiences themselves preventing a reversal of that progression such that one can imagine a universe of closeness without precedent nextness, and then nextness without precedent withness, etc. There can be unity first, with less connected forms of relations being *post hoc* projected onto them from a dualistically individualizing ontology.

Certainly, one might counter this by saying that the very notion of an *inception* point for relations means there must be a precedent separateness, or lack of relations. But, this still does not imply that the togetherness, at its inception, is inherently *relationally* distant. The phenomenon of closeness sometimes occurs instantaneously. For example, Ortigue, Bianchi-Demicheli, Patel, Frum, and Lewis (2010) discovered that falling in love can take approximately one-fifth of a second. Noddings (1984) describes "feeling with" (p. 30) another human being as having a quality of immediacy in its formation. As an example of this "feeling with," she speaks of having lunch with a group of people where, at some point one person for whom she did not have feelings of care suddenly opens up emotionally and through that she is "touched—not only by sentiment—but by something else. It is as though his eyes and mine have combined to look at the scene he describes…I feel what he says he felt. I have been invaded by this other" (p. 31). Here, closeness occurs spontaneously, and in a relationship that, if anything, worked *against* closeness prior to the events described. Therefore, I suggest that in James' relational spectrum, ontologically basic relations are as likely to start with closeness as they are with distance. I acknowledge that in a world of modern thought ruled by dualisms, such a thing is hard to accept. In such a worldview, I can imagine the one-fifth of a second to fall in love being portrayed as a very rapid progression through James' spectrum at least at a material and perhaps unconscious level. But, if one dispenses with such dualistically sponsored assumptions, it's possible to see that closeness may not be an agglutinative process but one immediately instantiated through such a thing as shared feeling in the Peircian (1960) sense that I'll discuss in the next chapter. Suffice it to say here that Peirce advocated for feelings suffusing those experiencing them such that they were sharing the *same existential feeling*. In the context of the horse and me, the feeling of territoriality I described is coming from the horse to me and I *perceive* it directly via the matrix of the existential encounter. I am invaded by the horse's feelings—we share them and have an instant closeness (or its relational antipode), or at least the grounds for one. Of course one does not have to read my encounter with the horse this way, my point here is that there is nothing *ontological* preventing such a reading from being valid.

Therefore, I conclude that while James is right that there is a wide range of possible ontological relations, I suggest that they can take on any of the forms that

James notices occurring in experience itself. Thus, *a priori* denial of the possibility of closeness between humans and more-than-human beings becomes a *post hoc* imposition at odds with the immediate and direct closeness of some human-nature relational experiences. This, in turn, violates the tenets of Occam's Razor by multiplying and unnecessarily adding the conceptual entity of relational *distance*.

That relational distance may be a *post hoc* conceptual element rooted in dualisms is supported by the explanation Rollin (1990) gives for the abatement of serious scientific inquiry into animal consciousness that occurred around the turn of the 20th century. He notes that most scientists attribute significant shifts in scientific thinking to either the discovery of new data refuting past claims or the discovery of logical flaws in those claims. But, he notes, in the case of animal consciousness neither of these occurred. Thus, he suggests, there is a third way that scientific thinking can shift, and that is through the *"rise of new values which usher in new philosophical commitments or new basic assumptions* [emphasis original]" (p. 380). The loss of scientific acceptance of animal consciousness, he says, is the result of this third element.

He goes on to explain that at the time this rejection of animal consciousness occurred, there was a shift across modern societies to a set of "reductive" social values, and with it a desire for "simplification and paring away of frills" (p. 382). Because, according to him, animal consciousness was falsely positioned by those criticizing it as unverifiable, it became a frill to be pared away. I suggest that something very like this occurred concomitantly in ontological theories of human-nature relations, where the more basic theories were the ones that comported best with an assumed substantivist or materialist ontology and any more complex, intimate, or "frilly" relations were discarded. Posthumanism does nothing to remedy this reductive trajectory because it accepts as foundational the same materialism responsible for igniting and/or underwriting the reduction. The addition of lively matter or vital elements does nothing to reverse this. It only serves to reinforce the purely material as the starting point for subsequent, more intimate or close relationships.

But, when reality reveals relational closeness, relational inception *qua* relational distance is revealed as the frill *it* is—as the *post hoc* conceptualized entity that has been unnecessarily multiplied or superadded. Thus, regardless of the hegemonically produced veneer of reality that has been laid over top of this distance-containing interpretation of primary experience, it ought to be treated with an amount of suspicion equal to that to which close human-nature relational experiences are subjected by dualist modern thought.

6.4 Human-Nature Relations as Ontologically Basic

So far I have argued for ontologically basic relations in general terms. Here, I'll extend the argument to include human-nature relations as ontologically basic. Pragmatism is one school of thought that offers a path to a human-nature relational ontology. For example, in discussing pragmatist John Dewey's perspective on human relations to the world, McDonald (2004) suggests that, to Dewey, "Humans are in and of nature…[and that b]eing in nature…involves a system of relations…to other natural events or processes" (pp. 68–69). Fesmire (2004), in speaking of one of pragmatism's founders, Charles Peirce, says, "as with Peirce's doctrine of synechism [i.e., continuity], there are no ontological barriers to continuity between human and other forms of life" (p. 58).

In addition to pragmatism, ecofeminism has much to offer the possibility of a relational ontology. For example, Warren (1990) suggests that

> [h]umans are who we are in large part by virtue of the historical and social contexts and the relationships we are in, including our relationships with nonhuman nature…Relationships of humans to the nonhuman environment are, in part, constitutive of what it is to be a human. (p. 143)

Whatmore (1997) says of "a relational understanding of political and moral agency…[that it] centres on a recognition of…environmental embeddedness" (p. 37–38). Plumwood (1991) also comments on this, for example, when articulating a self-in-relationship with more-than-human nature by saying that such a self "clearly recognizes the distinctness of nature but also our relationship and continuity with it…[a] self…embedded in a network of essential relationships with distinct others" (p. 20).

I note that in most of these discussions, however, there is little description of the particular qualities of that relationship or the various capacities of its participants to engage in them. Thus, how humans and more-than-human beings respond to each other (i.e., what qualities and capacities they employ in those responses) is either unknown or a matter of vigorous debate. At a material level, it is uncontroversial to hold that human-nature relations are ontologically basic. But, since I contend that human-nature relationships can be close just like interhuman ones, this book will press forward in exploring the possibility of humans and more-than-human beings having ontologically irreducible material and more-than-material qualities that contribute to closeness between them.

References

Barad, K. (2007). *Meeting the universe halfway: Quantum physics and the entanglement of matter and meaning*. Durham, NC: Duke University Press.

Buber, M. (1958). *I and thou*. (trans: Smith, R. G.). New York: Charles Scribner's Sons. (Original work published 1923).

Buber, M. (1970). *I and thou*. (trans: Kaufmann, W.). New York: Charles Scribner's Sons. (Original work published 1923).

Cohen, S. M. (2012). Aristotle's metaphysics. *The stanford encyclopedia of philosophy* (Fall 2012 Ed.). Retrieved from http://plato.stanford.edu/entries/aristotles-metaphysics

Exist. (2018, June). *Oxford English dictionary online*. Retrieved from http://www.oed.com

Fesmire, S. (2004). Dewey and animal ethics. In E. McKenna & A. Light (Eds.), *Animal pragmatism: Rethinking human-nonhuman relationships* (pp. 43–62). Bloomington, IN: Indiana University Press.

James, W. (1912). *Essays in Radical Empiricism.* New York: Longman, Green, and Co..

Lothstein, A. S. (1996). To be is to be relational: Martin Buber and John Dewey. In M. S. Friedman (Ed.), *Martin Buber and the human sciences* (pp. 33–50). Albany, NY: State University of New York Press.

McDonald, H. P. (2004). *John Dewey and environmental philosophy.* Albany, NY: State University of New York Press.

Noddings, N. (1984). *Caring: A feminine approach to ethics & moral education.* Berkeley, CA: University of California Press.

Ortigue, S., Bianchi-Demicheli, F., Patel, N., Frum, C., & Lewis, J. W. (2010). Neuroimaging of love: FMRI Meta-Analysis evidence toward new perspectives in sexual medicine. *The Journal of Sexual Medicine, 7*(11), 3541–3552.

Peirce, C. S. (1960). *Collected papers of Charles Sanders Peirce* (C. Hartshorne, P. Weiss, & A. W. Burks, Eds.). Cambridge, MA: Harvard University Press.

Pfuetze, P. E. (1967). Martin Buber and American pragmatism. In P. A. Schilpp & M. S. Friedman (Eds.), *The philosophy of Martin Buber* (pp. 511–542). London: Cambridge University Press.

Plumwood, V. (1991). Nature, self, and gender: Feminism, environmental philosophy, and the critique of rationalism. *Hypatia, 6*(1), 3–27.

Rollin, B. E. (1990). How the animals lost their minds: Animal mentation and scientific ideology. In M. Bekoff & D. Jamieson (Eds.), *Interpretation and explanation in the study of animal behavior* (pp. 375–393). Boulder, CO: Westview Press.

Smith, C. (2010). *What is a person?: Rethinking humanity, social life, and the moral good from the person up.* Chicago: University of Chicago Press.

Warren, K. J. (1990). The power and the promise of ecological feminism. *Environmental Ethics, 12*(2), 125–146.

Whatmore, S. (1997). Dissecting the autonomous self: Hybrid cartographies for a relational ethics. *Environment and Planning D: Society and Space, 15*(1), 37–53.

Chapter 7
Relational Perception and Knowledge

7.1 Perceiving Closeness in Human-Nature Relationships

In my discussion of James' range of relations in the previous chapter, I argue that elements of closeness in human-nature interchange may be perceived in primary experience. If that's the case, a major impediment to this perception is being able to detect which close relational elements are real and which are only *post hoc* conceived of by the human participant. In response to this problem, Dewey (1929) recommends treating all perceptual data as "undefinable and indescribable" (p. 56), and only allowing for *post hoc* conceptualization to determine their knowability. His justification for this position begins with a critique of Cartesian dualism, where he says, "some of the most cherished metaphysical distinctions... [are really just] learned counterparts" (p. 56). What he means by this is that while distinctions can be made in primary experience, treating them as metaphysically (or ontologically) real is incorrect. I agree with him that these counterparts are complementary and interdependent, forming a dyadic whole that cannot, in reality, be separated. By suggesting that these dyads are "learned," it's possible that Dewey means that they are *post hoc* conceptual overlays of a reality in which these elements carry no such ability to impress upon the experiencer that which distinguishes them from other elements of reality—and have what is impressed come through as an existential thing. But, I contend that one can equally perceive a thing such as love as love, as well as come to understand it as such through the process of *post hoc* reflection. Dewey might argue that one cannot know love *as love* without this *post hoc* processing. But, I suggest that such an objection is based on the belief that all knowledge is only possible as the result of some conceptual processing instead of knowledge itself being either an element of the experience, or an instantaneous response to the experience that is not cumulative or reflection-based, but perception-based.

   I explore *Peircian feelings* later in this chapter as one mechanism by which things can come to be both communicated and known immediately through experience.

© Springer Nature Switzerland AG 2019   195
N. H. Kessler, *Ontology and Closeness in Human-Nature Relationships*,
AESS Interdisciplinary Environmental Studies and Sciences Series,
https://doi.org/10.1007/978-3-319-99274-7_7

Another avenue I explore below is *poetic knowledge*, where knowledge is sensory-emotional, and comes through directly from the experience. Nobel Prize winning geneticist Barbara McClintock offers an excellent example of this kind of knowledge. When asked how she came to have knowledge of certain things about the genetics of corn, she responded by saying,

> *Why* do you know? Why were you so sure of something when you couldn't tell anyone else? You weren't sure in a boastful way; you were sure in what I call a completely internal way... you work with so-called scientific methods to put [that knowing] into their frame [but only] after you know. (Keller, 1983, p. 203)

Here we have McClintock suggesting that the genesis of some knowledge is not in the *post hoc* processing of it into more scientifically acceptable, intellectual frames of reference. The knowledge comes *before* that and is immediate—it comes through from the experience of relations with the corn plants themselves. If this is possible, then the distinguishable attributes of things in our experience may either be inherent in the things themselves or in the matrix of one's relations with them. Thus, they would be perceivable and knowable as a part of experience. Peirce reinforces such a perspective when he says that, since "feelings are communicated to the nerves by continuity...there must be something like them in the excitants themselves" (Peirce, 1960, 6.158, p. 111). In this context, Barad's (2007) posthumanist "agential realist" ontology also lends credence to my point when she says,

> In traditional humanist accounts, intelligibility requires an intellective agent (that to which something is intelligible), and intellection is framed as a specifically human capacity. But in my agential realist account, intelligibility is an ontological performance of the world in its ongoing articulation. It is not a human-dependent characteristic but a feature of the world in its differential becoming. The world articulates itself differently. (p. 149)

One such potentially perceivable thing in experience is relations themselves and their features. For example, relations between the horse and me were *in the experience* beyond any question in my mind. Of course, understanding this encounter as "territorial" instead of friendly could be seen as *post hoc* interpretation. But then, it could just as easily be a Peircian *feeling*—as that's certainly how I experienced it—and an external one communicated to its participants via the matrix of the experience itself (again, see "Peircian Feelings" section below for a more fulsome explanation). If, to counter such a claim, one were to suggest that I wouldn't know that feeling *as territorial* without some past conceptions that I applied to interpretation of the present feeling, I'd suggest that there could be something innate in horses and humans that recognizes and understands this relational element directly and immediately—that is predisposed to *recognize* it happening externally instead of only *conceptualizing* about it internally and after the fact. That's one explanation for why my immediate inclination was to move backward as the horse stood before me even before I had a chance to think.

It's also possible that Dewey's aim is less to suggest that things are inherently indistinguishable, but that they are unknowable in the sense that knowledge is a thing we must support and separate from belief, the latter having no inbuilt mechanism to legitimize itself as real and able to be experienced by others. The case of religious fanatics comes to mind as an example of experiences labeled as true by those having them that most others would dismiss as fiction. To support this interpretation of Dewey (1929), I note him saying, "Things in their immediacy are unknown and unknowable...because knowledge...is a memorandum of conditions of their appearance, concerned...with sequences, coexistences, relations. Immediate things may be pointed to by words, but not described or defined" (p. 86). In such a frame of reference, then, Dewey may be asking that experiences of feelings of territoriality between the horse and me, for example, be subjected to the *post hoc* process of verifying them as *true* in order to distinguish them from only a belief on my part that this is what the horse was feeling. But, there are two problems with this requirement.

First, the mechanics of the verification process have a great influence over what is ultimately taken to be true. In particular, subjecting a phenomenon to modernist truth verification in no way inoculates the subsequently produced "knowledge" against the heavy sway of belief, it only displaces that belief from the main lens under which we place our experiences for verification. For example, we take modern methods of analysis (i.e., objective, detached, and rational) as more accurate than one's immediate sense of feelings or intuition not because they are more accurate, but because we believe them to be. As Rollin (1990) points out, assumptions rooted in belief (which he calls "foundational presuppositions") underpin all "human cognitive enterprises" (p. 378), be they based in feeling or the intellect. As I discussed in relation to Rollin's take on animal consciousness above, it is beliefs rooted in foundational assumptions about the world-as-parsimonious that led to a largely inaccurate dismissal of animal consciousness. Reliance on *post hoc* conceptualizations, then, does not provide us with a better avenue to truth and knowledge, only one more acceptable to the very modern thinking that, I've suggested throughout the book thus far, is rife with the dualisms that marginalize and negate avenues of thinking, belief, and knowledge that don't comport with pre-existing ontological commitments that undergird the modernist view. The shared, Peircian feelings, poetic knowledge and personal acquaintance knowledge that I explore below (and then spiritual experience, intuition and other avenues to knowing as well) are just a few ways by which knowledge may be legitimately gained outside the dualist paradigm.

One last point on perceiving relations directly in experience is that, in perceiving particular kinds of relations in experience, "truth" may be less relevant than the recognition of the simple act of relational *participation*. For example, it may matter less that the horse and I agree to any "fact" that he was being territorial and more that we shared feelings across the liminal, relational space between us. This alone can lead to *genuine* closeness. Of course I don't support closeness of relations based on a sham, but the commitment to the encounter has a power of its own, divorced from any "facts" of the relational encounter. Noddings (1984) says of the encounter between two humans that may end up caring for each other, "Feeling is not *all* that is involved...but it is essentially involved" (p. 32). When falling in love, one does not refer to facts, one just lets oneself go into the relational maelstrom. The act of doing so *is* its own relational truth to be perceived in the experience of it and nowhere else. The sheer willingness of two relational Selves to participate in relations becomes an element in the establishment and quality of the relational ties.

To elaborate on this, I refer to Noddings at length, to whom I attribute the genesis of this point about perceiving relations in experience. In response to the question of how she can *know* that she is actually "receiving the other" in a caring encounter, she says,

> How can I know?...If I respond that I cannot be mistaken in a basic act of receptivity, I fall into the trap that has already snared the phenomenologist when he speaks of the infallibility of basic intuitions. He asserts his position and presents it as right by definition. Surely, I do not want to respond in this way...I am not claiming that I know either in my receptivity itself or in my description of it. It is not at bottom a matter of knowledge but one of feeling and sensitivity...When I receive the other, I am totally with the other. The relation is for the moment exactly as Buber has described it in *I and Thou*. The other "fills the firmament." I do not think the other, and I do not ask myself whether what I am feeling is correct in some way. When I have a sudden, severe pain in my mouth, for example, I may complain of a toothache. I cannot be wrong in responding to what I feel as a pain. It is not a matter of knowledge at all. Later, when the pain has gone and I think back on it, however, I may say, "Well, I guess it was not a toothache after all. It's gone. Perhaps it was bit of neuralgia caused by the cold or altitude." I do not say, "Well, I guess I did not have a pain." Of course I had a pain. My error, if one occurred, lay in assessing the pain as a toothache. Similarly, I may, in looking back, become aware that there was a failure somewhere in my movement from feeling to assessment. But in the receptive mode itself, I am not thinking the other as object. I am not making claims to knowledge. There can be failures to receive...but these are not matters of faulty claims to knowledge...What is offered is not a set of knowledge claims to be tested but an invitation to see things from an alternative perspective... (p. 32)

What Noddings describes is a wholly imperfect, yet relationally incontrovertible interpretation of encounter that leads to care and its ally closeness. In the context of the point I'm making here, whether accurate or not that the horse was feeling exactly territorial, I am not wrong about the fact that a charged exchange occurred, reverberating back and forth between us over that low fence like an arc of electricity—one by which I felt wholly subsumed. We co-responded, the horse and I, and I believe that this is what is necessary to establish the conditions for real closeness had the encounter continued along a certain relational path, or had we had many encounters over a period of time. Granted, the territoriality could also work against closeness, as it certainly did then in driving me away from the edge of the horse's field, but if I lived there? Some of the great friendships are formed between those that first encounter each other oppositionally. Besides, even if a territorial challenge works against, rather than for, closeness, between *this horse* and me, it is still having a relational effect—it is still *in the relationship*. That is the essential point to make. With the elements of reciprocal thoughts and feelings in place, upon which I elaborate in Chaps. 9 and 10, there is nothing ontological that impedes my ability to directly perceive and become part of the ontologically *relational* nature of this experience. It seems that posthumanist views align well here in the sense that relationships—the relational matrix—becomes a source for relational knowledge. The difference is that while posthumanists largely rely on material means of knowing and relating, here I am rejecting such means as the sole or best source of gleaning relational attributes. That, and I reject the subordination of the individual to the relationship that is the hallmark of such posthumanist thinking. My goal is to instantiate and defend the horse's Self-ness, not to deny it as a means to somehow account for the horse's influence.

Thus, I conclude here that things can very well be distinguishable, true and knowable in experience itself. This means that things are differentiable in immediate experience, and the concept of the indistinguishability of the flux of immediate experience must be rejected. Though we may not be able to think about or systematize all the information coming in at any moment via feelings, sight, or other senses—and though we can be wrong—if our subsequent undertaking of such cognitive processing correlates with the attributes of the experience itself and the other Selves in that experience, then we have as good a chance as through any other means to know what our relational partners and relations are like, and to participate in the latter as fully as their structure and dynamics allow.

I've argued that human-nature relations are ontologically basic, and that when dualisms are accounted for and discarded, that there is nothing yet preventing perception of human-nature relations and their structure and dynamics in primary experience. In the sections that follow I explore some means of perceiving and having knowledge of human and more-than-human Selves, and the relations between them, that would permit us to call such perceptions knowledge.

7.2 Peircian Feelings

One of the ways in which the horse and I related was through feelings. As I suggested above, I perceived through my own feelings a territoriality that the horse was also feeling. While I'll argue for the horse and all other beings having the capacity for feelings in Chap. 9, here I'll explore just how those feelings might be communicated in ways that circumvent the dualist, materialist view.

To pragmatist Charles Peirce (1960), one can have both an internal response of feelings to an external phenomenon or one can experience external feelings directly—the latter literally being shared between relational participants. His view of the latter kind of feelings is that they are external to any particular individual feeling or responding to them. They are first something *perceived*, not an internal response to a perception of a material nature. Peirce (1960) explains his view on external feelings in the context of experiencing the color red. He says, "[I]t is the psychic feeling of red without us which arouses a sympathetic feeling of red in our senses" (1.311, p. 155). Of the fact that these feelings are something that can be shared, or are communicable, Peirce finds support in the fact that a response to the color red by various humans is similar. He offers the example of how even a blind man knows what red is by hearing the descriptions of others talking about it and, in so doing, correctly analogizes it to the blare of a trumpet (1.314, p. 157). Such a sharing of feeling even crosses species lines, in his estimation, where he notes that "I am confident that a bull and I feel much alike at the sight of a red rag" (1.314, p. 157).

As an anecdotal example, my 6-year old son will not watch movies that have any "drama." It's because of the music. Whenever there is anything remotely tense conveyed by the building of a certain kind of music, he runs from the room and tells us to shut it off. He has always been this way. There is no negative association for him with the music. From the time he was even remotely aware of the television and before he understood anything of storylines or plots or even dialogue, as soon as any suspenseful music swelled, he fled. I attribute this reaction to something existential called *tension* or *suspense* that certain musical arrangements carry inherently. My son, being a human innately predisposed to perceive tension or suspense, feels it directly and knows what it is. There is nothing material about it. The soundwaves may be material, but the suspense they carry is not. He feels it directly and what's more, *he's right*. Just like Peirce's color red, my son experiences the suspense because there is something *in the music itself which carries it* and communicates it directly to those who hear it.

While not trying to be exhaustive in my treatment of Peirce's proof of how shared feelings are not internal to experiencers, I will now undertake a brief exploration of it as it relates to my purposes. For Peirce, feelings are things which totally occupy immediate consciousness. As such, they cannot be decomposed, nor are they resemblances or memories since they are experienced in the present, are self-contained, and thus are all that they are without being any other thing (1.310, pp. 153–154). According to Peirce, all memories, mental resemblances, or responses of internal feelings to these feelings are a diminished version and not the actual feeling itself. Thus, if two humans have a feeling, then both are experiencing the external psychic stimulus of the feeling itself, and nothing else. This, according to Peirce, is how feeling can be shared—through the capacity of more than one entity to be immersed in this external feeling and to experience it simultaneously. Of this sharing between humans and more-than-human beings, he says,

> You would never persuade me...that the canary bird that takes such delight in joking with me does not feel with me and I with him; and this instinctive confidence of mine that it is so, is to my mind evidence that it really is so. (1.314, p. 158)

In addition, he says of a friend who doubts "whether we can ever enter into one another's feelings" that this doubt is akin to asking "whether I am sure that red looked to me yesterday as it does today" (1.314, p. 158). For him, feelings are as real as the color red, and as communicable as the experience of seeing the color red. They are the same, not just in description, but in actual instantiation and shared experience for all with the faculties to perceive them. Of this, Peirce says, "I know experimentally that sensations do vary slightly even from hour to hour; but in the main the evidence is ample that they are common to all beings whose senses are sufficiently developed" (1.314, p. 158).

This Peircian notion of the communication of external feeling is what is meant, in my mind, by the common expression of "getting a feeling." As in, "I don't know, I just get the feeling that he doesn't like me." The attribution of origin most commonly made about such a feeling is that it is internal to the individual, and arises in response to some external visual or verbal, non-emotional, expression (such as body language). This attribution, I believe, is partly rooted in a substantivist ontology where all individuals are taken as ontologically distinct, therefore some physically perceivable "signal flare" must be sent up to which the individual can have an internally-sourced response. As I do, posthumanism would also reject such ontological individualism, but would still root the origin of feeling-getting in material elements that are, ultimately, obscure and do, ultimately, keep us materially separate. See my discussion of contagion as it relates to posthumanism in Chap. 3 above. By rejecting the materialist explanation as dualist assumption, relations can be *a priori* taken to exist, and subsume their participants. Thus, and according to Peirce, the feelings of the relationship can be external to any individual within them, and therefore can be felt by more than one being at a time. In this context, "getting a feeling" is a literal description of the experience and to be taken preliminarily as factually correct.

With regard to the human-nature relationship, for Peirce it is unproblematic that these external feelings can be shared across species boundaries. For example, Anderson (2004) says that for Peirce, "horses, dogs, and canaries reveal a continuity of feeling (as does protoplasm)" (p. 89). Another example of the possibility of sharing feelings across species boundaries lies in the research into whether plants respond to emotional states in certain human beings (Backster, 2003; Puthoff & Fontes, 1975). Granted, this research has long been regarded with an overwhelming degree of skepticism, but even some of the skeptics have found things about plant responses that are not easily dismissed. I'll address the work of Backster and others in detail in my discussion of the "Kook Fringe" in Chap. 9, but conclude here by stating that if feelings operate in a Peircian way, and there is a continuity of them across the boundaries of beings as different as humans and plants, then it's possible that this is a pathway along which the feeling element necessary for close relationships can develop between human and more-than-human. This pathway is also a means by which a thing can be known by humans about more-than-human beings that bypasses the dualistic *post hoc* conceptualization processes that so often distort the interpretation of closely relational primary experiences humans have with those in the more-than-human world.

At this juncture, one might ask how it would be possible to differentiate Peirce's external feelings from the subsequent, internal feeling responses to them that he mentions. As a possibility here, I'll give an example from my own life experience. When I was in my twenties I took part in a 10-day Vipassana Buddhist meditation retreat. Unlike other meditation practices that focus the mind on breath or mantra, Vipassana focuses it on sensation—on physical feelings in one's body. It trains the mind to doggedly follow sensation without thought, judgment or other mental response or interruption. The work was quite intense, and at some point during my stay, like a flood, the world of sensation opened to me. I could feel it passing through me in a kind of vibrational shimmering that was constant, though not uniform. Like wind over water these sensations ebbed and flowed, grew more and less intense, and spread and moved in irregular, constantly shifting shapes all over and through me. Some parts were filled with sensation, like my fingers, while others were more subtle or hard to feel, such as my lower back. At some point about halfway through my time there, I had a realization that all these sensations were bedrock familiar to me, and what's more, that I had *always been feeling them*. The only difference was, at the retreat I was being trained to give them my clear, conscious attention. Having that attention honed in such a way was like "coming home" to myself in a way that I've never felt as strongly in any other circumstance.

After that experience, I began to attend closely to these feelings and found that they routinely corresponded to my external experiences with others. For example, when someone isn't telling the truth, it *feels* different to me from when they are. And so, over the years since that retreat, I've gotten good at being able to differentiate when someone is being dishonest from when they're being truthful. I can't always tell the reasons behind the dishonesty I seem to detect, but I can feel its presence as certainly as I can feel the sun on my skin when my eyes are closed. While materialists might point out that the feelings are still physical, and thus material, I suggest that they are the physical signatures, or counterparts, to the feelings that are being shared. Because of these experiences, I now treat this kind of feeling as a sense as real as hearing or touch, and take it to be the communication and experience of something from outside of myself. In reading Peirce's treatment of shared feelings, while it stands in contrast to materialist and individualistic interpretations of the phenomena he describes, when considering experiences such as mine within a relational ontological context, his descriptions taken literally are unproblematic. Peircian feelings offer an avenue along which things in the world beyond the self and beyond the human may be experienced and known directly without the interposing of dualistic ontological commitments that obliterate their legitimacy as external or subsuming elements of primary experience. In the following sections of this chapter and in the chapters that follow, I will periodically allude to their capacity to provide this pathway of knowing.

In this section I discussed Peircian feeling as a means of relational perception and communication that crosses species lines to contribute to potential human-nature relational closeness. Though an in-depth exploration of the epistemological implications of my relational ontology is beyond the scope of this book, in the three sections that follow I'll discuss what it means to have relational knowledge rooted in this type of feeling, and also through other means. In total I'll discuss three types of knowing that comport with the possibility of human-nature relationships being close ones.

7.3 Poetic Knowledge

The first type of knowledge under consideration is what Taylor (1998) calls "poetic knowledge" (p. 5). He explains that this kind of knowledge is not a knowledge of poetry, but "rather a poetic (a sensory-emotional) experience of reality" such that learning and the resultant knowledge become a "poetic impulse to reflect what is already there" (p. 5). According to Taylor, this is not a new form of knowledge, but instead find its roots in "ancient, classical and medieval times" (p. 5), with it only finally giving way to "less intuitive, less integrated" forms of knowledge during the time of "the Renaissance and Cartesian revolution in philosophy" (p. 5).

According to Taylor, poetic knowledge is not rooted in the rational, intellectual capacities that humans have, and that underpin current academic and scientific ways of knowing, but instead in "the intuitive nature of human beings who are able to know reality in a profound and intimate way that is prior to and, in a certain sense, superior to scientific knowledge" (p. 4). Since this kind of knowledge is not formed through *post hoc* construction inside of a human Self, but instead as a spontaneous, direct reflection of (and response to) that which one experiences, this sort of knowledge fits well within pragmatist and ecofeminist views that also position experience as the source and arbiter of knowledge. This circumstance is reflected in Taylor's quote of Aristotle saying, "intuition will be the originative source of scientific knowledge" (p. 60) and his comment that "first knowledge of being (that things are as they are) is not only intuitive, but is never surpassed in its initial importance" (p. 61). Such views of knowledge and human capacities for it are exemplified today in statements such as the one from Barbara McClintock to which I referred earlier in this chapter. It's worth repeating as it pertains to my point here. She says,

> *Why* do you know? Why were you so sure of something when you couldn't tell anyone else? You weren't sure in a boastful way; you were sure in what I call a completely internal way... you work with so-called scientific methods to put [that knowing] into their frame *after* you know. (Keller, 1983, p. 203)

This is poetic knowledge.

If I'm right that the dualistic view of the relational qualities and capacities of more-than-human beings as largely passive and material is the result of *post hoc* dualistic reductions of the closeness-capable, relational more-than-human Self, then getting to know a more-than-human being in the poetic sense ought to allow one to bypass this reductive filtering. If the poetic knowledge that results from human-nature encounters yields knowledge of a closeness in human-nature relationship—between me and the horse—then this must be taken not only as true, but as *more* accurate than anything resulting from *post hoc* intellectual processing. Of course one must ensure that this poetic knowledge is arrived at through a diligence of mind, but the same stipulation applies to *post hoc,* intellectual knowledge formation as well. Of course even with poetic ways of knowing, there will always be *post hoc* conceptualization, where things experienced and known in the moment are contemplated or "thought through" in order to reach other and deeper understandings. But, these understandings must be consistent with the poetic knowledge of the experience that forms first. Thus, when one understands a human-nature relational experience as close through poetic means, any dualistic interpretations that result in radically different conclusions ought to be treated with suspicion.

If, as I contend, more-than-human beings have the qualities and capacities to enter into close relationships, then poetic knowledge gained by being in relationship with them means that the relationship itself is of a fairly personal, or close, nature. What's more, since Aristotle suggests that poetic knowledge is "the originative source of scientific knowledge" then by necessity, these personal elements both come before the development of scientific knowledge, and if awareness of them is in place, they become science's constraining context. In light of the potentially personal nature of poetic knowledge, the Menominee's more-than-material relationship with the forest discussed in Chap. 5 cannot be separated from a "more relevant" material/ecological one. The personal knowledge, with its material and more-than-material elements, must be understood to be the origin, and thus the indispensable context, of any ecological knowledge that follows.

Ultimately in poetic knowledge, our intuition, senses and emotion become not only acceptable media through which we can gain knowledge of more-than-human beings and our relationships with them, but if poetic knowledge is the basis for any scientific knowledge that comes from this source, these faculties become *the* basis for knowing more-than-human beings at all. If our poetic knowledge of more-than-human beings, and of the relationship we're having with them, is the means by which we know both that more-than-human beings are capable of entering into close relationships with us, and that our present experience of our relationship with them is close, then knowledge that we can, and/or have, entered into close relationship with a more-than-human being becomes the foundation of any other sort of investigation of this relationship.

7.4 Personal Acquaintance Knowledge

The second type of relational knowledge I'll discuss is what I'll term *personal acquaintance knowledge*. To begin this discussion, I note that one of the key elements of traditional scientific approaches to knowledge acquisition is that of repeatability (e.g., Stodden, Leisch & Peng, 2014). If the more-than-human world is to be dualistically taken as a passive, material backdrop, then repeatability of result is not problematic on its face. If we push the button of the soda machine each time after inserting our money, we expect that each time, a soda will emerge. This is a reliable result from a machine designed to dispense soda, and that we know to have this purpose. A researcher might believe that a bird, a Loggerhead Shrike for example, is similarly mechanistic—operating largely on instinct, lacking any teleology, intelligence, feelings or other qualities and capacities that would enable him to enter into close relationships with human beings. This researcher would then also believe that her observations or manipulations of the bird would engage the full range of the bird's behavior and purpose in relation to the manipulation every time, and that any of the researcher's unaccounted for influences—her presence, or thoughts and feelings about the bird—would not affect the Shrike in any appreciable way. Let's say further that this researcher hypothesizes that the bird's various food caching strategies, like hanging frogs on barbed wire or snapped-off branches, is a means for food consumption such that the bird would use the branch to hold the frog and, with the resistance of the branch, pull the frog apart to eat it. In the field observing from a blind, if on enough occasions the researcher fails to see this behavior, she'd conclude that it likely doesn't exist.

But, if more-than-human beings like Shrikes are relational Selves, then not only are they not machine-like in the way that humans are not, the relationship between Shrikes and humans is not mechanical or controllable in this way either. In such a scenario, the kind of knowledge it's possible to gather from the Shrike, and the means by which it's gathered, are foundationally relational, and thus radically different from what the researcher believes. In this example, if the Shrike is aware of the observer, let's say through Peircian feeling, and detects some derision, say, as the observer snorts to herself about how stupid the bird is not to use the branch as a lever, then this will have an effect on the bird. Instead of it being a case of Shrikes never using such a technique to eat, perhaps the bird simply *chooses* not to show the behavior in question to *this observer*. Obviously, I've not yet made the case that the bird is capable of such perceptions and exercise of free will, as I will undertake more of that in the chapters on Thoughts and Feelings in Part IV below. But if it's possible, here one sees that the means by which the attempt to gather knowledge is undertaken affect the knowledge obtained. In this case, the researcher's information will be false and her knowledge faulty. Backster, who gained infamy in scientific circles for arguing that plants can detect harm intended for them and other life forms, holds a view of them as relational Selves akin to my suggestion about the Shrike. He says of this possibility that, "[t]he…problem in [traditional scientific] research is that Mother Nature does not want to jump through the hoop ten times in a row, simply because someone wants her to" (Hunter, 1973, p. 9). I'll further discuss Backster's work, and the valuable elements I believe it does contain, in the Chap. 9.

To take a step back and talk about types of knowledge, Western epistemological theories generally offer three basic kinds (Fantl, 2014). The first, propositional knowledge is explained as "knowledge *that*." The statement, "I know *that* Loggerhead Shrikes use caching techniques to help them consume food" is an example. The second, "knowledge how," can be seen in a statement like, "I know *how* Loggerhead Shrikes consume their food." Lastly there is "acquaintance knowledge" with the most common example being that of knowing another person. In my example, acquaintance knowledge would be exemplified by the statement, "I know a Loggerhead Shrike." In regard to my suggestion about the possible influence of the researcher on the Shrike, such a suggestion comports best with acquaintance knowledge since the response is of the particular Shrike to the relational approach of the particular human researcher.

But, while acquaintance knowledge might be seen as a good explanation for this knowledge relationship, when one delves a bit more deeply into the literature on acquaintance knowledge, one finds that theorists tend to hold the Known (the Shrike, in my example) as a fixed object and not as a relational Self with influence over what *is* known. In other words, though the example of knowing another person is often given for acquaintance knowledge, that is not really what is meant by the term. Instead, Russell (1912) defines the acquaintance portion of acquaintance knowledge by saying, "We shall say that we have acquaintance with anything of which we are directly aware, without the intermediary of any process of inference or any knowledge of truths" (p. 78). The key to acquaintance here, then, is the direct experience of the Knower before *post hoc* conceptualization, judgment, or possible truth derivation—without any consideration of what, or whom, is being experienced. It may be noted that this definition has close similarity to the pragmatists' "primary experience," or ecofeminists' "particular experience." But, just as I've suggested in response to those theories, the qualities and capacities of *both* relational partners is foundationally influential over the experience of the relationship and knowledge obtained from it. In the knowledge relationships under consideration here I suggest that it is no different.

To wit: when one says, "I know my wife" I suggest that one does not really mean the same thing as when one says, "I know my pencil." It's not just the amount of knowledge, but the quality of it as well. In acquaintance knowledge theory in general, however, acquaintance knowledge of humans, Shrikes or pencils are taken to be qualitatively similar. Such a definition accounts to some degree for the orientation of the Knower to the Known—but misses the qualitative differences in the qualities and capacities of the Known almost entirely. In addition, it fails to account for the fact that the qualities and the capacities of the Known do not just dictate the Known's behavior toward the Knower, but make demands upon the qualities and capacities of the Knower in response—ones that might not otherwise be engaged when forming knowledge. I mention this because, while within traditional scientific methods the intellect is the faculty of the Knower believed to be predominantly engaged, in the kinds of knowledge relationships I'm discussing—such as poetic knowledge—many other faculties can be, and are, employed. The qualities and capacities of the Known will, then (and to a substantial degree) exert determinative influence over what those faculties turn out to be and in what ways they are engaged.

If the Shrike is a machine, the intellect will be engaged with perhaps a small emotional response. If the Shrike is a relational Self like a human is a relational Self, then the sensory, emotional and intuitive capacities that are involved in poetic knowledge will be called upon in the Knower. As Taylor (1998) quotes Renard, knowing another in this way means that "one must somehow *become* another, for to know is to be another. It is a sort of participation in the 'to be' of another" (p. 61). There appears to be then, a direct communication between Knower and Known that is akin to the conflation of Selves that Noddings' describes as "feeling with" when it occurs between humans. I explore this phenomenon in detail in Part IV. Taylor uses the term "sympathy" (p. 9) to describe this, and quotes Renard as calling it an "immaterial...union" (p. 63). Things like love, care and other relationally reinforcing elements can and most likely will develop with such a close knowing. I call this kind of acquaintance knowledge "personal acquaintance knowledge" to distinguish it from the more common epistemological form of acquaintance knowledge I just described, the latter of which places its focus only on the orientation of the Knower to an undefined (and thus abstracted and relationally irrelevant) Known.

It's commonly understood in interhuman relationships that people rarely reveal to strangers the deepest, most meaningful things about themselves. The willingness to do that builds relationally over time as personal acquaintance knowledge, trust, care and even love develop. If a more-than-human being like the Loggerhead Shrike has the thoughts, feelings and ability to act on them that contribute to close relationships, then he has the capacity to enter into relationship with a human observer in the same way. Knowledge gained of the Shrike, then, is not something taken from an object, but learned almost as a byproduct of the more central, and necessarily genuine, relationship the human has with him. This means that the researcher must step out from behind the blind and join the relationship she's having with the Shrike whether she thinks she's having one or not. Only then will she come to truly *know* the Shrike—to have him show her that, after pulling on the frog and having it fall off the branch upon which he'd impaled it, he re-secures it on the branch and tugs on it again until a piece small enough to eat has come free.

In this context, while Goodall's methods of living with and interacting with her chimpanzee "subjects" has received its share of criticism as unscientific (interview with Goodall by Riefe, 2017), it is still knowledge she has gathered. It's just rooted in personal acquaintance. Given the possibility of personal acquaintance knowledge with more-than-human beings, and the depth of influence Goodall's knowledge has had since its collection, I must ask which knowledge, hers or the mainstream scientist's, is more fruitful, real, or better? I suppose the answer hinges upon the ontological context within which such things are determined, and to what real world contexts each is applied. Loosed from any relational, contextual constraint, the atom bomb traveled its path from imagination to reality. Knowledge of the more-than-human world gathered from detached, objectifying, material, and otherwise dualistically reduced and anthropocentrically co-opted human-nature relationships may yet prove to be as destructive. How much longer can we have knowledge *that* material natural gas can be extracted from the ground via the "fracking" process before realizing that detachment from, and ignorance of, the more-than-human beings affected makes such a knowledge relationship nothing short of abusive? Ultimately, personal acquaintance knowledge of more-than-human beings is not only possible and needed, when one's ontology is stripped of the dualisms that obfuscate our view of any other relational possibilities, it becomes both primary and unavoidable.

7.5 Propositional Knowledge

Of the three types of knowledge common to epistemological discussions (that I articulate above), propositional knowledge is the one which receives the most use and attention in modern approaches to generating knowledge of more-than-human beings. It is this type of knowledge that is the last which I'll discuss. I believe that the focus on propositional knowledge in mainstream epistemology is due, in part, to its orientation around discreet, rather than relationally-embedded, facts. As such, propositional knowledge appears to not lend itself to the kinds of knowledge that can be gained between two relational Selves. But, if one knows one's wife in the personal acquaintance sense I've defined in the section above, it is also possible to know *that* she has the ability to run a full marathon in the propositional sense. Therefore, there is nothing inherently problematic with coupling propositional knowledge with relational knowledge, it's just that the former cannot be thought of as the entirety of possible knowledge, nor its exclusive foundation.

Yet, this is precisely how propositional knowledge is taken within modern humanist paradigms when it comes to seeking knowledge of the more-than-human world. It is seen as the most important kind of knowledge, and one that need not be governed by or reside within the context of relational knowledge. If there is nothing inherently antagonistic toward relational knowledge within propositional knowledge itself, then the only source of the near-exclusive focus on propositional knowledge must be the assumption that relational knowledge such as personal acquaintance knowledge is simply unavailable in the context of seeking knowledge about the more-than-human world. As I've argued throughout this volume, that assumption is a dualist one, and not supported by evidence. Even in posthumanism, where the foundational project is to account non-anthropocentrically for the more-than-human contribution to human-nature relationships, its materialist ontological underpinnings allows the dualism that elevates humans to the status of having the *most* relational capability to marginalize, reduce and/or negate any substantive more-than-human contributions to the kinds of relationships which generate personal acquaintance knowledge. As I suggested earlier in this chapter, when scientific knowledge flows from the primary, relational knowledge to which personal acquaintance and poetic knowledge contribute, such knowledge is properly situated. Since propositional knowledge is the predominant form of scientific knowledge, then it appears there is nothing problematic about propositional knowledge of more-than-human beings within a relational context.

I will now examine two hallmarks of the scientific method—objectivity and reliability—to show that it is dualisms within such ways of acquiring propositional knowledge, rather than any innate quality of propositional knowledge itself, that is largely responsible for the lack of relational grounding for this type of knowledge.

7.5.1   Objectivity

The OED states that a thing is objective when it is "external to or independent of the mind" ("Objective", 2018). In this scientific use of the word, objectivity is an attempt to "produce knowledge independent of the particular people who make it" (Porter, 1995, p. ix) or that has "immunity to worldview differences" (Gauch, 2003, p. 36). Since the object of scientific inquiry in this context is held to be independent of those making such inquiries, it is generally thought that the scientist ought to be "detached" (Porter, 1995, p. 86) from the object, have an "impersonal" (p. 229) attitude toward it and to his or others' worldviews concerning it, and be "disinterested" (p. 4) in the result. This approach is sought in order for the scientist not to betray her own stance or worldview as it relates to the object, or whatever knowledge is obtained from or about that object.

In posthumanist thinking, true detachment is called out for what it is: a fiction. Knowledge, and those seeking it, are situated (Barad, 2007; Braidotti, 2006; Haraway, 2008; etc.), and each relational participant influences the knowledge gained. But if, as I suggest, the relational knowledge I discuss in previous sections is, and ought to be, the seat of propositional knowledge, posthumanism's new materialist ontology works against this by obscuring relational possibilities and therefore distorting or downplaying knowledge of such relations. As discussed in Chap. 3, new materialist ontology tends to reinforce, rather than subvert, the dualist superiorization of humans in relationship with the more-than-human world. It does so by leaving unexamined the dualistically rooted notion that the material evidence at modern humans' disposal supports their having the most developed relational qualities and capacities amongst all beings. Posthumanists speak often to an asymmetry or inequality in human-nature relationships (Barad, 2007; Haraway, 2008; Taylor & Giugni, 2012) that, because such inequality is rooted in the material, is supposed to bypass dualism with its "nonhierarchical multiplicity" (Haraway, 2008, p. 77). Such a material focus is also supposed to avoid the "logic of domination" (Warren, 1990, p. 128) that ecofeminism identifies as the enabling mechanism of dualism. But, if the logic of domination and its concomitant dualism *has already done its work* to distort the equality or lack thereof in human-nature relationships—either because such inequality does not exist, or because it is being portrayed through dualist lenses as substantively different than it actually is—then dualism has not been avoided, it's been further entrenched.

This materialist ontology aligns with modernist beliefs about the material bases of relational qualities and capacities, and so positions more-than-human beings as generally not having the kinds of thoughts and feelings necessary to form close relationships with one another or with humans. Under such circumstances, not only does relational knowledge cease to be the context or root of any propositional knowledge, it leaves the disconnected material roots of some forms of propositional knowledge as by far the most accessible. This ensures a maximized relational detachment, where whatever influence the human relational partner does exert, and whatever response the more-than-human has to such influences, is either ignored or understood only in material, and therefore reduced relational terms. Thus, while posthumanist theory has correctly characterized ontological objectivity as nonexistent, the material basis of their analysis of human-nature interactions leaves objectivity and relational detachment as the destination, albeit unreachable, toward which the path of least epistemological resistance leads.

As an example of the effects of relational knowledge's absence in the production of propositional knowledge in posthumanist theory, I note that Barad's (2007) more-than-human beings makes their central appearance in human-nature relationships not in a whale, chimpanzee, insect or houseplant, but in a scanning tunneling microscope. And even in the relationship with that microscope, nothing is mentioned about the atoms being probed by it, nor how those atoms might feel about such perturbations. Even less relational knowledge is produced about the place and more-than-human beings from which such atoms were extracted, and the human-nature relationships (and possibilities thereof) that were damaged or ignored in the process. That is the danger of a purely materialist ontology, it tends to subordinate, ignore, or work to make unreal, any substantive relational connections and possibilities for them between human and more-than-human. Even when posthumanism makes room for them, it seems that there is a cold distance—an antiseptic, relationally vacuous treatment of it. For example, as I discussed in Chap. 3, Haraway (2008) describes her relationships with her dogs in ways that seem at odds with true closeness. There is something missing from the relationship. Is that thing care? She says they share love. Is it a protective impulse? Whatever it is, it is why on the one hand, in her relationships with her dogs she can profess to share the "contagion" of love and on the other, she can recommend as anti-humanist a dog trainer who demands total obedience from her dogs and subjects them to all manner of harsh manipulation to get it (see Weisberg's, 2009, critique). It's because she still views the dogs, even her own, objectively. Objectivity requires an object. Materialism, and also the displacement of agency and other attributes like *telos* from Selves-in-relation (again, see Chap. 3), facilitate the creation of such objects, especially out of more-than-human beings seen as relationally less capable. This allows dualist objectification and instrumentalization to continue to operate, even in supposedly posthumanistically sanitized relationships.

One might counter my claim that agency and *telos* are displaced from more-than-human beings by pointing out that in posthumanism, it has also been displaced from humans. To this I would respond that, though this very thing has been stated by the likes of Barad and Haraway, in reality such individually possessed qualities are alive and well in humans within these theories. For example, Weisberg notes Haraway's reliance on the existence of "human 'curiosity'" (p. 34) as a justification for animal experimentation. This means that humans *possess* the curiosity. One could counter that supposition by pointing out that in posthumanism, even though curiosity is enacted by the human-in-relation, it is not hers, but instead the product of the human-nature relationship in the laboratory and not to be taken as human trait in an essentialist sense. But, because *animal* curiosity is never mentioned in Haraway's' construction, one must assume either that more-than-human curiosity does not rise to the level of relational relevance for Haraway, or that humans enact the matter's curiosity but more-than-human beings do not. Either way, this is an example of the relational "inequality" or "asymmetry" between human and more-than-human whose "recognition" is an important part of posthumanist thinking. Yet nothing, in my estimation, could be a better example of objectification of more-than-human beings than to suppose that more-than-human curiosity, or any other expression of agency, such as Weisberg's notion of more-than-human "freedom *to* fulfill [their] potentialities" (p. 35), is so insignificant that it doesn't warrant consideration. This is humanist. It is also dualist. Thus, while human experimenters are able to detach because of the lack of substantive relational connection, it's not actually because it doesn't exist.

What Haraway does do in the face of this reduction of more-than-human agency is attempt to reintroduce its presence in the intra-action by metamorphosing the subordinated more-than-human beings' behavioral responses to an experimenter's oppression and torture into, for example, what Weisberg says. Weisberg notes Haraway's suggestion that animals, in such situations,

> "have many degrees of freedom." An example of this "freedom," Haraway writes, is *"the inability of experiments to work if animals and other organisms do not cooperate"* (my italics) (ibid.). If there is a more perverse conception of freedom than this, however, it is hard to imagine. (p. 35)

When a materialist ontology forces even posthumanist theorists to see more-than-humans objectively, then to avoid the charge of humanism, it is precisely such "perversions" of the actual more-than-human contribution to human-nature relationships that must inevitably be invoked.

To pursue some of the ramifications of attempts at objectivity and detachment both in humanist and posthumanist thinking, I'll consider Brown, Davidson, Munger and Inouye's (1986) work on ecological interactions between rodents, ants, birds and plants in the Sonoran and Chihuahuan deserts and attempt to put their work in a relational epistemological context. Over more than a decade, Brown et al. applied various manipulations to desert granivore communities in order to observe how various species altered food consumption and reproductive patterns in response to these manipulations. Their main method of manipulation was to eradicate a certain species within a study plot and observe changes in behavior of the remaining ones. For example, after exterminating the Banner-tailed kangaroo rat from the community, they found that while other species of kangaroo rats responded by shifting their microhabitat use into the vacated areas, the populations of those other rats didn't increase to fill the density void the experimenters created. The experimenters attributed this to the short duration of their experiment, elaborating later by saying that the cause for the delay was either the "failure of the rodents to perceive a significant change in the availability of resources or their inability to respond quickly to detected change" (p. 48).

The authors clearly attempt to be objective in this work since explanations like the one above do not consider their own, human impact on the desert community other than as an invisible arm manipulating material rodent variables. And while posthumanist thinkers might suggest that these experimenters are wrong to not account for themselves within this arrangement, by allowing their predominant materialist ontology to hold sway so that they ignore the relational elements of such "manipulations," they still fail to account for what I suggest must be the governor on such interactions with the more-than-human world.

The experimenters' attempts at objectivity leave unexamined several variables that may significantly alter the community's response to these manipulations. Thus, the propositional knowledge they gain may be faulty. For example, the authors are attempting to observe and explain normal responses of granivore ecosystem members to disturbances, yet their manipulations artificially create highly *abnormal* circumstances. There is only one instance I can think of where there would be a sudden and total loss of a community member species that doesn't originate in the idiosyncratic characteristics of anthropogenic perturbation. It would have to be some sort of disease only deadly to a particular species. This is highly unlikely in itself. Further, because there are several species of kangaroo rats in the study plots, one would also think that if a disease were to afflict one species of kangaroo rat, that other, similar species would also be affected. Thus, the likelihood of such an event is exceedingly rare. The result is that the response of the other species to these events sheds far less light on their "normal" behavior in granivore communities than the authors seem to suppose.

I suggest that most of the researchers' failures to see how abnormal their manipulations are in this community can be attributed to their failure to see themselves as having joined the community. By objectifying the community, they've failed to notice their integration into it as *whole individuals* instead of disembodied arms waving magic wands and "poof," the rats disappear. Again, while my critique aligns with posthumanist views superficially, the materialist ontology posthumanists employ does little to account for the kinds of *relational* perturbations I am noting here as ones that must be accounted for in any formation of propositional knowledge. No disturbance stands alone, nor does it travel unidirectionally from human to more-than-human community members. In a human-nature relational reality, any connection is reciprocal and cyclical, and like a live wire held by the two individuals, touches and affects human and more-than-human alike. Further, such a connection is a conduit through which human and more-than-human communicate (both intentionally and unintentionally—and through verbal and nonverbal means), and becomes the foundation of a connection between them, whatever that connection may be.

What is the nature of the connection between human and more-than-human in the Brown et al. studies? To put it plainly, it is a murderous one by the humans in the name of their carrying off knowledge. Through a relational lens, their research activity is an exploration of what happens to more-than-human community (including humans) when the humans within it eradicate certain species. Describing it in such terms means I've both moved the boundary of the granivore ecosystem outward to include the humans involved, and I've called out as relevant the more-than-material elements of human intent to kill, and through that, extract knowledge. Seen in this light, it becomes clearer that the study is a microcosm of the modern human-nature relationship as a whole. This should not be surprising given that the human-nature relationship in such a set of studies is very much sponsored by the dualistic ontologies that underpin the more general modern relationship humans generally have with more-than-human beings. Every time a housing development is built into a tract of woods this dynamic occurs. It occurs every time beavers are trapped out of an area for the profit their fur brings their human eradicators. Every time I see ants on my kitchen counter and set out poison, it occurs. It is *this* kind of study that the authors are unwittingly undertaking.

In addition, and as I mention above when speaking of intent to kill and desire for knowledge, in a relational ontology the effects of the more-than-material must be taken into account. For instance, if Peircian feelings are a means of relational interchange, then the feelings of the researchers both before and after they eradicate a community member will affect the other more-than-human members in perceivable and potentially substantial ways. Perhaps the other kangaroo rats that didn't immediately reproduce to fill the void left by the Banner-tailed Kangaroo Rat didn't do so because they sensed some danger in the space that the Banner-tailed Kangaroo Rats had occupied. Perhaps they felt grief at the loss of a community member that they knew well. Perhaps they felt anger toward the humans in their midst. I'll discuss animal emotions extensively in Chap. 9, but state here that all manner of animals have feelings. What if an individual Ord's Kangaroo rat *liked* seeing one particular Banner-tailed Kangaroo Rat every day? What if she was *upset* by his disappearance? As I will discuss in detail in Chap. 9, even honeybees have been shown to have emotions (Bateson, Desire, Gartside & Wright, 2011). Why not these Kangaroo rats?

It's interesting to note that in other portions of Brown et al.'s (1986) study the authors also found a "failure" to immediately fill the void left behind by whatever species had been killed off by them. For example, the authors found that birds foraged less on plots where rats had been exterminated. They suggest that this result "may well represent the outcome of a long-term indirect mutualistic interaction mediated through the direct effects of rodents and birds as selective predators on different but competing species of annual plants" (p. 50). This translates to a suggestion by the authors that the extermination of Kangaroo Rats, and with them the controls on population densities of plants whose seeds they eat, might mean that those plants would have a competitive advantage over plants whose seeds the birds eat. Thus, the food source for the birds would be reduced, which in turn would reduce their populations. But, what if instead the birds took the sudden disappearance of their fellow community members as an indication of danger? What if they sensed malice and/or danger from the humans there? What if they were sad?! A posture of objectivity obscures these very real possibilities because, in some way, it allows the human researcher to not *feel* or intuit it happening—or to discount as irrelevant or internal-only any feelings about it that they might have.

And if the possible alternative explanations I've articulated above are real, then objectivity can be called out as a fiction. At a more basic level I ask: how else *can* we describe a so-called "object's" existence without describing that from which it "stands forth" as viewed from our vantage point in relation with it? To try to be wholly objective with respect to the granivore community is to remove oneself from a relationship in which one is inextricably, willfully, and profoundly involving oneself. Posthumanists, while attempting to reinstall the same unavoidable interconnectedness that I am, fail when they see that interconnectedness first and primarily as a physical, non-relational one. When one's ontology is no longer individualistic and material-only, the human in all aspects of her participation must be accounted for within the community. In terms of the knowledge relationship, the Known is not how he is without the Knower being how she is. One cannot know without, in some sense, *being* known. Just as the human researcher hopes to know the rats, the rats know the human researcher. That is the reciprocity of the kind of ontologically basic relations for which I make the case in this book and, through it, the structure and reciprocal dynamic that underpins relational knowledge. Attempting to separate oneself from those who one is attempting to know is an attempt to remove oneself from the relationships upon which, unavoidably, that which can be known rests.

Of course we don't want the prejudice and petty political views that spurred the original drive to objectivity (Porter, 1995) to color what we think we know. But, relational interconnectedness can only be put in the same category as those prejudices if relations of any sort—and the idiosyncratic vantage point they unavoidably provide each individual—are taken to be prejudicial and irrelevant to knowing. If they are, this must mean that they are taken to be *post hoc* human creations, and not part of relational reality already existing "out there." That's how the act of *accounting for* relational elements gets to be equated with their *removal* in present day efforts at objectivity. As I've said many times in this book, however, such views of relations are only possible if one's uses one's ontological commitments to *a priori* deny relations a foundational role. To invert the widespread inclination toward doing so, in a relational ontology, objectivity is a *relational* stance, just one that tries to both attenuate and "unbend" the intermediary structure and dynamics of an ontologically basic relationalism, the latter of which tends to move individuals toward conjunction directly in primary experience. Through this relational lens, objectivity does nothing less than *distort* the truth "out there" and any knowledge of it. Thus, it ought to be replaced by a more accurate and relationally grounded knowledge that accounts for its particular perspective, but does not do so by trying to treat that perspective as separable from knowledge or as purely or primarily material.

7.5.2   Reliability

The second notion I'll consider is reliability. The OED defines scientific reliability as the "degree to which repeated measurements of the same subject under identical conditions yield consistent results" ("Reliability", 2018). Kirk and Miller (1986) define reliability as the "the extent to which a measurement procedure yields the same answer however and whenever it is carried out" (p. 19). As can be seen, repetition is the cornerstone of reliability. In dualistic human-nature relational ontologies, however, this repetition is equated with repeatability, and though the difference between the two may appear subtle, in a relational ontology, they are two quite distinct ideas.

In repeatability, if one's ontology dualistically positions more-than-human beings as either passive and/or relationally limited, material objects, then the human is at the controls, and according to her individual human will and capacities she tries to see the more-than-human world behave in a certain way as many times as is her preference. The desire is for repetition of *result*, and the only way to get passive objects to yield this is to initiate it unilaterally. Thus, the "ability" in repeat-ability is largely the researcher's, and depends almost entirely on her will and capacities. In the more general term "repetition," however, *how* the repetition is carried out goes unspecified. In an ontology where human and more-than-human are relational Selves, repetition is partly the result of the more-than-human being's will and capacity and partly that of the human's, but more importantly and centrally, it is a result of the relationship between them. Just like I don't tell my wife to "Dance, Rummy," in a relational ontology where more-than-human beings are relational Selves, I can't tell the Red maple in a study plot in the Hubbard Brook Experimental Forest in New Hampshire to do so either.

Given the distinction I'm making between repeatability and repetition, repetition cannot be "carried out," as Miller suggests, but instead it is what *happens* when two relational Selves interact. What repeats is a byproduct of the relationship, which is primary and determinative. Thus, even when a researcher believes more-than-human beings to be relationally irrelevant, the relevance of phenomena begins with the material and more-than-material interaction, not with what occurs after the unexamined relationship is established and the human carries out whatever manipulations she chooses. Whether or not researchers *post hoc* conceptually shield themselves from this relations-as-source reality does not lessen in any way its formative influence.

Further, in an ontology where the relationship is the origin of any phenomena that repeat, the general tack that researchers take of selecting a few key variables, some to be controlled and others allowed to vary, becomes problematic as well. That's because in a relational context there are variables that, regardless of anyone's desire to account for them, cannot be controlled. Yet such variables figure prominently in what is repeated. In my example of Brown et al.'s (1986) work in the "Objectivity" discussion just above, the desire for knowledge and the intent to kill are variables that have an effect on the results whether the researchers care to consider them, or believe them to be controlled. When repetition is the result of human-nature encounter, it is the whole human—her conscious and unconscious thoughts, her like or dislike for her work, etc.—that influences what is repeated. The same can be said for the nonhuman if my argument that more-than-human beings have thoughts and feelings enabling their participation in close human-nature relationships is true. Even whether one considers more-than-human beings to be relational Selves or the passive, material objects *is itself a variable*. When the relationship becomes the vehicle of repetition, instead of the human in command of a biotic machine—or at least in command of her attention to certain features of its functioning such that they come to be called an experiment—there are a great many relevant, often difficult to measure and control, variables. Thus the human is not in charge of repeatability, but at best chooses to immerse herself in a roil of relationally relevant variables that may lead to some consistency in certain features of the interaction over time.

In my earlier example about the Shrike, an unaccounted for variable could be my return to the place where I saw the Shrike using a branch to help him tear off pieces of food *with* the desire to see the behavior repeated. This introduces the relevant, more-than-material variable of my desire for repetition into the second meeting that did not exist in the first. In a relational ontology where more-than-material elements are accounted for, this desire is influential over my own approach to the Shrike, a change which may be detected by the Shrike, or the desire may be felt by him directly through Peircian feeling. In response, the Shrike's behavior might be different. Even the *cause* of my desire can become a relevant, more-than-material variable. If I want to see the behavior again so I can make myself important by publishing my findings in some journal, then the results might be different than if I simply sat quietly for hours, so enamored of the Shrike that I was glad to see whatever the Shrike chose to show me, hoping that I got to see that amazing thing he did before with his food and the broken branch.

But what if the Shrike does do it again with my desire for personal notoriety intact? Some with materialist ontological commitments might take that as evidence that my more-than-material variables didn't affect the Shrike's behavior, meaning that a) those variables don't really exist or b) if they did, that the Shrike as a passive and/or relationally limited, material object wouldn't be influenced by them. But again, such disregard of more-than-material variables is injected *a priori* from ontological commitments, and is not founded in any evidence showing my suggestions to be untrue. Even were someone with a modern approach to human-nature relationships to devise an experiment to study the difference in the Shrike's response, I think it reasonable to assume the experiment has a high likelihood of being confounded by human attitudes and experimental construction that would betray the human's disregard for the more-than-human's standing as a relational Self. In a relational ontology, any desire to manipulate the Shrike will affect the Shrike differently than if one seeks a genuine, reciprocal relationship as an end in itself. Until dualistic ontologies are out in the open, discussed, and remediated for *as variables* in any research meant to build knowledge, they will continue to work their self-reinforcing, largely invisible (to those carrying them), effects on what is known.

As an informal, anecdotal example of what I mean, a few years ago when I was visiting Masada in Israel with my wife and her cousin, my wife kept trying to take a photograph of some beautiful birds. But, whenever she'd lift the camera to take the picture, the birds would fly further away. At some point she and her cousin were yelling at the birds, "Hey, stay still!" and when I suggested that they ask the birds to sit still instead of telling them to do so, they tried it and it worked! For me, it was confirmation that the birds understood the difference between the two approaches. To a materialist, it could have been a soothing tone in their vocalizations. It could've been pure, dumb luck. But in a relational world, the thing about repetition is that if one employs and experiences these subtle gradations of approach and response between humans and more-than-human beings, one begins to build a body of experiential evidence for their reality that is not derived from any scenario with a set of "controlled" variables. Again, in a relational world one cannot build a cage and then ask the birds at Masada to "Dance, Rummy." Instead, one just *has* human-nature relationships, and when certain things start to consistently happen in them, to occur over and over in some discernible pattern against the backdrop of in-the-present, fluid, relationally genuine and relevant variables that are numerous, sometimes impossible to discern, and material and more-than-material in origin, then one has a relational repetition. It is from this repetition that relationally grounded reliability and knowledge can be formed.

Even were we to build a cage and get consistent behavior from the Shrike, this is not confirmation for the dualistic ontological view of human-nature relationships. Even humans enslaved will, at times, capitulate to the unilateral desires of another (and at other times rather die than do so). But what knowledge is really gained by forcing repeated performances from the slave? It is not that this slave can work 14 hours without rest if that's the experimental question. That is *information* divorced almost entirely from its relational source. In a relational ontology, that does not constitute knowable truth. That the slave can do this is a *capacity*. And I suggest that truth, as some approximation of the indispensable elements of the reality of any phenomenon, must have both capacity/information and its relational matrix in order to be truth. Putting it differently, one cannot "know" that the slave can work 14 hours a day. One can know that under certain circumstances, that is what he will do. Thus the slave's capacity for work is only one informational element in a larger, irreducible truth. The slave's capacity to work an element of truth, but not a truth unto itself. Further, if this element is taken as truth, it immediately becomes part of a less accurate, abstracted knowledge which, when applied in reality as it must be, will often have unexpected and uncontrollable results. In response, I can imagine a materialist asking, "What about steel production?" They have the formula down to a science. It's the same every time and the quality of the steel is the same. But, what I'm suggesting is that the real, relevant human-nature relationships which produce that particular batch of steel are lost. On one end of the relationally relevant variables is the destruction of the latest mountain to retrieve iron ore. What is the material and more-than-material response of that particular place to such an extraction? At the other end, after the steel is produced, one relationally relevant variable is climate change.

Moving back to my hypothetical of slavery, I suggest that the *least* we can *know* as *true* is that when one human compels another to work 14 hours per day, under the "right" circumstances the slave will comply. The source of this knowledge is the understanding of how the phenomenon occurs relationally first. And so what is true is that a slave can do this when another is willing to force him to do it. There is no slave to be known without a slaver. Just as there is no repeated phenomena without those that participate in its repetition, be it through force as modern humans too often employ with more-than-human beings, or through the close relational approach for which I advocate. As in my discussion of Brown et al. above, by forcing repetition upon more-than-human Selves, we learn as much about the one compelling as those compelled.

My portrait of fluidity and reciprocity of relations means that we don't need to rid ourselves of reliability's cornerstone of repetition, but instead its dualistically underwritten and artificial narrowness. To offer a revised definition of reliability, then, I say that it is *the extent to which a human-nature relationship as an end unto itself produces some phenomenon in the present similar to one that it, or similar ones, produced in the past under similar, relevant material and more-than-material conditions.* In this definition, both human and more-than-human contribute, as do the circumstances surrounding and suffusing their relationship. Here it is the relationship, and the contributions of each to it, that will or will not produce the same phenomenon repeatedly.

In this relational light, propositional knowledge becomes a byproduct of good relations. This means that researchers that get personally involved with their subjects, like Goodall that I reference above, or Timothy Treadwell, the latter of whom lived in close, personal contact with the Grizzly bears of Alaska's Katmai National Park for 13 years before being attacked and killed by a young male (Associated Press, 2003), are producing results and propositional knowledge that are *more* reliable than objective forms of it could ever be. I am aware that more than one critic quietly chortled over Treadwell's demise at the hands of the bears with whom he was "supposedly" close (Marquez, 2004; Medred, 2003). But that sort of derision, too, only shows an ignorance of what it might mean to be a member of a bear community—to be in position to know a bear best and most accurately. In the latter context, the attack and death can be construed as a strong form of acceptance, as bears are innately aggressive, territorial, and immensely powerful animals who routinely challenge each other, kill cubs, and otherwise behave with a level of physical intensity that is far more easily fatal to a human being. Perhaps the young male, so fully accepting of Treadwell *as a bear*, challenged him like he would any other rival and, Treadwell, being much less powerful, succumbed easily.

When researchers like Treadwell and Goodall experience things over and over, the repetition grows from a mutually contributed-to desire for the knowing and is a product of the intimacy established, instead of that intimacy being a means to the end of human-desired knowledge. Not stopping at animals, I'd suggest that geneticist Barbara McClintock did the same thing with corn, as I refer to in several areas of this book. If there is manipulation for selfish ends, it only takes away from what may be true about chimpanzees, bears or corn and adds to what is true about the selfishness of modern humans in their relationships with more-than-human beings. While many criticized Treadwell for hopelessly tainting bear behavior and any knowledge of it with his personal closeness with them, it's only because the dualist ontology dominating scientific thought allows those who secret themselves in a blind to conceive of themselves as being foundationally apart from those whom they are observing. In a relational ontology, such researchers are simply ignoring their effects on more-than-human beings, and in treating the more-than-human beings as largely passive and material, ignoring the capacity of those more-than-human beings to know about it—to understand perfectly well their presence, their behavior, and their intentions.

Ultimately, reliability is not an issue of repeatability, but repetition-in-relationship. The means by which a thing comes to repeat itself tells the story of the ontology that underpins it, and gives form to whatever knowledge flows from it. If the means of knowledge cultivation are unilaterally determined by the human and counted only when materially expressed and observable, in a relational ontology the knowledge produced will be abstracted, reduced and of generally poor quality. In saying this, I'm reminded of an exchange I had with a congenial critic of this suggestion who said something in the vein of, "Yes, but in general, ecology *works*." At the time, I had no real answer for him, but upon reflection, my answer to him here is, "Works for *what?*" Objectively measuring and manipulating nature in largely material ways produces some level of repeated results, but ultimately, the material knowledge that emerges is only part of harmonious human-nature relationships and knowledge thereof. It will never on its own *solve* environmental problems. As I took up in my discussion of Menominee forest management practices, and as the likes of Leopold (1949/1989) suggest when he says, "It is inconceivable to me that an ethical relation to land can exist without love, respect, and admiration for land…" (p. 223) something more is required of us if we want to lay claim to knowing a place and what it's like. Without personal acquaintance knowledge and any propositional knowledge situated within it, our knowledge is incomplete, and will never drive a human to really work to apply that knowledge such that both human and more-than-human in the relationship benefit and flourish. Perhaps some radically expanded ecology (e.g., as truly the study of *home*) could accommodate it, but such an expansion, I am arguing in this book, would have to include the more-than-material and the possibilities for closeness that such ontologically irreducible elements enable.

If good human-nature relations are entered into for their own sake, and from that, knowledge flows back and forth between both relational participants, the knowledge can be said to be true for this particular experience in this particular time, which ecofeminists argue is the only kind of experience there is. This is not to make reliability relativist, since under similar conditions with similarly oriented participants, human-nature relationships will produce similar knowledge. Thus, when I finally say I know that a Shrike uses food caching locations such as broken-off branches as tools, and base that on repeated observations of the behavior, those observations can be said to be reliable if they are repeated by anyone who approaches the Shrike with patience and a relational openness, and whom the Shrike accepts as a relational partner in his turn.

References

Anderson, D. R. (2004). Peirce's horse: A sympathetic and semeiotic bond. In E. McKenna & A. Light (Eds.), *Animal pragmatism: Rethinking human-nonhuman relationships* (pp. 86–96). Bloomington, IN: Indiana University Press.

Associated Press. (2003, October 7). Grizzly mauls, kills a bear 'expert'. *Seattle Post Intelligencer.*

Backster, C. (2003). *Primary perception: Biocommunication with plants, living foods, and human cells*. Anza, CA: White Rose Millennium Press.

Barad, K. (2007). *Meeting the universe halfway: Quantum physics and the entanglement of matter and meaning*. Durham, NC: Duke University Press.

Bateson, M., Desire, S., Gartside, S. E., & Wright, G. A. (2011). Agitated honeybees exhibit pessimistic cognitive biases. *Current Biology, 21*(12), 1070–1073.

Braidotti, R. (2006). Posthuman, all too human: Towards a new process ontology. *Theory, Culture & Society, 23*(7–8), 197–208.

Brown, J. H., Davidson, D. W., Munger, J. C., & Inouye, R. S. (1986). Experimental community ecology: The desert granivore system. In J. M. Diamond & T. J. Case (Eds.), *Community ecology* (pp. 41–61). New York: Harper and Row.

Dewey, J. (1929). *Experience and nature*. London: George Allen & Unwin, Ltd..

Fantl, J. (2014). Knowledge how. *The Stanford Encyclopedia of Philosophy* (Fall 2014 ed.). Retrieved from http://plato.stanford.edu/entries/knowledge-how

Gauch, H. G. (2003). *Scientific method in practice*. Cambridge, UK: Cambridge University Press.

Haraway, D. J. (2008). *When species meet*. Minneapolis, MN: University of Minnesota Press.

Hunter, F. (1973, December 11). Are plants 'people'? *Christian Science Monitor.*

Keller, E. F. (1983). *A feeling for the organism, 10th anniversary edition: The life and work of Barbara McClintock*. San Francisco: W. H. Freeman and Company.

Kirk, J., & Miller, M. L. (1986). *Reliability and validity in qualitative research*. Newbury Park, CA: Sage Publications.

Leopold, A. (1989). *A sand county almanac, and sketches here and there*. New York: Oxford University Press (Original work published 1949).

Marquez, J. (2004, January 5). Was California bear advocate naturalist or 'con man?'. *Juneau Empire.*

Medred, C. (2003, December 14). Man, martyr, myth. *LA Times.*

Noddings, N. (1984). *Caring: A feminine approach to ethics & moral education*. Berkeley, CA: University of California Press.

Objective. (2018, June). *Oxford English dictionary online*. Retrieved from http://www.oed.com

Peirce, C. S. (1960). *Collected papers of Charles Sanders Peirce* (C. Hartshorne, P. Weiss, & A. W. Burks, Eds.). Cambridge, MA: Harvard University Press.

Porter, T. (1995). *Trust in numbers: The pursuit of objectivity in science and public life*. Princeton, NJ: Princeton University Press.

Puthoff, H., & Fontes, R. G. (1975). *Organic biofield sensor*. Menlo Park, CA: Stanford Research Institute.

Reliability. (2018, June). *Oxford English dictionary online*. Retrieved from http://www.oed.com

Riefe, J. (2017, November 1). Jane Goodall on sexism, controversial feeding stations and science deniers. *The Hollywood Reporter*. Retrieved from https://www.hollywoodreporter.com/news/jane-goodall-sexism-controversial-feeding-stations-science-deniers-1055417

Rollin, B. E. (1990). How the animals lost their minds: Animal mentation and scientific ideology. In M. Bekoff & D. Jamieson (Eds.), *Interpretation and explanation in the study of animal behavior* (pp. 375–393). Boulder, CO: Westview Press.

Russell, B. (1912). *Problems of philosophy*. New York: Longman, Green and Company.

Stodden, V., Leisch, F., & Peng, R. D. (Eds.). (2014). *Implementing reproducible research*. Boca Raton, FL: CRC Press.

Taylor, A., & Giugni, M. (2012). Common worlds: Reconceptualising inclusion in early childhood communities. *Contemporary Issues in Early Childhood, 13*(2), 108–119.

Taylor, J. S. (1998). *Poetic knowledge: The recovery of education*. Albany, NY: State University of New York Press.

Warren, K. J. (1990). The power and the promise of ecological feminism. *Environmental Ethics, 12*(2), 125–146.

Weisberg, Z. (2009). The broken promises of monsters: Haraway, animals and the humanist legacy. *Journal for Critical Animal Studies, 7*(2), 22–62.

Chapter 8
Material and More-than-Material Considerations

Thus far, the elements of reality that I've suggested are more-than-material in origin, such as the feeling of territoriality shared between the horse and me, can still be argued by some to be material. For example, most modern scientists suggest that emotions have a material origin (Panksepp, 1998; Ramachandran, Blakeslee & Sacks, 1998). While I'll explore this topic in greater depth in Chap. 9, here I will contend that an ontology that admits closeness in human-nature relationships requires that these and other seemingly more-than-material elements be more-than-material in origin. I say this for two reasons. First, as I will argue in subsequent chapters, some of the things noted about the behavior of plants, humans, more-than-human beings, and even inanimate beings cannot be explained by current, material understandings. Second, because materialism has been so dualistically wielded to conceptually extirpate the possibility of closeness with more-than-human beings, an alternative view that undercuts such a practice seems at least prudent, if not necessary. Should someone desire to fashion my claims into a greatly expanded or more "forgiving" notion of materialism, that work may yet be of value. My suspicion, however, is that because materialism will have no truck with spirituality or intuition as ontological elements, a true reconciliation is not possible.

I begin by noting that both Peircian feeling and relational knowledge comport fairly "naturally" with an ontology that allows for more-than-material elements. While I made no explicit appeal to more-than-material ontological elements in discussing poetic knowledge above, Taylor (1998), who I take as an authority on this kind of knowledge, does. For example, he says that

> since the mind as the immaterial power of knowledge must, in order to know, be able to receive something likewise immaterial from the objects of reality, it follows that there be a *form* of the thing, its immaterial substance, obviously not its physical presence, that is impressed into the mind... (p. 62)

The rest of this chapter contains a further consideration of some important points about the material/more-than-material distinction.

© Springer Nature Switzerland AG 2019
N. H. Kessler, *Ontology and Closeness in Human-Nature Relationships*,
AESS Interdisciplinary Environmental Studies and Sciences Series,
https://doi.org/10.1007/978-3-319-99274-7_8

8.1 Materialism's Support for More-than-Human Relational Qualities and Capacities

So far in this book I have critiqued materialism's dualistic marginalization and nega-tion of more-than-human beings as relational Selves. But, it would be inaccurate to suggest that all materialist stances work against the suggestion that more-than-human beings have the necessary qualities and capacities for close relationships. For example, Rollin (1990) notes that consciousness in animals was an accepted fact for Darwin and other 19th century biologists—most of whom were thoroughgoing materialists in their ontological commitments. To them, animal mentation and con-sciousness were "an inevitable consequence of phylogenic continuity…[where i]f morphological and physiological traits were evolutionarily continuous, so too were psychological ones" (p. 376). The result of such views is that instead of theories of animal consciousness being anthropomorphic—which positions humans as the stan-dard of consciousness against which all more-than-human beings are measured—consciousness itself becomes the measure, with human and more-than-human alike sharing (or not) in such qualities. Another materialist whose work lends support to elements of my own is Sheets-Johnstone (2011), who accuses materialists who deny the existential nature of consciousness of reductively ignoring the evidence for a "knowing subject" with "corporeal consciousness" (p. 69). Posthumanism also claims to support this, but my exploration of posthumanism in both Chap. 3 and the latter portion of Chap. 5 show that while their ontology may make room for some form of materially-rooted consciousness, too often the old dualistic assumptions remain unchallenged, and consciousness is only thought to manifest in the same hierarchical way as it does in humanistic views.

Ultimately, the weakness of broader-based forms of materialism is twofold. First, they are still reductive and ultimately dualist, since they've really only moved the boundary for inclusion in capacity for feeling and consciousness outward from what remains a human center. As I'll address more in Part IV, in an expanded mate-rialism human physiology is still the dualist standard by which more-than-human feelings and consciousness are measured. This is anthropocentric, and tends to cause theorists to either overlook or discount entirely evidence showing that beings with physiology vastly different from humans have feelings, thoughts and con-sciousness. As I'll discuss in Chap. 10, if a plant can be shown to have a mind, materialism either has to be radically altered as it pertains to the understanding of what physiology is required for a mind, or the commitment to a monist material view has to be discarded entirely. The latter point dovetails with the second weak-ness of even a broadened form of materialism. That is, there appears to be *no* evi-dence that is sufficient to sway the materialist from his or her material ontological commitments. If a position such as this isn't falsifiable—falsifiability being a requirement of any scientifically valid theory or hypothesis—then it's not really a theory, but a belief closed off to alternatives. I'll address this in more detail in the next two sections, so conclude here by again rejecting any wholly materialist bases for the qualities and capacities necessary for close human-nature relationships.

8.2 Evidence for the More-than-Material?

More-than-material ontological elements may, at first, seem harder to find than material ones. This is mainly because, by definition, suggesting something is more-than-material implies it is intangible. In materialist-dominated modern societies where the ability to touch something has been equated with its reality, it makes sense that the intangibility of more-than-material ontological elements would come to be thought of as unreal.

I acknowledge that here is a certain comfort from the solidity of repeatedly see-ing a mosquito lay viable eggs after a blood meal, and failing to do so when the mosquito lacks one, that on the face of it supports a material view of the world. But is even this an originally material event? Where does the mosquito get her blood, the human or the deer? How does she decide? Certainly, there may be material factors (Hess, Hayes & Tempelis, 1968), but that only primes the mosquito to choose, it doesn't *determine* the choice unless one assumes that mosquitos operate purely on instinct. This is often exactly what is assumed, but recent findings suggesting that honeybees, a fellow insect to the mosquito, can have a pessimistic bias (Bateson et al., 2011), call into question such materially facile conclusions. For honeybees, past negative events such as an attack on the hive can color the choices they make in the future. Such experimental results led Bateson et al. to wonder whether honey-bees have minds. Thus, either the notion of what makes up a material mind has to expand radically, or there must be an admission that minds are not wholly material. A commitment to the former may be based in evidence, but as I've pointed out thus far, it is just as likely based in a pre-existing assumption of a materialist reality.

If one were to entertain that more-than-material ontological elements do exist, this brings us to the question of how evidence for them might be found. This may appear at first daunting given that the most accepted modern means of finding evi-dence are material in nature. Thus, one might assume, evidence of this sort is sup-portive only of a materialist ontology. But this itself is a faulty assumption for a number of reasons. First, material evidence is only "better" and more supportive of a materialist ontology if one ignores the influence of the more-than-material on that material evidence. In the case of the mosquito, depending on one's ontology the mosquito's more-than-materially original choice of blood sources may be the deter-minative influence over whether and how she reproduces. And again, *the choice itself* is only originally material in an assumed materialist ontology. I'll discuss how this operates at material and more-than-material levels in more depth in Part IV.

Second, material evidence can only be taken as proof *against* the reality of the more-than-material in an approach that *a priori* holds reality to be material only. This means that the arguments supporting the claim are circular. In the relational ontology for which I've sketched portions in Chaps. 9 and 10, materially observable phenomena could, and I suggest should, be taken as evidence *to support* the exis-tence of the more-than-material. For example, with respect to emotions, renowned neuroscientist Candace Pert (2006), says that

> the movement of energy in the body was possibly due to the flow of emotions going back
> and forth between the physical and the spiritual, existing on the physical level as peptides

and receptors, but also in the spiritual realm as information. Chi, prana, and meridians—
these phenomena could possibly be explained by the flow of information and emotions.
(p. 253)

In this view, emotions are both material and more-than-material, and the existence of each is reflected in, and evidence for, the other. To underscore my point that the seeking of academic or scientific evidence for the more-than-material is "rigged" to support pre-existing materialist ontological views, I'd ask of those holding a materialist ontology what *would* constitute evidence for the existence of ontologically more-than-material elements. I think it can be fairly uncontroversially asserted that the answer would be, "not much." Thus, the question has been closed off dualistically by a materialist ontology before it can even be diligently considered.

In the scenario where the material and more-than-material influence but are irreducible to each other, material evidence becomes the "signature" or evidence for the more-than-material. For example, everyone knows that stress has many physiological effects (US Department of Health and Human Services, 2012). In a relational ontological orientation admitting material and more-than-material elements, suggesting that chemicals change and *then* one "feels" stress is a case of putting the cart before the horse. You feel stress from an outside or inside more-than-material "force," and this changes things like the chemistry in your brain. You can give a prisoner-of-war all the anti-depressants in the world and the more-than-material stress and pressure will keep him depressed. It is a more-than-material problem with a more-than-material solution—in the latter of this example, the solution is his freedom. And again, one cannot suggest that the stress originated in some antecedent material events unless one holds a sponsoring materialist ontology—clinging tenaciously to one's egg over the other's chicken. I suggest instead that there is a back and forth flow between the material and more-than-material. Thus, some uplifting of emotions from antidepressants given to such a person can result in an influence of the material, but the influence is *post hoc*, and does not go to what I see, in this example, as the more-than-material root of the emotionally problematic reality. In my portrait of a relational ontology, the more-than-material can easily be seen as the source of the emotions in this situation. I will discuss this dynamic in greater detail in Part IV, so conclude here by saying that the material results of hypertension, heart disease, and altered serotonin or dopamine levels from stress is evidence that supports stress's position as a more-than-material source.

So far, my discussion has centered on material evidence for more-than-material ontological elements. But what of more-than-material evidence? I argue in the previous chapter that Peircian feelings, for example, can be shared directly amongst individuals. If one takes the medium of this sharing to be more-than-material, this means that one can experience more-than-material feelings directly, and no material markers for such foundationally more-than-material phenomena are necessary. While some argue that ultimately, even feelings and their sharing are materially based such as in brain-injured humans whose emotional lives change radically after the injury (Ramachandran et al., 1998), I suggest that such conclusions are a case of mistaking the material apparatus with which feelings operate for their source. I suggest that it is areas of the human brain that aid in the *perception* of feelings that are injured, and that it is akin to the destruction of the antenna of a radio. The radio will no longer receive the signal clearly if at all, but it doesn't mean that the antenna created the signals, nor does it mean that the signals aren't still being broadcast. For example, Freedman (2007) quotes Pert as saying, "We're not just little hunks of meat. We're vibrating like a tuning fork—we send out a vibration to other people. We broadcast and receive" (para. 5). Thus, while such material phenomena may shed some light on the contributions the material makes to the ability to perceive more-than-material feelings, these material elements are not necessary to support the reality of the irreducibly more-than-material nature of these feelings.

Poetic knowledge, gathered as a result of repeated direct experiences and "transmitted" through the medium of the more-than-material, if taken to function in this way, offer support for the reality of the more-than-material. Thus, if I'm having the experience of being in love in the Buberian sense of actually being *inside* the phenomenon of love with another Self, that constitutes evidence for the more-than-material nature of that love. I am not arguing here that it is conclusive proof, since some of the principles of good evidence gathering (e.g., repeated observations) should still be followed, but I am saying that the category into which one fits each instance of such phenomena can no longer be uncritically assumed to be material.

To the suggestion that such a perception about the experience of love is the stuff of fantasy, I suggest that without an unsupported, materialist ontology as backing, the exact same charge can be made against the argument that love is ultimately materially based. The only reason the latter is less controversial is because of the hegemonic force with which materialist ontologies have operated in modern societies to reinforce their own positions while undercutting the more-than-material. To the suggestion that one cannot rely on self-reports of being in more-than-material love as proof of a more-than-material experience I'd respond by noting that self-report is often deemed as quite reliable in scientific circles (Stone et al., 2000). Again, if the report is based on well-formed relational knowledge, whether the knowledge is of the material or more-than-material is, to put tongue firmly in cheek, immaterial. More importantly, however, I'd respond to the questioning of more-than-material self-reports by wondering what *would* constitute better proof of the existence of the more-than-material than this kind of information. I assume that a materialist might resort to some "independent" means of verification such as, in the case of reports of feelings of love, elevated heartrate, loss of sleep, etc. But, by doing so, we have come full circle, and back to seeking material verification of more-than-material reality, and for the materialist, this will never do. If the equivalent of valid proof is *material* proof, then as long as the ontology is materialist, *no* amount of the latter will do anything other than bolster the materialist position. In the end, by already holding a materialist ontology, and through that, preordaining only material proof as valid—and as proof only of a materialist ontology—one has dualistically eliminated the possibility of any evidence to the contrary, regardless of what's actually experienced and how. I will explore other self-reinforcing forms of materialist argument in Chap. 10 when I critique the notion of emergence.

8.3 Ruminations on More-than-Material Relational Elements

In the chapter thus far, I've attempted to "make room" for the more-than-material by pushing aside the dualism inherent in a monist material position. In this section I'll explore some features of human-nature relationships in order to further explore their whether their best fit is within material or more-than-material ontological categories.

The predominant, modern perspective on human-nature relationships tends to portray more-than-human beings as possessing material-only attributes that interact with the attributes of humans. Though in purely materialist perspectives, all human attributes are ultimately also taken to be materially sourced, whenever substantive relational elements are under discussion, it is usually human or human-like animals that are seen to be in possession of the material structures and processes necessary for the development of close relationships. For example, at times environmental education programs seek to develop in humans an awareness of various material aspects of more-than-human beings so that within the human care, stewardship or any number of other benevolent, or at least benign, human relational overtures toward more-than-human beings are inspired. Such programs are attempting to inspire wonder, curiosity, and a subsequent sense of responsibility (e.g., Chawla & Cushing, 2007), thus the human-nature relationship is improved Though wonder can inspire better human-nature relationships, I question whether that wonder is really of a material origin. That is, is it really the *material* that, observed or interacted with by the child, inspires in that child an upwelling of wonder—an affective response rooted itself in some purely internal-to-human, material affective structures and processes? I admit to being of two minds on this topic, so will explore both.

My first inclination is that the material can have this direct effect. James (1907) says it best when he suggests, in somewhat macabre fashion, that it's possible that

> one's opposition to materialism springs from one's disdain of matter as something 'crass'... [but instead is] infinitely and incredibly refined. To anyone who has ever looked on the face of a dead child or parent the mere fact that matter *could* have taken for a time that precious form, ought to make matter sacred ever after. (p. 95)

I believe he's saying that matter itself can inspire "more-than-material" response—it does not need any more-than-material qualities of its own. As such, a material-only conception of more-than-human beings can suffice to explain the strong more-than-material response of humans to them. At one level, I'm inclined to leave it at that. After all, anyone who has seen glaciers calving in Alaskan fjords, or the flick of a hummingbird's tongue as it hangs three feet in front of one's face, can understand that matter itself is at times achingly wondrous and beautiful. Yet I hesitate. And do so because I cannot leave these material things alone without what I see as their inextricable, more-than-material complements.

Would these dead faces be so precious if not for the more-than-material counterpart of the love we feel to ones so close? That's why in Chap. 3, in my discussion of

Haraway's description of love between her and her dog, I note that there is something discordant and even distasteful in her material explanation of the origins of love. It just doesn't "ring true." While there has to be a grouping of physical individuals for love to have anything upon which to operate, that they are the source of love instead of its correlated experiencers can only be definitively claimed if one assumes a materialist ontology. Such a construction is not evidence for such an ontology. As my critique of notions like the material mechanism of "emergence" show in Chap. 10, things like love and consciousness do not comport easily if at all with material explanations. Who is it, after all, that quickens the tongue of the hummingbird except the hummingbird as a relational Self akin to me as a relational Self? It is the spark of that more-than-material encounter between Selves, just like between the horse and me, that is in part responsible for my amazement. All of these things—love, Selves, sparks—seem to defy facile material compartmentalization.

The glacier I mention, being "inanimate," is potentially harder to argue for those skeptical about my claims thus far. Because I'm disinclined to broaden my references to spirituality at this juncture, I instead use the glacier as a departure point for contemplating just exactly what the nature of beauty, wonder, or "amazingness," are. For example, if I see a glacier, or a river stone, and am amazed at the river stone's bands of white then is the beauty and amazement in me as the beholder alone or in the stone or the glacier? Or, is it something that inheres in our relationship? At some level, my response *is* mine, but I wonder what exactly it is I am responding to in the other. Is it the material banding, or is it beauty in the thing itself along with the capacity of the stone to invoke wonder over that beauty in a creature such as me? As such, it is not the material of the other that brings this about, but beauty and its capacity to invoke amazement. If true, not only can we be *in* a relationship in the Buberian sense, we can be *in* an amazement-producing one as well.

In a relational world with relational interactions such as these, there cannot be a one-way interaction, nor can that which occurs as a result of an encounter (e.g., wonder) be something that is only in me. Otherwise, we are all hopelessly disconnected, and no quality at all of the stone could evoke such a thing on its own. In that scenario, it is utterly random and according to some isolated nature of mine (apart from the nature of the stone or other more-than-human being) that the stone invokes this. It could just as easily be a landfill. There is no external, qualitative difference between the two. And yet wouldn't common sense dictate that there is some beauty in the stone that the landfill does not possess—and that this is foundationally, or ontologically, intrinsic to the stone? If in response one were to suggest that it is just our nature to adjudge river stones as beautiful and landfills ugly—a random byproduct of an evolutionarily material necessity perhaps—then I believe we've again arrived at the unfounded assumption of the primacy of the material. This over the possibility of there being an equally plausible, existential thing that is more-than-material *beauty*.

That this beauty inheres in the material and is experienced in pairing with it does not mean it *is* material or that it originates there, however. Again, in an ontology that holds the material and more-than-material to be co-existing and complementary, the two can be, no *are*, different and irreducible, yet interdependent.

Therefore, I suggest that there is something *in the river stone* that is capable of being seen as beautiful. At which point the difference between the capacity to be seen as beautiful and actually having the existential quality of beauty borders on the semantic.

But, though I suggest the difference borders on the semantic, there is still an important difference that must be resolved. Is this beauty a more-than-material quality of the thing thought beautiful, or is the beauty just a way for humans to conceptually, and internally-to-themselves, encapsulate something material that we deem 'beautiful' instead of 'ugly?' I'll press this further. If I suggest that a thing in nature is astonishing (i.e., has the internal, more-than-material quality of being astonishing), the argument against this is that some are not astonished by it. But, there are *two* explanations for that lack of an astonishment response. One is what the materialist counterargument assumes, that astonishment is wholly inside the one astonished, and so a lack of astonishment can be accounted for by some difference in personality, brain chemistry or the like. But, a second explanation is that the person who is not astonished is unusual in some way—one might even say flawed—and *ontologically ought to be* astonished because the thing to which one reacts with astonishment is inherently astonishing. To support such a notion, I again quote Peirce (1960) as saying of external feelings that, since "feelings are communicated to the nerves by continuity...there must be something like them in the excitants themselves" (6.158, p. 111). Thus, as Noddings (1984) says of the occurrence of this phenomenon between humans, here one ought to "feel with" the astonishing thing by feeling the irreducible astonishment conveyed.

Let's forget about the banding of river stone for a moment to consider a more obvious example of something externally, and largely universally, thought to be astonishing such as the Grand Canyon in Arizona. Of a person unmoved by such a sight, we would think there was something a little "off" with them, so powerful is the capacity of the Grand Canyon to move humans in general to astonishment. We would say of the one un-astonished that he is not perceiving correctly—that he doesn't have the capacity to *recognize* the obvious magnificence of the place. It would stretch the limits of common sense to say that there is *nothing* astonishing and the person is merely unlike most others who respond to it by constructing a conception of that astonishing place inside themselves. Of course I recognize that a staunch materialism not uncommonly found in modern discourse would hold to just such a position. In response I can only suggest again that such a position is sponsored by an *a priori*, not evidentiary, conceptualization of such an experience, and is driven by the theory, not the experience. Eschewing such dualist assumptions, it is as likely that there is an astonishing quality that is at least in part external to the one astonished. In this scenario, the problem is not lacking an internal sense of astonishment, but failing to recognize that which *is* astonishing.

And if the argument against such a claim is that different things are astonishing to different people (therefore a universally agreed upon astonishing quality cannot be truly external to humans), I'd counter that by arguing that in reality *everything is astonishing in its own way*. According to the particular natures, socialization and human-nature relational experiences of particular humans, each is receptive to discerning a particular kind of astonishment-inducing thing. For one it is the trail of an ant across the sand, for another it is the eddies in a river. For yet another, the wood grain of oak or the banding of a stone. To each according to her own predisposition to wonder at the great diversity of wonders in the world. For every Green Peace warrior in love with the Right Whales off the Atlantic Coast of North America there is the entomologist. How different would the entomologist's relationship be with bugs if the astonishment that drove him to his career was loosed from its materialist/scientific constraints to be something founded in, and openly driven by, the innate thing in him that loves what's innately wondrous in bugs? In a modern, materialist world, his love can only be his unilaterally for a dualistically reduced, material and largely passive other. Thus, he has a career in which, at least in professionally published work, he maintains an emotional distance that won't betray his love, and that portrays bugs as "really" just material others. But that is only his experience of the bugs if his ontology suggests that his astonishment is his, and not the bug's, or not communicated and/or shared in some way between them.

Yet another way to see astonishment is by taking a pragmatist approach of letting the primary experience of astonishment dictate the truth of the relationship. In such a context, the experience of astonishment does not align well with a view of it as a *post hoc* product of internal-to-human processing. Like the shock of delight of a hummingbird circling my head beneath a spruce tree in South San Francisco, astonishment seems, more than anything, to *bypass* any internal "processing" and come straight to one's "I" from Buber's "Thou" in the experience. The explanation of astonishment as an internal construction has always seemed to me a weak one given the power and immediacy of such relational phenomena. Thus, astonishment, wonder, or beauty seem much more in keeping with Peirce's external feelings than any subsequent, internal responses. Such an observation is unproblematic when one discards the dualist ontology that is monist materialism. Besides, the root meaning of amazement is that which stupefies ("Amaze", 2018). If so, to be amazed means to be *stopped* from thinking. It means to have one's thinking outflanked by one's experience such that one only feels that which there is to feel from the other as a relational Self, and to respond to that feeling before having had a chance to think or construct anything.

References

Amaze. (2018, June). *Oxford English dictionary online*. Retrieved from http://www.oed.com

Bateson, M., Desire, S., Gartside, S. E., & Wright, G. A. (2011). Agitated honeybees exhibit pessimistic cognitive biases. *Current Biology, 21*(12), 1070–1073.

Chawla, L., & Cushing, D. F. (2007). Education for strategic environmental behavior. *Environmental Education Research, 13*(4), 437–452.

Freedman, J. (2007, January 26). The physics of emotion: Candace pert on feeling go(o)d. Retrieved from http://www.6seconds.org/2007/01/26/the-physics-of-emotion-candace-pert-on-feeling-good

Hess, A., Hayes, R. O., & Tempelis, C. (1968). The use of the forage ratio technique in mosquito host preference studies. *Mosquito News, 28*(3), 386–389.

James, W. (1907). *Pragmatism: A new name for some old ways of thinking*. New York: Longman, Green, and Co..

Noddings, N. (1984). *Caring: A feminine approach to ethics & moral education*. Berkeley, CA: University of California Press.

Panksepp, J. (1998). *Affective neuroscience: The foundations of human and animal emotions*. Oxford: Oxford University Press.

Peirce, C. S. (1960). *Collected papers of Charles Sanders Peirce* (C. Hartshorne, P. Weiss, & A. W. Burks, Eds.). Cambridge, MA: Harvard University Press.

Pert, C. B. (2006). *Everything you need to know to feel go(o)d*. Carlsbad, CA: Hay House, Inc..

Ramachandran, V. S., Blakeslee, S., & Sacks, O. W. (1998). *Phantoms in the brain: Probing the mysteries of the human mind*. New York: Quill.

Rollin, B. E. (1990). How the animals lost their minds: Animal mentation and scientific ideology. In M. Bekoff & D. Jamieson (Eds.), *Interpretation and explanation in the study of animal behavior* (pp. 375–393). Boulder, CO: Westview Press.

Sheets-Johnstone, M. (2011). *The primacy of movement*. Amsterdam: John Benjamins Publishing.

Stone, A. A., Turkaan, J. S., Bachrach, C. A., Jobe, J. B., Kurtzman, H. S., & Cain, V. S. (Eds.). (2000). *The science of self-report: Implications for research and practice*. Mahwah, NJ: Lawrence Erlbaum Assoc.

Taylor, J. S. (1998). *Poetic knowledge: The recovery of education*. Albany, NY: State University of New York Press.

US Department of Health and Human Services. (2012). Stress and your health fact sheet. Retrieved from http://womenshealth.gov/publications/our-publications/fact-sheet/stress-your-health.html

Part IV
Vectors of Interdependence

To this point in the book, I have focused on explicating dualisms in modern thinking, and on showing how they operate in largely *post hoc* fashion to define more-than-human beings as inferior in their capacity to engage in close relationships with humans. By supplanting such dualistic views with a foundationally relational ontology, I've also attempted to show that there is nothing standing in the way of the possibility for human-nature closeness. Having laid this ontological groundwork, the remainder of the book focuses on exploring the two main vectors of interdependent closeness that I articulated in Chap. 4—feelings and thoughts. In the next two chapters, I'll show in more depth how dualisms operate in distorting the possibilities for each of these elements, and offer alternative views such that they become components of the possibility for closeness.

Part IV
Vectors of interdependence

Chapter 9
Feelings

Feelings, or emotions, figure centrally in any discussion of close relationships. Human interdependence theorists Kelley et al. (1983) suggest that emotions are so strongly tied to the notion of close relationships that "[m]any do not consider a relationship between two people to be close unless there exist strong positive affective [i.e., emotional] ties between the participants" (p. 110). Similarly, attachment theorists Hazan and Shaver (1994) note that the attachment that later leads to close relationships is primarily an emotional bond (p. 3).

Given the essentiality of these feelings, it might be supposed that, since beings in the more-than-human world are generally not taken to have emotions that can contribute substantively to close relationships, the likelihood of their ability to enter into close relationships is minimal. This view, however, is rooted in three erroneous assumptions. The first is that emotions can only be experienced through, sourced in, and/or are equivalent to, material emotional structures and processes. The second is that emotions are something wholly contained within an individual—they have no externality and thus no way of directly encompassing relational participants. The third is that only humans or human-like animals can experience emotions—and we can only know that human or human-like animals do so through analogy with our human selves. This last assumption is partly predicated on the first, and also on the concomitant belief that only *human or human-like* material structures and processes can originate emotions.

If these three assumptions are true, then the denial of the feelings necessary to forming close relationships to all but humans and human-like animals is justified. It is my intention in this chapter to challenge these assumptions on the grounds that they are rooted in dualisms instead of in anything empirical. I'll pause here to note that while posthumanism itself appears at first to expand the field for inclusion of any more-than-human beings whose intra-action with humans generate some feelings that contribute to closeness, as I discussed in Chap. 3, notions of asymmetry operate as a Trojan horse for a relational hierarchy that dualistically leaves humans at the apex. Thus, I find the material basis for their arguments to be insufficient.

© Springer Nature Switzerland AG 2019
N. H. Kessler, *Ontology and Closeness in Human-Nature Relationships*,
AESS Interdisciplinary Environmental Studies and Sciences Series,
https://doi.org/10.1007/978-3-319-99274-7_9

My hope here is to push beyond material constraints and to make conceptual "room" for feelings in all beings—from a stone to a Killer Whale—and thus to allow for the potential capacity for all beings to participate in close relationships. This chapter will take up discussion of assumptions that feelings are purely material, purely internal-to-a-self, and also almost exclusively human, in order to move beyond them.

9.1 Feelings As Purely Material

Researchers have linked material changes in humans and animals to various emotional states. Most often, such material evidence of emotion is taken as sufficient to conclude that emotions are materially based. The only way that this evidence can be considered sufficient to draw such a conclusion, however, is if one already holds an assumption of a materialist ontology. If, instead, one believes that feelings have more-than-material elements as well, then the very same material evidence can be interpreted as being supportive of the existence of those more-than-material elements. This is especially the case if the material evidence is seen to occur subsequent to the more-than-material emotional stimuli.

As an example, one phenomenon that can be interpreted as having more-than-material origins is that of antidepressant "tachyphylaxis," which is the loss of efficacy of anti-depressant drugs. From within a materialist ontology, authors such as Byrne and Rothschild (1998) nominate material explanations such as what the drug does to the body, changes in the material source of the disease, the accumulation of metabolites that block the drug's efficacy, etc. From within an ontological frame of reference that includes irreducible more-than-material elements, however, just as plausible is the suggestion that depression has a more-than-material origin that material therapies can never wholly address. From within the latter frame of reference, an explanation for tachyphylaxis is that while initially the anti-depressant has an effect, it leaves intact the more-than-material causes that eventually reassert their influence. In such a view, the cure for depression is happiness. I don't mean that glibly, as I'm well aware that diseases such as major depressive disorder are best treated by psychotherapy and pharmacotherapy (Pampallona, Bollini, Tibaldi, Kupelnick, & Munizza, 2004). What I do mean, though, is that even in a person suffering from major depressive disorder, it's still possible that the problem is at root more-than-material. To support this conjecture, Midgley (2004) says that there is a "reductive and atomistic picture [of reality that] now leads many enquirers to propose biochemical solutions to today's social and psychological problems, offering each citizen more and better Prozac rather than asking what made them unhappy in the first place" (p. 2). Servan-Schreiber (2004), a physician, neuroscientist and clinical Professor of Psychiatry at the University of Pittsburgh School of Medicine, also echoes this point of view when relating a personal experience:

> During a visit to France a very close childhood friend told me about her recovery from a serious depression. She had refused the medications that her doctor had offered and she had

sought the care of a sort of healer. She was treated with 'sophrology,' a technique that involves deep relaxation and reexperiencing of old, buried emotions. She had come out of this treatment 'better than normal.' Not only was she no longer depressed, she was also freed from the weight of 30 years of unexpressed grief over the loss of her father, who had died when she was 6 years old. (p. 6)

Even a mental illness such as schizophrenia, held by most to be material in origin and best controlled (note I didn't say "cured") by drug therapies, may not necessarily be caused (and thus radically affected) by material interventions. Phenomenological psychotherapist R.D. Laing (1971/2001), in his book *The Divided Self*, suggests that at root a schizoid personality is not experiencing problems due to any physiological cause. Instead, Laing says of such a person that his "experience is split in two main ways...in his relation with his world and...his relation with himself" (p. 15). Laing's cure for this person is to follow her psychologically through the mental labyrinths she constructs as a result of those splits in order to help her find a way back "out" to a feeling of coherence (in the sense of internal, mental unity). In this case, while schizophrenia has a material "signature" in altered serotonin levels, etc., it is a more-than-material problem at its source, requiring understanding of it in these terms and the grounding of its most complete cure in approaches aimed at addressing them.

Those adhering to materialist ontological commitments can always counter my categorization of these phenomena by saying they are preceded by the material. One likely tack their rationale might take is that psychotherapy is still conducted by originally material beings, and that unhappiness-producing circumstances are created by beings or a world that is ultimately material in its origins as well. The soup of cells that made a father that the eventually schizophrenic girl lost at 6 years old is still, and originally, a soup of cells. The problem with such an argument, however, lies in the fact that it is exactly examples like those of clinical depression and schizophrenia affected by material interventions that are offered as *support* for such materialist interpretations. This means that the examples offered as support for a materialist interpretation of reality are wholly dependent on a pre-existing supposition of that interpretation for their validity. This is circular, and thus largely empty. By no means am I denying that there are material components to feelings, I just don't see where their discovery and elucidation stands as proof that they are feelings' *originators*. As Pert (2006) suggests, material elements are affected by emotional experience and, as I suggest, with an adjusted ontology they can just as easily be interpreted as having more-than-material components to their origin.

To underscore the difference in material and more-than-material origins of emotion, I refer to an example from Panksepp (1998), who holds that feelings are material in origin. One of his supporting experiments involves the insertion of an electrical probe into a certain area in the brain of a cat. When electricity is delivered to that area of the brain, the cat becomes enraged. Thus, he concludes, rage is material in origin. But, this conclusion is flawed for two reasons. First, the rage he induces in the cat shows marked differences from rage when it is "naturally" produced. Second, Panksepp and others like him appear *a priori* committed to a materialist ontology, allowing it to artificially constrain their definition of the feeling phenomenon. Thus, they miss or ignore what a feeling *experience* actually is.

My first suggestion regarding Panksepp's experiment is that the rage the cat feels is not the same as rage the cat might feel if driven to it by more "natural" means, such as the introduction of a territorial rival. One element of the experiment that hints at their not being the same is when Panksepp describes the aftermath of the rage. He says, "Within a fraction of a minute after terminating the stimulation, the cat was again relaxed and peaceful, and could be petted without further retribution" (p. 194). Panksepp sees this as proof that rage is material in origin, and does so because once the material trigger was terminated, the anger was terminated. But, I suggest that in a *real* anger experience, the cat would not so quickly, easily, and, I suggest, unnaturally, be able to move back into a peaceful state. In a real anger experience, humans, cats and other creatures have to take some time to "come down" from the feeling. What's more, in a real rage experience, an association is often created in the mind of the one experiencing it that the target of the rage is to remain a target if encountered again. But, as Panksepp notes in his description of the experiment, while experiencing the "rage" the cat leapt at Panksepp's head, but in the aftermath he allowed Panksepp to pet him as if nothing adverse had occurred. I suggest that this is because, in the context of real emotional experiences, nothing adverse *did* occur.

The difference I'm pointing out is akin to the difference between a smile being produced by providing an individual with a happiness-inducing experience or by pulling up the corners of that individual's mouth. In Panksepp's experiment, has he engaged in the former or the latter? I admit that my parallel is oversimplified, but it is an oversimplification of degree, not kind, and illustrates my point that if Panksepp has discovered the origin of rage (e.g., material electrical activity), then one would expect it to behave like rage in all its facets, not just in a narrow moment of its materially observable expression. In a smiling event, an individual would likely not feel happy when the corners of her mouth were lifted, and while a probe delivering electricity into the brain may more closely stimulate the expression of happiness or rage, given the differences in Panksepp's version of induced rage, I think it's legitimate to raise questions as to whether it is truly emotion or feeling experiences whose origins have been discovered.

In response, a materialist might note that for the *particular* emotion of rage stimulated by the probe, the material is clearly being shown to precede it, thus it has to be its originator, even if the electrical probe does not stimulate the entirety of the complexity of such a material emotional event. But this logic is faulty as well. If the rage is not the same rage, then what Panksepp has shown is that one element of rage—its observable, end-product expression—can be induced through material means. Panksepp is confusing the observable, material action *in response* to the rage with the rage itself.

In my second suggestion above, I note that Panksepp and others are dualistically allowing a materialist ontological commitment to artificially constrain what is considered part of a feeling experience or phenomenon. Under normal, healthy and intact physiological circumstances how does one become enraged? Each instance is preceded by something more-than-material. A thought. Another feeling. Some physical injury *the experience of which* leads to thoughts that lead to rage or leads directly to rage. What materialist experimenters like Panksepp are forgetting is that by inserting a probe into the brain, they're not going far enough "upstream" in an intact feeling phenomenon. They're bisecting the process and then suggesting that by achieving some approximation of the same end result that they've happened upon the same cause. Posthumanist Barad (2007) can help us here with some terminology. In her ontology, she'd say that the experimenters are not accounting for all components of the experimental *intra-action*.

This discussion obviously raises the issue of just what "a feeling" is—where it begins and ends. If, as I've suggested in earlier chapters, it can be communicated directly between those experiencing it, or can suffuse a relationship from outside both participants in Peircian fashion, then where the feeling starts and ends certainly escapes the boundaries of the individual. I will address this in more detail in the next section, but here, I think, there are really only two possibilities for where feelings start and end, and both pose problems for a purely materialist interpretation. First, it could be supposed by the materialist that the electrical stimulation of the cat's brain is part of the feeling itself. But one can always critique such a position as having dualistically drawn too confining a boundary around the material elements of the feeling intra-action. To its furthest extent in this example, one could charge that such a boundary has failed to include the origin of the feeling in a human researcher whose more-than-material *desire* to understand the nature and origin of rage has caused the cat's rage. In a close relational ontology, this could certainly be the cause of the cat leaping at the *researcher's* head during the event instead of at some random object.

To counter this, the materialist might adopt a second, inverse stance: that of conceptually isolating the electrical stimulation from the feeling and saying that the experience of rage and its simultaneous expression equate to a "feeling." But this releases the feeling from any dependence on a material origin which, ontologically, has any number of precedent, more-than-material origins such as those to which I've pointed above. In this latter possibility, feelings have no exclusive ontological continuity with material influences on them. I don't believe the loss of this material association is what those seeking a purely material ontological explanation for feelings of emotions have in mind, so this tactic fails as well.

But, there is yet a third way out for the materialist. One can adopt a strict materialism such as eliminative materialism, where all experiences as distinguishable from material structures and processes are a form of fiction. And while there are some who adhere to such ontological positions, they are generally seen as being more ontologically extreme, and run counter to the more widely accepted notion that consciousness, and the complex feelings and thoughts that go with them, are more than the sum of their material parts (see Chap. 10 for more on this in the discussion of Emergence theory).

To conclude, there is no real way to argue that feelings are definitively material or more-than-material. One's ontology, in this case, is determinative. And if the only real, bedrock support for a purely materialist explanation for the origin of feelings is the pre-existing assumption of its exclusive legitimacy, this cannot stand.

9.2   Feelings As Purely Internal

The second assumption about feelings under consideration here is that they are wholly internal to the individual experiencing them. While posthumanist conceptualizations are one way to argue against this humanist position, because of their largely materialist ontological commitments, the feeling intra-action itself must still emerge from whichever individual-in-relation is expressing it. Some posthumanists have attempted to explain this mechanism through appeals to contagion or emergence, but as I discussed in Chap. 3 and will take up in more detail as it relates to emergence theories in Chap. 10, these explanations are ultimately insufficient. Therefore, even within much of the posthumanist literature, at their ontological roots feelings are still quarantined to the individual-in-relation experiencing them. In Part III above I discussed more-than-material alternatives to purely internal experiences of feeling, and in this section will explore the implications of such a possibility.

If one assumes, or is forced to concede due to one's ontology, that feelings must be purely internal, it seems sensible to ask how it is possible to share them, or to have the feelings of one relational partner mirror or influence the feelings of another. Posed differently, what basis exists for the continuity of feeling between those sharing them, or is if that is even possible? In a commonsense take on feelings, most would agree that feelings are shared *somehow*. Any group of people who lose a loved one would certainly attest to the fact that the grief that each individual experiences is one that is also shared amongst the group. But, those theorizing about feelings hold that such a sharing is not direct. A typical example of this type of explanation comes from the psychologist, Bucci (2001), who suggests that the

> affective communication of one individual—in sensory and motoric as well as verbal form—is received and known through the [five] sensory systems of the other, as well as through feedback from the motoric systems that are activated in response. (p. 56)

This means that the observer's emotional response is actually not to the other directly, but to his own response to his perception through the five senses. Figure 9.1 depicts this mechanism, where there is continuity of materially perceivable phenomena, but no continuity of the feeling itself. In sharing a feeling such as love in such a construct, one can only trade representations of internal emotional states and hope the other receives the transmission. Cast in satirical terms, most theories equate emotional communication to the functioning of the transporter beam in the various permutations of the television franchise, *Star Trek*. Emotion is broken down into the medium of sensory and motor expression and perception, transported via this medium, and then reconstituted as emotion at the other end. In the television show, there are always some who are mistrustful of whether what comes out the other end of the transporter is what went into it, and I am similarly mistrustful of emotional communication theorized to work in this manner. In such a system, how could two people ever really *know* that they're in love? Yet those that do know that they are probably do more certainly than anything they've ever known.

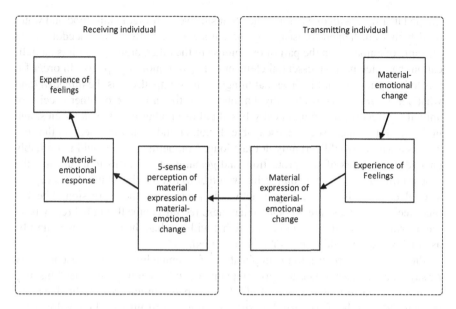

Fig. 9.1 Materialist ontological representation of the sharing of feelings

I suggest first that the *experience* of intense, shared emotion such as grief, love or worry is that they are immediate and direct, and that no such intervening mechanism is present. In Chap. 6, I spoke of Noddings' (1984) sense of "feeling with" a co-worker who reveals something personal, and in the moment of doing so dissolves the barrier between him and Noddings that prevented her caring for him in some way. Another example Noddings gives along these lines is that of a crying infant, of which she says "we react with the infant and feel that something is wrong. *Something is wrong.* This is the infant's feeling, and it is ours. We receive it and share it" (p. 31). While at first it's not clear if Noddings feels that this is an example of the kind of direct communication of feeling to which I refer in Chap. 6 and Chap. 7, her argument eventually positions the crying infant as Buber's (1923/1970) "Thou," and when he cries she quotes Buber in saying that the baby "fills the firmament" (p. 32). In Buber there is no equivocation regarding the ontological sharing of these feelings. They are direct in the Peircian sense. They are not a facsimile of the feelings the other feels, but *the same feeling*. Thus, in referring to Buber, when Noddings' infant cries, his feelings are not internal to him, displayed to another via sound, sight and touch, processed by that other via the five senses, triggering in the other her own sensory-motor response, and only then felt as a mirror of what the infant feels. Intuitively, such materialist characterizations appear to be labored, *post hoc* conceptualizations, failing to mirror the immediacy, power, directness, and continuity of such experiences as they subsume their participants.

Having suggested this alternative to the materialist explanation, I also suggest that it not be confined to the sharing of feelings between humans. Pierce (2008), in discussing the possibility of animal empathy, relates an anecdote of two mice trapped in a sink, with one helping the other survive by bringing him food and water until both were helped to escape by the author. Pierce offers this anecdote as an example of empathy on the part of one mouse to the other, arguing that it is an indication that mice have an essential element to forming morality (p. 76). In order for this to happen, though, I believe that things such as empathy must be *intended* (and what's more *received*) as empathy. On one level, that a mouse or other species of animal can receive the empathically-based aid of another of his own species, and understand it as such, seems inescapable. If not, an individual mouse, and the species as a whole, would certainly not last long. Competitive acts would not reliably engender responses of self-protection, and gestures of an altruistic nature like the mouse bringing food to her compatriot, would be as likely to be rebuffed as accepted. Though empathy is not a feeling, exactly, it is a vehicle for the expression of benevolent feelings such as care. I contend that these feelings come through directly from the provider of care to the recipient, and that without that, the empathy wouldn't be as effective, and wouldn't be as readily accepted.

Pierce pushes further into the implication of an empathy-guided mouse morality by saying that since humans also use empathy to guide their morality, that "this may set in motion a serious reappraisal of…how empathy in animals might shape our behavior toward them" (p. 76). Her suggestion is that, if mice feel empathy as we do, maybe they ought to be thought of as more like us, and thus admitted, somehow, into our moral sphere. But, I believe that what she observed carries a far more radical implication. That is, if mice feel empathy as we do, then mice and humans have a *common ground* by which they might communicate their feelings directly with each other. Thus, feelings might be a relational *lingua franca* through which humans and mice might converse. If so, this could certainly be a means by which closeness between humans and beings in the more-than-human world might develop.

What's also interesting in the relation of this anecdote is that Pierce's discussion centers around the empathy expressed by one mouse to the other, and though she abstractly ponders how, in seeing such a thing, a human might be motivated to treat mice differently, what she fails to see is that in the event itself she already had. Didn't she place a piece of wood in the sink to allow them to climb out and escape? Wasn't this participation in her relationship with the mice a form of empathy and altruism? Pierce doesn't say, so we're left to wonder. If she placed the piece of wood there out of care for the well-being of the mice, then one might say it is motivated by empathy and care. If, instead, the human adopted more of a detached experimenter role and was only intellectually curious as to what the mice would do if an avenue of escape were provided, then the feeling of care would be absent. To be able to tell if emotional communication can happen in human-nature relationships, the real question is, would the mice have been able to tell the difference? In a dualistic interpretation, the mice using the piece of wood to escape would most likely be read as responding in largely bio-mechanical ways to the sudden, mysterious and unattached presence of a path of escape. But, can we write the mice off so easily, especially if we already see them as able to receive empathy as such from one another?

Nollman (1987) would likely answer in the negative. In an experience interacting with Howler monkeys in Panama, he wonders about the role played by a human's intentions/feelings toward a more-than-human being in the type and level of close-ness of their interactions. Nollman is well known for engaging in dialogues with various animals through music. In Panama, he had gone out and done just this with one group of monkeys. During one particular session, the monkeys joined him with their own vocalizations. As he tells it, he was relating this experience to a zoologist also there to study the monkeys, and though the zoologist attributed this rare occur-rence of Howler musical participation to Nollman having a "very special rapport with animals" Nollman suggests otherwise. He says that "anyone with basic musi-cal skills, *and a benign intent toward the monkeys* [emphasis added], would achieve the same result" (p. 96). He believes that the monkeys had "'read' [him] as being benign" (p. 97). In the language of emotional communication, this means that Nollman expressed positive feelings toward the monkeys and the monkeys received it as such, responding reciprocally with a move toward physical proximity and, at least potentially, emotional closeness as well. If true, how could the monkeys have received it if not for it having been shared directly across the physical distance between them?

For all the weight that Noddings (1984) has given to direct communication of feelings between humans, she isn't as sure as Nollman that this can occur between humans and more-than-human beings. Specifically, she doesn't think that her "feel-ing with" can extend to human-nature relationships because she believes there is something innate in humans in terms of caring for one another that is too variable when it comes to humans caring for more-than-human beings (much less more-than-human beings caring for humans). Though I believe the variability she notes has more to do with what modern societies encourage and discourage in human-nature relationships, Noddings does admit the potential for the existence of a mech-anism of "feeling with" in human-nature relationships. She says that "there may be some feeling with respect to animals that is universally accessible or nearly so" (p. 149). So, while she believes other elements are necessary to support an ethic of care, and thus only interhuman caring is sure to occur, the element of being able to feel with another in reciprocal and direct ways is unproblematic in her ontology.

The implication here is that if I were to get to know a dog or bird well enough, just as with a human, I could reciprocate feelings with him or her, and be confident that the communication and reciprocation was occurring. While it's certainly pos-sible to be mistaken about what feelings are coming through to you, that they *are* coming through can no longer be taken as *prima facie* absurd. We could even poten-tially feel what a particular *stone* feels if we rid our ontologies of the dualisms that both obliterate our ability to accept the possibility of co-occupying, with any rela-tional partner, an existentially subsuming feeling, and enable the belief that feelings are only material and thus able to be experienced and shared only by those with similar emotional apparati as us. Perhaps, after having hiked up a hill and finding yourself standing in your favorite forest tract, when you are overwhelmed with a feeling of kinship with a deer or meadow vole you spy, it's because *they* are feeling this way, too, and you are perceiving it and responding to it without even knowing you are doing so. How this would influence a land manager's decision about how to "use" with that hillside is inestimable.

To the possibility of shared feeling and its potential outcome: closeness, Bekoff (2007) relates a story of friendship between a rat snake and dwarf hamster in a zoo. In witnessing it he asks, "If a snake and hamster can become friends, then why not humans and other animals?" (p. xix). With the dualistically sponsored strictures of a materialist and internal-only basis for feelings removed, the answer is *Why not, indeed?*

In the remainder of this chapter, I'll explore what I take to be the erroneous assumption that only humans, or those physiologically similar to them, can experience close relational feelings. I begin by considering animals partly because their having feelings is least controversial within a mainstream context, and partly because exposing the conceptual weaknesses in a stance of near-human-exclusivity will serve as the basis for my eventual assertion that all things in existence, from plants to "inanimate" beings, can have feelings.

9.3   Animal Feelings

9.3.1   The Denial of Feelings to Animals Is Historically Recent

Rollin (1990) suggests that most shifts in scientific thinking are precipitated by either an accumulation of evidence that stands in refutation of former claims or beliefs or the discovery of internal logical flaws within claims or beliefs (p. 377). But, according to Rollin, the denial of emotions and cognition to more-than-human beings follows neither of these paths. Since his suggestion is that scientific claims or beliefs are rooted in foundational assumptions to which all scientists must eventually commit, he believes that a third way that claims or beliefs change is when their foundational assumptions change (p. 378).

To begin to trace the denial of thought, feeling and consciousness to nonhuman animals, Rollin notes that before the turn of the 20th century, there was a strong belief in both animal cognition and emotion by scientists and lay people alike. As an example, he says that Darwin believed in an evolutionary continuity with animals and humans for both emotions and cognition (differences were differences in degree, not kind). Rollin believes that the ontological shift away from this knowledge or belief was due to a reductionist trend in society in general, with a new "emphasis on sweeping away frills, excesses, superfluities" (p. 387). This resulted in the emergence of a positivism in science that denied the knowability of animal consciousness, which in turn led to its outright dismissal. He notes the self-serving nature of such a dismissal when he says that this positivist ontology "exonerat[ed]… scientists from having to think about the morality of the animal use which came increasingly to be seen as presupposition to scientific progress" (p. 389). It also reinforced the lack of "moral shock" in "ordinary [non-scientific] common sense" at the treatment of animals whether or not they could think or feel. To the latter point, Rollin says that the lack of shock that preceded positivism probably emerged in the 19th century as a result of "so much of ordinary life depend[ing] on exploiting animals and causing them pain and suffering" (p. 390). I note here that in my discussion of Weisberg's (2009) critique of Haraway earlier in the book, the failure of Haraway to express shock or horror when justifying some forms of animal experimentation is an indicator that this positivism or humanism still undergirds materialist theorizing about animals even in posthumanist theories.

While I agree with Rollin's move to correlate the shifts in ordinary 19th century life that increased exploitation of animals with a shift in ontology, I believe that they did not precede the moral disregard of scientists, but instead were an earlier form of the same ontological shift, just played out in everyday, non-scientific circles. The key point I'm making is that the ontological shift preceded both of these trends. Merchant (1980) traces this shift to the middle ages. Plumwood (1993) to the Greeks. I agree with Plumwood's assessment that the Greeks are a very early influence on the ontology of what has become modern science, but I also think that Merchant's focus on the middle ages, and the Enlightenment, Scientific Revolution and Industrial Revolution that followed them is more useful in that this is where modern, dualistic ontological stances took the form in which we largely see them today.

I believe that philosophers such as Bacon, Descartes and other prominent Enlightenment and Scientific Revolution thinkers were attempting to wrest power from two dominant societal forces of the time. First, it was from the religious leaders who authoritatively wielded the power afforded them by their anthropocentric brand of Divine nature. Foster (2000) details this form of religious totalitarianism extensively when discussing materialism as an ontological response to it. Second, it was from more-than-human beings themselves and the power they were seen to have by human societies of that era. I discuss this in Chap. 3 when tracing posthumanism's version of the history of humanism. As I did there, I note here that Merchant's (1980) exploration of the details of this process is magnificent. I take from her exploration that these philosophers wanted to control nature for their own ends, so they set about both vilifying both women and nature, and dualistically reducing more-than-human beings as relational Selves as a means of establishing this control.

For example, Merchant mentions how Bacon "sets forth the need for prying into nature's nooks and crannies in searching out her secrets for human improvement" (p. 33). These philosophers embarked on a course of denying more-than-human beings everything from feelings to existence as Selves so that they could advance their desire to take from more-than-human nature whatever they wanted without obligation to those who'd given it, or acknowledgement of what had been given. To free themselves of the burden of anything but a mechanical, material human-nature relationship they went about denying the possibility of any other kind of relational reality by *post hoc* conceptually elevating humans to uniqueness on various fronts, and then setting them apart as atomistically individualized. To humans thus was granted exclusive domain over things humans find of most value: an emotional life, *telos*, happiness, etc. Thus today it is largely unproblematic even for a conservation-minded land manager to think of herself as having the right to "manage" land without the input, *at a relational level*, of the more-than-human beings affected.

This is because in the earlier era, more-than-human beings were conceptually recast as devoid of most if not all qualities of a Self. They became passive and/or relationally limited material objects incapable of entering into relationships of any substantial kind with human-beings, least of all close ones.

Posthumanism has made some strides to reclaim the agency and other attributes that more-than-human beings possess, but have faltered embracing materialism as the mechanism by which to redistribute them. In their theories, while more-than-human beings no longer lack agency, neither do they possess it. It's been installed as a disembodied phenomenon—which is a funny trick to try to pull off as a materialist. If it's not *in* an individual-in-relation, then where is it spatially? If the material intra-action is larger than the traditionally individual human or more-than-human beings within its bounds—if it is its own, larger "body"—the question still remains: How do, I as a member of that intra-action, embody or enact it? Note also that *telos* is largely rejected within posthumanist materialist theories. This is, in part, due to the rejection of spiritual/religious notions of an atomistically-rooted "purpose" that is discussed at length in Foster (2000). It is also a byproduct of materialism. In a materialist ontology, there is no purpose. Instead, a being's existence and path through life is the product of the universe's chaos in never-ending permutations that miraculously and dynamically organizes itself into, at times, beings-in-relation with consciousness to ponder the origin of such things. See Kessler's *Chaos or Relationalism? A Pragmatist Metaphysical Foundation for Human-Nature Relationships* (2012) for a discussion of the ways in which chaos is an insufficient ontological grounding for a relational reality.

Not only did earlier, humanist work destroy whatever was left of the popular, common concepts of close human-nature relationships, but it put in place concepts of ontological foundations that would prevent them from reforming. Put another way, these philosophers and scientists "salted the ontological earth" in which human-nature relationships normally grow. If a material reality for more-than-human beings is made the starting point, then at the outset, any observations of more-than-material qualities and capacities has to pass through its reductive filter. Once through, they are no longer the elements as experienced, but have been reduced to just more evidence for the materialist claim. The denial of feelings to more-than-human beings is self-fulfilling, and at some point in more recent centuries, became hegemonic to the point that to suggest that all animals, and beyond them, plants and stones, might have feelings is to commit an "easily" dismissible heresy. If an ornithologist has the experience of falling in love with a particular species of bird he is studying, he best leave that out of any paper he writes about their life cycle—because that love has been *post hoc* recast as a fantasy projected outward onto a non-reciprocating, material object. That, or the category into which his topic would be moved would be poetry, fiction or something else safely outside serious scholarship about the "reality" of birds—outside of truth as modern societies have come to define it. It is only recently that a toehold has once more been regained for some human-nature relational experiences, with research showing that at least some non-human-like animals do have emotions (see Sect. 9.3.2 discussion next), and that animals have consciousness just as humans do (Allen & Bekoff, 1999; Griffin &

Speck, 2004; Low, Panksepp, Reiss, Edelman, & Van Swinderen, 2012; Radner & Radner, 1989).

Ultimately, what Rollin's history lesson shows us is what I've suggested throughout this book, that human beliefs about what more-than-human beings are, and what they are capable of, is as often the product of unexamined ontology as it is any evidence. Sheets-Johnstone (2011) speaks directly to this point when, in her critique of materialist reduction of notions of consciousness in nonhuman animals, she says, "[M]aterialists offer a metaphysics in advance of an epistemology and a natural history that support it. Their metaphysics is in advance of a supportive epistemology in that both experience and meticulous study belie [their] theory" (p. 44). Essentially, what she means is that materialists discount more-than-human feeling and consciousness by simply assuming there isn't any, and do so in the face of more than ample evidence to the contrary. Sheets-Johnstone is still a materialist, however, advancing an evolutionary/corporeal explanation of consciousness rather than aligning with my assertion that feelings can have more-than-material origins, but my conflict with her model doesn't affect the agreement between our suppositions that more dualistically reductive materialists have not supported their denial of animal feelings or cognition with any substantive evidence.

9.3.2 Examples of Animal Feelings

The first, specific emotion I'll explore is that of *grief*. Bekoff (2000) notes that "many animals display grief at the loss or absence of a close friend or loved one" (p. 865). Bekoff notes this feeling in descriptions of chimpanzees from Goodall, in geese from Lorenz, and in his own work observing sea lions, dolphins, and elephants. He quotes Goodall describing the grief of a chimpanzee named Flint whose mother, Flo, had died, saying,

> [I]n the presence of his big brother [Figan], [Flint] had seemed to shake off a little of his depression. But then he suddenly left the group and raced back to the place where Flo had died and there sank into ever deeper depression...Flint became increasingly lethargic, refused food and, with his immune system thus weakened, fell sick. The last time I saw him alive, he was hollow-eyed, gaunt and utterly depressed, huddled in the vegetation close to where Flo had died...the last short journey he made, pausing to rest every few feet, was to the very place where Flo's body had lain...He struggled on a little further, then curled up—and never moved again. (p. 865)

For evidence from non-primates, Bekoff says of elephants that "Orphan elephants who have seen their mothers being killed often wake up screaming" (p. 866). Pierce (2008) notes that elephants grieve openly for their dead, such as when a baby elephant was "killed by a lioness; [and] over the course of the day, elephants from the herd gathered in a rough circle around the remains of the baby. Many of them touched the body with their trunks" (p. 85).

In my own experience, I'll never forget the time, when I was a boy, that I ended up watching a nature show on the television about Harp Seal slaughter. You'll have to take issue with my mother for letting me watch such a thing, but watch it I did. And even at that age, the reaction of the mother seal as they clubbed her baby and dragged him away was unmistakable. The memory of her slipping over the ice in pursuit, nuzzling her dead baby to see if he'd wake up as the man dragged his small body away, her eyes radiating desperation and anguish in equal measures, makes me well up to this day. To suggest that this is anything other than grief on her part at the loss of her child, and to suggest that my emotional response to it is anything short of some form of me "feeling with" her in the sense Noddings uses it—of it being the sharpest perception of an existential thing—flies in the face of common sense.

Skutch (1996) notes that feelings of grief and the impetus to act upon it in birds is also purely emotional, having no material survival component to it. He relates a story of a female Scarlet-rumped Tanager who he found

> hanging by one wing from the jaws of a snake she had vainly tried to drive from her eggs. Such superogatory zeal for progeny could hardly be promoted by natural selection, for the loss of the parent would in most cases be followed by the loss of the brood. Likewise, the futile cries that many birds utter while fluttering around a nest that is being pillaged by a predator or inspected by a naturalist are expressions of parental distress without survival value. (p. 41)

Bekoff (2007) gives another example of bird grief in relating a personal experience of his:

> A few years ago my friend Rod and I were riding our bicycles around Boulder, Colorado, when we witnessed a very interesting encounter among five magpies...One magpie had obviously been hit by a car and was lying dead on the side of the road. The four other magpies were standing around him. One approached the corpse, gently pecked at it...and stepped back. Another magpie did the same thing. Next, one of the magpies flew off, brought back some grass, and laid it by the corpse. Another magpie did the same. Then, all four magpies stood vigil for a few seconds and one by one flew off...We can't know that what they were actually thinking or feeling, but reading their actions there's no reason not to believe these birds were saying a magpie farewell to their friend. (pp. 1–2)

As a last example, I quote at length from Ken Kesey's (1987) description of the death of a bull amongst his herd:

> ...that night we called Sam's [the butcher], and the next morning John came turning in the drive before I'd even had coffee. Riding the running board, I directed him out to where the herd was bedded in the green clover around the main irrigation pipe. Ebenezer commenced bellering a warning...but she was too late. John was already out, walking around toward the target I had pointed out...Abdul was just blinking awake when the shot exploded in his brow. He fell over the pipe without a sound.

> As the herd bucked and bawled John hooked his winch cable to Abdul's hind feet and dragged the carcass away about fifty yards. I used to insist that he drag them clear from the field out of sight, so the herd wouldn't have to watch the gory peeling and gutting of their fallen relative, but John'd shown me it wasn't necessary. They don't follow the carcass; they stay to circle the spot where the actual death occurred, keening around the taking-off place though the hoisted husk is in full view mere yards away. As time passes, this circle spreads larger. If one were to hang overhead in a balloon and take hourly photos of this outline of mourning, I believe it would describe the diffusing energy field of the dead animal.

Abdul was the biggest animal we'd ever killed, and this mourning lasted the longest. Off and on between grazing, the herd returned to the dented pipe and stood in a lowing circle that was a tight ten feet in diameter the first day, and the next day fifteen feet, and the next day twenty. For a full week they grieved. It was fitting: he'd been their old man and a great one, and it was only right that the funeral last until a great circle had been observed, only natural—with the proper period of respect fading naturally toward forgetting... (pp. 51–52)

These are all examples from more and less scientific or at least modern sources, giving diverse voice to the particular experience of humans having witnessed grief in more-than-human beings and knowing it as grief.

Since one strong element of close relationships is altruism, then the *empathy* defined by Bekoff and Pierce (2009) as the "the capacity to read and understand the emotions of others and respond in a sensitive and helpful manner." (p. 138), is another emotion-related element worth considering. While empathy may not be an emotion, precisely, it is an important framework within which emotions of care, concern and love—amongst other generally benevolent feelings—find expression.

Pierce (2008) relates several examples of empathy, one of which (about two mice trapped in a sink) I explored above. Another lies in her reference to the research of Langford, who found that mice who witness the pain of fellow mice become more sensitive to pain in general themselves (p. 76). She discusses Rhesus monkeys refusing food or refusing to engage in a trained behavior to acquire food if it subjected another monkey to a painful experience (p. 81). Finally, she speaks of a pod of False Killer Whales who stayed with an injured pod member, at risk to themselves, for days until the companion died (p. 84). She notes that elephants show empathy in a variety of ways, from helping and comforting distressed calves to helping wounded companions (pp. 84–85). Bates et al. (2008), in their elephant research, describe

an adult...pulling a tranquillizing dart out of another male...that had been darted by a vet prior to treatment for a spear wound...The actor...dropped the dart as soon as he had pulled it out, suggesting he was not interested in the dart itself, but rather in removing it from the other elephant. (p. 218)

In Preston and de Waal's (2002) account of animal empathy, they suggest what I have about the sharing of emotions between individuals when they say that "empathy is by definition a shared state phenomenon. As such, one cannot experience it without to some extent feeling the distress of another" (p. 287). Though they don't source that sharing in more-than-material elements, they do suggest that there is a lack of hard boundary between the two individuals sharing, in this case, the distress that promotes an empathetic response. They cite research showing empathetic responses in rats, pigeons, monkeys and apes and convincingly refute quite a bit of the research that tries to characterize this evidence for animal empathy as something else.

In addition to these examples of intraspecific empathy, there are lay accounts of assistive actions by animals toward humans in need. There's the story of surfer Todd Endris who was being attacked by a great white shark and a "pod of bottlenose dolphins intervened, forming a protective ring around Endris, allowing him to get to shore" (Celizic, 2007, para. 2). According to the article, "stories of the marine mammals rescuing humans go back to ancient Greece" ("Our Finned Friends", para. 1). There are domestic dogs who, when a "stranger pretended to cry, rather than approaching their usual source of comfort, their owner, dogs sniffed, nuzzled and licked the stranger instead" (Custance & Mayer, 2012, p. 851). And there are domestic cats, one of whom, freshly adopted, sensed her new owner was in trouble when the owner collapsed in diabetic shock. The cat hopped onto her chest and nipped her face until she regained enough consciousness to yell for her sleeping son, but when the son didn't awaken, the cat then ran into the son's room and jumped on his bed until he woke up and helped the woman (Hernandez, 2012).

The last feeling I'll examine is *romantic love*. In making a case for romantic love in animals, Fisher (2004) says that

> Darwin is among very few scientists who have maintained that animals feel love for one another...[others] rarely say that animals love, even though descriptions of animal courtship are filled with references to behaviors that are akin to human romantic passion. (p. 28)

In fact, she says, "I have read about the amorous lives of some hundred different species, and in every animal society, courting males and females display traits that are central components of human romantic love" (p. 32). She notes that animals of various sorts feel "mad with delight" (p. 32), nervous, lose their appetite, are choosy about and possessive of mates, and display "love at first sight." Of beavers, she quotes Wilson observing a pair of courting beavers who

> sleep curled up close together during the daytime and at night they seek each other out at regular intervals to groom one another or just simply to sit close side by side and 'talk' for a little while in special contact sounds, the tones and nuances of which seem to a human expressive of nothing but intimacy and affection. (pp. 37–38)

Others, like Skutch (1996), also note such occurrences. In his studies of birds, he notes many instances of romantic love or affection. He says that the

> prolonged nuptial fidelity of many birds suggests they are affectionately attached to their partners...[and that t]he many tanagers and other birds that travel in inseparable pairs through the months when their sexual urges are dormant appear to be held together by mutual affection. (p. 35)

Perhaps his strongest example is of another's experiment on duck mating pairs. In the experiment, male-female pairs of ducks were sequestered with mates they hadn't chosen (having not chosen any or having chosen another and being separated from them). And, "...although there was apparently no physiological impediment to breeding...psychic factors inhibited mating and reproduction in all except...spontaneously formed pairs [the ones that had chosen their own mates]" (p. 35). In addition, for those who had chosen a mate but who were separated from that mate, the female of the pair "became extremely aggressive toward their new [forced] companions...Five of [the forced companions] succumbed to [i.e., died from] this harsh treatment [from the females]" (p. 35). Would these ducks have needed to die so that humans could learn this? Did the researchers feel the pain of the ducks? If they had simply trusted their own emotional response to witnessing the natural mating selection of ducks, could they have perceived the feeling of love directly as it was broadcast outward from the duck pairs? I believe the answers to all these questions are "Yes," and further, that experiments which account for the human observer, and in which the feelings of all involved are of primary importance, will go a long way toward helping establish close(r) human-nature relationships.

To return to Fisher's (2004) work, I suggest that while her examples and discussion are absolutely right, I believe her to be mistaken in attributing love to purely material causes as she does, saying, "there is chemistry to animal [including human] attraction. And this chemistry must be the precursor of human romantic love" (p. 47). Concluding this from her work is unwarranted and, I suggest, supported more by a dualist ontology rather than her results. For example, Fisher says of laboratory rats that "female laboratory rats express their amorous intentions by hopping and darting, behaviors associated with increased levels of dopamine. And in prairie voles...elevated levels of dopamine in the brain are directly associated with the *preference* for a particular mating partner" (p. 47). From these two "associations," or correlations, she incorrectly assumes causation. Of course to counter my point, one might suggest that causation is shown when Fisher refers to another element of the vole study, where "when scientists injected a specific region of the female prairie vole's brain with a substance that reduced levels of dopamine, she no longer preferred this partner over other males" (p. 48). But, as a converse, more-than-material interpretation I suggest that if one took away any male prairie voles that the female might actually prefer—eliminating her more-than-material attraction to any particular one—dopamine levels would drop as well. Thus, the more-than-material preference or its lack would precede and thus *determine* any chemical changes. What might be a much more effective tactic would be to *inject* dopamine or a chemical that would stimulate its production (though I don't advocate this since, relationally speaking, it is cruel to the vole), to see if the vole falls in love *with the same mate*, preferring him over others. If she did, maybe the material is more of a source than I think it is. If she didn't, then love's origin is more-than-material. Hence my suggestion that the material and more-than-material elements in any emotions ought to be taken as complementary, and that the more-than-material elements can precede and determine the material.

9.3.3 Insect Feelings

Further evidence to support my contention that feelings are not strictly human nor most likely to occur in those with human-like physiology, can be found in a study by Bateson, Desire, Gartside, and Wright (2011). In it, the authors suggest that not only do certain "higher" animals have emotions, but so does one very well-known insect: the honeybee. Their study falls into the category of research into "cognitive bias" in nonhuman animals, defined as the tendency of individuals who have been exposed to a negative or anxiety-producing stimulus to process future information more negatively than those that have not. These negative experiences "are associated with increased expectation of punishment, greater attention to potential threats, and a tendency to interpret ambiguous stimuli as if they were threats (i.e., a "glass-half-empty" or pessimistic bias)" (p. 1070).

According to Bateson et al., the success of correlating positive and negative cognitive bias with emotional states in human beings led researchers to attempt this with animals. The result is that emotions of both optimism and pessimism have been identified in rats, sheep, dogs, birds...and now honeybees. In their study, Bateson et al. took two groups of honeybees, habituated them to both positive and negative food stimuli (sweet and bitter scents, respectively) and exposed one group to a simulated attack on the hive by shaking it for 60 seconds. After ruling out any physiological damage from the simulated attack, they found that when exposed to a negative food stimulus again, the shaken bees "were more likely to withhold their mouthparts" from it (an indication of negative response) and from the most similar (next most bitter) novel odor. Thus, the shaken bees exhibited an "increased expectation of punishment" (p. 1071). The authors suggest that these results indicate that "agitated bees display a negative emotional state" (p. 1072).

Of cognitive bias research in general, the authors say that "pessimistic judgment biases are likely to be a good measure of negative emotional states across species" (p. 1071). To address the significance of finding such a cognitive bias in an insect, they more broadly suggest:

> Although our results do not allow us to make any claims about the presence of negative subjective feelings in honeybees, they call into question how we identify emotions in any nonhuman animal. It is logically inconsistent to claim that the presence of pessimistic cognitive biases should be taken as confirmation that dogs or rats are anxious but to deny the same conclusion in the case of honeybees. (p. 1072)

Bateson et al.'s explanation for the presence of this emotion in honeybees is materially based. For example, in response to their own findings they say, "The physiological mechanisms that produce this change are poorly understood" (p. 1072). In the ontological context under discussion here, however, such an explanation seems at least slightly revisionist. Confidence levels have generally been very *high* when material explanations for emotions have been offered for beings with physiologies similar to humans. To suggest otherwise only when faced with a materially very different kind of being strikes this author as a case of altering the explanation to fit the ontological assumption of a material basis for feelings. From a materialist perspective, the authors' findings suggest that either material evidence for the origins of feelings lies somewhere as yet undiscovered, or the confidence with which previous material claims have been lodged is misplaced. In the end, this kind of conceptual sleight-of-hand underscores my point that the foundation upon which the confidence of these kinds of materialist conclusions stands is installed as *a priori* ontological assumption, not the product of any kind of definitive proof.

As another example of this kind of unjustified confidence, I refer to Bekoff (2007), who is very supportive of the claim that "higher" nonhuman animals have emotions. Four years prior to the publication of Bateson et al.'s research, Bekoff (2007) asks: "Do even mosquitoes have emotional lives? Of course, mosquitoes have tiny brains and lack the neural apparatus *necessary for the evolution of emotions* [emphasis added], so it's doubtful they do. But in truth, we just don't know"

(p. 2). While I take his qualification of the ultimate unknowability of mosquito emotions as a nod to the possibility of their existence, it's fairly clear that he finds the possibility remote given his "knowledge" that brains of a certain size and a certain neural apparatus are necessary for the presence of emotions. I suggest that such "knowledge" is due to his deep commitment to the ontological *gravitas* that evolution's material directives possess more than anything else. In light of Bates et al.'s (2011) work four years later, those suggestions can now more easily be seen as founded at least partially in ontological assumption.

If Bekoff (2007) and most other materialists were this confident in human-like neurophysiology in 2007, what must change to accommodate the honeybee's pessimism in 2011? If one's ontology is material, then the material foundation thought to sponsor such feelings must be broadened radically to accommodate the physiology of bees. But, if one is inclined to consider more-than-material origins, the honeybee certainly brightens one's prospects. Given these possibilities, it seems reasonable at this juncture to ask: When is there *enough* experience falling outside accepted material explanations to shake the confidence materialists have in their ontology? At times it seems that the answer is that there will never be enough. Regardless, as I've stated throughout this book, from the outside one can at least see that the confidence materialists have in their own ontology is often the product of a circular logic that first requires the assumption of a material reality as a basis to interpret observed phenomena such as feelings as material in origin. This is a far weaker position than the proponents of a materialist ontology attempt to occupy in determining who does and does not have the capacity for feelings. This, in turn, opens the door for a far greater number of beings to potentially have them.

9.4 Plant Feelings?

As I state above, the finding of pessimistic feelings in an insect forces the modern conceptual understanding of what is necessary for feelings to change radically. This means a redefinition both of the material structures and processes thought to co-occur with feelings and the admission that, by expunging one's ontological positions of dualisms, there is no inherent impediment to more-than-material elements playing a role. Given the now greater uncertainty as to the origins of and nature of feelings, it's useful to explore whom else besides humans and honeybees might have them.

9.4.1 Feeling and Electricity

Chemical processes are not the only material elements that correlate with experiences of feelings. For example, Schmidt and Trainor (2001), provide evidence that electrical activity in the human brain, measured by the electroencephalogram (EEG), is associated with positive feelings as they relate to music. In addition, these authors found that the pattern of electrical activation has been seen to correlate with the intensity of emotion felt. Other studies have used the EEG to measure the presence of happiness, relaxation and sadness (Li, Chai, Kaixiang, Wahab, & Abut, 2009), to measure differences in location of electrical activity in the brain for positive and negative emotions (Ahern & Schwartz, 1985), and to measure location and changes in electrical activity when humans are in relaxed, meditative states (Aftanas & Golocheikine, 2001). Other ways of gauging electrical activity, such as brain electromagnetism, have been used to find evidence for romantic love in humans (Bartels & Zeki, 2000; Zeki & Romaya, 2010), emotions such as happiness, sadness, anger, fear, and disgust (Esslen, Pascual-Marqui, Hell, Kochi, & Lehmann, 2004), the anticipation of reward in dogs (Berns, Brooks, & Spivak, 2012), etc. Electrical activity as measured by electrical resistance in the skin (galvanic skin response, or GSR), has also been correlated well with emotions such as fear and anger in human beings (Kreibig, Wilhelm, Roth, & Gross, 2007; Sinha & Parsons, 1996), sadness (Kreibig et al., 2007), detachment as part of coping (Pecchinenda & Smith, 1996), and romantic love (Bartels & Zeki, 2000; Zeki & Romaya, 2010).

As can be seen, there is a strong correlation of various forms of electrical activity with emotions. Freedman (2007) says in response to Candace Pert's work that, "Emotions…are not simply a chemical in the brain. They are electrochemical signals that affect the chemistry and electricity of every cell in the body" (para. 1). He quotes Pert as saying, "As our feelings change, this mixture of peptides travels throughout your body and your brain" (para. 6). In such a statement, I note that Pert *could* be read as indicating that changes in feeling *precede* and precipitate chemical changes, they do not follow them. In addressing how emotions like love can be shared amongst humans, Freedman quotes Pert as saying, "We're not just little hunks of meat. We're vibrating like a tuning fork—we send out a vibration to other people. We broadcast and receive" (para. 5). He says that "Feelings literally alter the electrical frequencies generated by our bodies, producing a nonverbal communication" (caption 1). Freedman's then asks Pert about "how one person's emotions can affect another person" and describes her response as follows:

> "You're still thinking about this as chemistry," Pert chides. "Of course it is chemistry, but it's also physics and vibrations." Neurotransmitters are chemicals, but they carry an electrical charge. The electrical signals in our brains and bodies affect the way cells interact and function. "You have receptors on every cell in your body. They actually are little mini electrical pumps." When the receptor is activated by a matching "molecule of emotion" the receptor passes a charge into the cell changing the cell's electrical frequency as well as its chemistry. Pert says that just as our individual cells carry an electrical charge, so does the body as a whole. Like an electromagnet generating a field, Pert says that people have a positive charge above their heads and a negative charge below. "So we're actually sending out various electrical signals – vibrations." "We're all familiar with one kind of vibration: When we talk, we send a vibration through the air that someone else perceives as sound…we're also sending out other kinds of vibrations…It is a whole new paradigm shift that basically leads you to realize you're not alone. You are connected to everybody else. Your emotions are key. And you are leaving a wake…" (para. 7)

Thus far, Pert's work appears to parallel, and be a cogent *material* explanation for, the continuity of external, Peircian feelings, and the internal and external emotions I have described. To Pert, the shared electrical fields come before chemistry—

they have to as they exist externally first—and are followed by individual chemical responses to being within such a field of emotions. As to my further contention that both of these material elements can be preceded by more-than-material stimuli, Pert (2006) says that

> the movement of energy in the body was possibly due to the flow of emotions going back and forth between the physical and the spiritual, existing on the physical level as peptides and receptors, but also in the spiritual realm as information. Chi, prana, and meridians—these phenomena could possibly be explained by the flow of information and emotions... (p. 253)

If one finds Pert's move to the more-than-material here a bit jarring, I'd suggest it's because of engrained material ontological expectations, not because the move is unwarranted. After all, Pert appears more than well-qualified to render an opinion on the subject. Since she had been speaking about physical phenomena, and because of her high position in mainstream science, one might expect her to remain within its preferred materialist ontological boundaries. Given this expectation (a thing whose weight she herself must certainly have felt) think of how strong her belief or knowledge must be that more-than-material elements such as spiritual information—a belief *based on*, not in contradiction with, her close study of the phenomena—are real and relevant. As I have, Pert suggests that feelings have both material and more-than-material underpinnings, that there is an interplay between the two, and that neither is the clearcut originator of the other. If what Pert and I suggest is correct, then we certainly have one form of what the honeybee research forces upon us: a radically revised notion of the material elements involved, and the entrance of the possibility of more-than-material elements as counterpart. For Pert, electricity gains precedence, and thus importance, in explaining the origin and nature of emotions. More-than-material origins of feelings that may precede this electricity also appear essential.

Given such expanded ontological possibilities for the origin of feelings, I cannot help but begin to wonder about plants. Do they have the necessary material and more-than-material capabilities for feelings? If electricity is a significant indicator of the presence of feelings, since it's commonly held that plants don't have the chemical structure and processes to experience feelings, I wonder if they have the electrical structures and processes to do so. I'll explore the possibility of this next.

9.4.2   Electrical Plants

Knowledge of electrical activity and communication in plants has become widespread. Cifra, Fields, and Farhadi (2011) cite examples of intercellular EMF communication (as well as ultraviolet and infrared light). These include growth rates in yeast cells (Musumeci, Scordino, Triglia, Blandino, & Milazzo, 1999), germination rates in pollen grains (Budagovskii, Turovtseva, & Budagovskii, 2001), bacterial spore germination (Nikolaev, 2000) and plant seed damage from low dose gamma radiation (Kuzin, Surkenova, Budagovskii, & Gudi, 1997). Fromm and Lautner (2007) note a wide variety of physiological changes in plants that are induced by electrical activity and communication.

In Keller's (1983) book about Nobel Prize winning geneticist Barbara McClintock, Keller quotes McClintock as saying, "Plants are extraordinary. For instance...if you pinch a leaf of a plant you set off electric pulses. You can't touch a plant without setting off an electric pulse" (p. 387). Brenner et al. (2006) speak to electrical signaling in plants when they tell us that "Electrical signals have been linked with changes in rates of respiration and photosynthesis, observed in response to pollination, phloem transport, and the rapid, systemic deployment of plant defenses" (p. 415). Simons (1992) notes that for flowering plants, electricity in the form of action potentials has been linked with "feeling touch, wound and tempera-ture stresses" (p. 232). Shepherd (2006) says that "it is now fully accepted that plants employ electrical signals in the integration of their responses to the world" (p. 7) and quotes Bose as holding that "[p]lants possessed a[n electrical] means for rapidly transmitting information about an urgent or injurious event as well as for navigating the physical aspects of the world through subtle and exploratory growth movements" (p. 24). Shepherd also notes,

> Bose compared a plant to a bar magnet, its two poles located at root and shoot...[and] as the two parts of a divided bar magnet both then show a north and a south pole, so it was with the plant, all the way down to the individual pulsating cell. (p. 24)

I note here the similarity between Bose's description of plant electrical fields and Pert's description of the human being's electrical field, the latter in relation to emo-tion. In referring to plant electrical activity research overall, Shepherd (2012) notes the discovery of a

> 'plant brain'...[in] the transition zone of roots, where actin-enriched fields of cell-to-cell communication channels (plasmodesmata) at the end-poles of cells act as synaptic connec-tions. Synapses in this region confer on the root apex the properties of a 'brain', or com-mand centre, where incoming sensory signals are processed (Baluska et al., 2004; Barlow, 2006, 2008). 'Higher plants show neuronal-like features in that the end-poles of elongating plant cells resemble chemical synapses' (Baluska et al., 2003a, b). It is here that synchro-nised electrical 'spikes' are measured, which are proposed to reflect integration of internal and external signals (Masi et al., 2009). (p. 32)

Beyond communication within the plant, some have speculated that the inter-plant communication phenomenon of self/non-self plant recognition, with discern-ment and relaxed competition with a plant's own roots and those of its conspecific neighbors, could be explained electrically. For example, Falik, Reides, Gersani, and Novoplansky (2003) suggest the possibility that the plant's roots have

> electrical oscillations...[that] might be perceived by neighbouring roots without direct con-tact. The perception of 'self' signals might be based on resonant amplification of oscillatory signals in the vicinity of other roots of the same individual plant. Such resonant amplifica-tion could not occur in roots that are not oscillating at the same rhythm. (p. 529)

Here again, I note the similarity of their description of the function of electrical activity within the organism with that of Dr. Pert's notion of the electrical commu-nication via overlap of emotions between humans.

Finally, not only do root electrical fields potentially form the basis of interplant-communication, but plant-animal communication as well. For example, Van West et al. (2002) found evidence that electrical fields generated by the plant's roots could electrically communicate with and attract certain zoomorphic life forms, literally causing them to swim in the direction of the roots. And, the authors theorize, the electrical fields that caused this could override any chemical cues emitted for the same purpose. This again positions electrical activity as precedent to chemical activity.

9.4.3   Plant-Animal Electrical Parallels

Of this work on electricity in plants, Shepherd (2012) says that "Bose interpreted his results as constituting evidence that plants possess the equivalent of a well-defined nervous system" (p. 19). Bose found that "sensitiveness of Mimosa to electrical stimulation is high and may exceed that of a human subject. (Bose, 1913, p. 51)" (p. 18). Shepherd points out Bose's finding that electrical "action potentials travelled at similar rates to...the nerve impulse [in animals] with Bose concluding that '[t]ransmission of excitation in the plant is a process fundamentally similar to that which takes place in the animal'" (p. 20) and that the "...physiological mechanism of the plant is identical with that of the animal ..." (p.19). Many other researchers believe that there is a parallel in animal and plant electrical signaling as well, so much so that a recent field of study has been born called "plant neurobiology."

I must note that not all plant biologists are sanguine with the parallels being drawn between plant and animal electrical activity, however. For example, Brenner et al. (2006) state that "the concept of a plant nervous-analog system lost popularity in the scientific community in favor of a chemical diffusion mechanism of signaling coinciding with the discovery and effects of plant hormones" (p. 414). They trace the cause for this shift to, amongst other factors, "[negative] publicity... generated by the controversial book 'The Secret Life of Plants' [that] stigmatized any possible similarities between plant signaling and animal neurobiology" (p. 414). This stigmatization, according to Brenner et al., induced "a form of self-censorship [in plant biologists] in thought, discussion and research that [has] inhibited asking relevant questions of possible homologies [i.e., parallels] between neurobiology and phytobiology" (p. 415). They note that the result has been a de facto "prohibition against anthropomorphizing plant function" (p. 415).

Shepherd (2006) traces this shift away from study of plant electricity to an earlier time, when in "the mid-nineteenth century...German mechanistic materialist philosophies had begun to influence the science of physiology...[where, a]s a tenet of Descartes' philosophy, cells, tissues and organisms [were seen to] respond passively to the physical and chemical features of their environments" (p. 9). As to the effect this had on parallels being drawn between plant and animal electrical activity at that time, Shepherd says,

> This concept of nerve-like electrical signaling in [Bose's research into the plant] Mimosa was unpopular at a time when many scientists sought to construct a purely physico-chemical theory of life (Agutter et al., 2000), exorcising...interpretations of plant response that were reminiscent of animal behaviour. (p. 10)

But, Yoon (2008) suggests that defenders of these plant-animal parallels are

well aware that plants do not have exact copies of animal nervous systems. For example, "No one proposes that we literally look for a walnut-shaped little brain in the root or shoot tip," five authors wrote in defense of the new group. Instead, the researchers say, they are asking that scientists be open to the possibility that plants may have their own system, perhaps analogous to an animal's nervous system, to transfer information around the body. (para. 30)

The evidence for drawing such a parallel between plant and animal electrical signaling is robust. Less well known experiments by Kaznacheev, Mikhailova, and Kartashov (1980) show that not only are plant cells damaged and communicating the nature of that damage to optically connected, unmanipulated cells, but so are animal cells. Cifra et al. (2011) also cite this electromagnetic communication in growth rates in yeast cells, bacterial spore germination and induced respiratory bursts in human inducer neutrophil cells. They also refer to the work of Albrecht-Buehler on interactions of baby hamster kidney cells separated by glass. That work showed that "Cells on one side tended to orient (traverse) themselves based on the orientation of the cells on the other side of a separator made of glass" (p. 234).

In referring to dinoflagellates, Simons (1992) says they "come in two types—the plant-like (photosynthetic) and animal-like—but they both behave electrically... This sort of sophisticated electrical behaviour in such 'simple' creatures is another measure of the common inheritance of plants and animals" (pp. 231–232). Cifra et al. (2011) also note that communication occurs similarly across species lines when discussing Budagovsky, et al.'s work finding "that human whole blood has a stimulatory effect on germination of radish seeds when those are optically coupled" (p. 234). They also note that there is continuity at the organismal level between plants and animals when considering Popp and Chang's research finding "synchronization of flashes of fireflies and dinoflagellates when cultures were connected optically. From these experiments, the authors suggested that electromagnetic bio-communication plays an important role in the interactions of whole organisms" (p. 234). If these authors are right that whole organism communication can occur electrically, and between plants and animals—and then Pert (2006) is also right that emotional communication occurs electrically between humans—then it is not far-fetched to suggest that if plants did feel, that humans and plants could communicate emotionally as well.

In answer to "how plants could have nerve-like signals when they don't have nerves" (p. 233), Simons (1992) paraphrases the response of late 19th century researcher Sir John Burdon-Sanderson as saying, "quite simply...they don't need nerves" (p. 233). To invert the zoocentric orientation of the discourse overall, let's consider Simons when he says, "what's interesting from an evolutionary point of view is that animals also appear to have something akin to [those plant structures for channeling electricity]" (p. 234). This shows that not only are plants similar to animals in their electrical signaling, but animals are now being discovered to be similar to plants, *with plants as the standard against which animals are measured.* That's hardly a "step down." If it were taken as such, it would betray a dualistic, anthropocentric and/or zoocentric orientation in judging electrical activity. Simons' statement merely means that plants are far more complex than generally supposed over the last 150 years of modern human study of them.

9.4.4 Possibility of Plant Feelings

It is instructive to note that Simons does not ultimately believe that the continuity he notes means that plants are animal-like. For example, he says, "Not even the sophistication of a...Venus's flytrap comes anywhere near the complexity of an earthworm, with its well-coordinated movements, feeding and sexual behaviour" (p. 233). Given the context of the discussion here, it's sensible to ask whether he is right. I suggest that his point is only valid if one is considering physical locomotion, and over a short temporal and relatively large spatial scale. We must note, though, that such parameters center the measurement around the human or human-like attributes of locomotion and relatively short temporal spans of consideration. Using a much finer frame of spatial reference, Bose (Shepherd, 2006) noticed that plants moved quite a lot. A long-time native plant nursery owner (personal communication), who's spent 30 years living in one place, told me all about how apple trees can walk over distances, the same as Trewavas (2003) points out about stilt palm plants (p. 15).

Simons' (1992) comparison of plants to earthworms illuminates a more general modernist ontological trend as well: that most researchers and theorists employ the dualism of fixing the reference point for "complexity" on human physiology. By doing so, any deviation from this fixed point will necessarily be categorized as out of center (and by extension, not as important, valuable or complex). Terms like "lower" or "simpler" make their appearances in these discussions to suggest some lack—that what more-than-human beings have is not what's best for "the job." That plants are more competent plants than humans are is not a point raised by many. I believe Simons is guilty of employing this technique in discussing electrical activity when he cautions us not to get too "carried away on the idea that plants are evolving into some sort of green animal" (p. 233). One notes first that he is not saying we humans shouldn't get carried away in *equating* plants and animals, which would be a bit less of a zoocentric statement. But in his statement, the bar has been set by animals and the plants are trying to get "high" enough to rub elbows with them. For Simons, as well as most others, animals are the pacesetter, with plants lagging behind in their travels down the evolutionary path. Of the higher value Simons places on animals, it is clearly visible when he says that the "most *flattering* [emphasis added] comparison we could make [for plants] is with jellyfish or sea anemones" (p. 233). Such a comparison can only be called "flattery" if it is thought to *be* flattering to be thought of as animal-like. What Simons fails to note is the irony such a statement evokes when, elsewhere, he notes that plants were mobile in an *earlier* phase of their evolution (p. 233). Some might reasonably infer from such a statement that animals are *less* evolved than plants, plants having evolutionarily "arrived" at the benefits of sessility that animals are lagging far behind in adopting.

My discussion of defining complexity anthropocentrically raises a more general issue in the logic of argumentation around which beings are acknowledged as having emotions and which are not. The general path of that logic is that because humans experience emotions, they must be the beings that experience them most thoroughly and best. In concluding such a thing, modern humans have taken a material representation of their species' emotional capacity and used it to create the standard by which to judge emotional capacity in all beings. Granted, if another being is materially similar to humans, one could reasonably infer that he or she would experience emotions, but to infer the converse from the same logic is falla-cious. Specifically, if another being is dissimilar from humans it is *not* reasonable to infer that they don't have emotions...unless we're *sure* the source of emotions is the material structures and processes that humans have (and have in the ideal pro-portions), only then moving out concentrically to less and less similar beings until emotional capacity is lost altogether. This is the current modern worldview in research into more-than-human emotions, but it is based almost entirely in what I've referred to in Sheets-Johnstone (2011), a "metaphysics in advance of an epistemol-ogy" (p. 44). This is especially the case given the foundationally contentious nature of the discourse on the source of emotions. Are they chemical? Certainly many working in the biochemistry field would have us think so. Are they electrical? Other researchers think this is plausible. Are both these material explanations of a *source,* or are they just descriptions of the material portion of reality—one among many possible descriptions—in which the feelings that I take to be more-than-materially sourced are complementarily manifested? With dualisms moved aside, there is no clear answer. Even if one were to hold that emotions are materially sourced, the logic that those who experience emotions ought to be the ones by which the stan-dard is set for emotional capacity would just as easily support the honeybee physiol-ogy as the standard by which human capacity for feelings ought to be judged.

Ultimately, the nature of plant electrical activity is not a matter of complexity or its lack because, in truth, there is no absolute "right" apparatus against which to judge it. Are plants less complex in their electrical activity than humans or more complex? It depends entirely on what is thought important and differentiable in the measurement. That dualistic ontology factors into such determinations is undeni-able, therefore humans as assumed to be "complex" is immediately suspect. Pragmatically speaking, what is and is not complex in the realm of organismal elec-trical activity depends entirely upon the purpose of the electrical activity. Certainly the type of electrical activity in a human would utterly fail the plant in her endeavors to excel as a plant, just as a plant's would fail a human in his endeavors to excel as a member of his own species.

Ultimately, this is not about complexity; it is about *fit*. As I mention above, at one juncture Simons notes the strangeness of plant evolution because plants used to be mobile but evolved to become sessile. Perhaps the electrical activity associated with locomotion is not a more complex or "higher" form of activity, but instead an earlier way of being that plants evolved away from to their now more "elegant" and "com-plex" electrical systems. They don't move over extended distances, at least not quickly, because they've evolved a means to get all that they need from one place—a far more "complex" thing to carry off, if one looked at it from a phytocentric standpoint. To move humans even further from the center of importance, Simons (1992) notes that some of the electrical elements previously held to be exclusive to the plant have now been found in animals, including humans, so here again the human is trying to "catch up" to the plant from an evolutionary perspective.

With regard to the term I used above, "fit," plant electrical systems are the perfect complexity and elegance for plant lives. They lack nothing in this, the only relevant context by which to judge their nature and capacities. Thus, any research into electrical activity in plants that can be correlated to experience of feelings is no longer a matter of some abstracted, absolute complexity or its lack, or of finding similarity to human material structures and processes. Instead, it's a matter of finding correlative evidence, through scientific, relational and poetic means, of the occurrence of emotions only within the plant's own electrical structure and processes. In fact, if feelings are ontologically more-than-material, then any material evidence found for their occurrence ought to *vary* according to the varying physiology of the entities experiencing them. Thus, the question becomes: What material changes does any entity undergo before, during and subsequent to an emotional experience given its physiology? Of course where physiologies are similar, there will be similarities. For example, there is similar electromagnetic activity in chimpanzees and humans experiencing emotion (Hirata et al., 2013).

There can also be chemical similarities, such as dopamine and serotonin in humans, animals and...plants? Brenner et al. (2006) talks about the most common

> metabolic neurotransmitters, acetylcholine, catecholamines, histamines, serotonin, dopamine, melatonin, GABA (g-aminobutyric acid) and glutamates...in the animal nervous system, playing roles in sensing, locomotion, vision, information processing and development. It has long been noted by scientists that each of these compounds are present in plants, often at relatively high concentrations [13]. However, it is unclear whether these compounds play a metabolic or a signaling role in plants despite numerous studies [1,58,59]. (p. 415)

Could these elements be evidence of emotional experience in plants? I make no claims that they *are* evidence of such a thing, but I do suggest that, given the strength of my exploration above, it is perfectly sensible to wonder. This is especially so given the complexity of plant electrical activity, its parallels with electrical activity in animals in several other areas, and the fact that the material aspects of animal emotions are, in part, reflected not just in chemical but in electrical activity. Therefore, it is simply inaccurate for those like Galston (quoted in Backster (2003)) to critique the possibility of plant sensations by saying, "There's no nervous system in a plant. There are no means by which sensation can be transferred" (p. 59).

Given all of this, however, the question of plant emotions is still taboo in traditional, modern plant research. Brenner et al. (2006) worries about his career prospects just talking about plant electrical activity (p. 415). How much worse, then, to discuss plant emotion? When the subject is raised, it is routinely dismissed as the work of the "kook-fringe" doing "decidedly nonscientific" (Yoon, 2008, para. 21) things. It seems there is particular venom aimed at the 1973 book *The Secret Life of Plants* (noted in Brenner et al. (2006), Davies (2004), Yoon (2008), etc.). And while much of the research regarding plant emotion presented in the book *is* questionable (and so, necessarily, are any conclusions based on it), it doesn't logically follow that because the methodology is suspect, the question it was designed to answer is as well. At the least, the question most certainly ought not to be denigrated with *ad hominem* statements from a Dr. Dudley, quoted in Yoon (2008) as saying, "Plants are not 'sensitive new age guys who cringe when something around them gets hurt and who love classical music and hate rock'" (para. 22).

At this juncture, it seems fair to ask if the invective coming from established researchers is due to the absurdity of the question or the threat such a question poses to the foundational worldview of the researchers. I note that the latter is at least possible. For example, similar treatment was given to Bose 100 years ago when he first suggested that the electrical activity he found in plants was also present in "inanimate" matter. In response to the experience of his rough treatment by the establishment at that time, Shepherd (2006) quotes Bose as saying,

I had...unwittingly strayed into the domain of a new and unfamiliar caste system [physics] and so offended its etiquette. An unconscious theological bias was also present...To the theological bias was added the misgivings about the inherent bent of the Indian mind towards mysticism and unchecked imagination... (p. 14)

Before the work on honeybee emotions, I'm sure similar treatment would've been meted out to anyone suggesting that insects could feel pessimistic. As to my point that the question of plant emotion ought to not be problematic, I note that Solfvin (2009), in reviewing Backster's (2003) oft-vilified book on plant communication, says, "Just because Backster's beliefs are not yet scientifically validated (and they are not) certainly does not rule out their potential veracity" (para. 5). Thus we stand on the threshold of the question: Can plants feel? The wonderfully provocative answer, based on available material *and* more-than-material evidence, is: We don't yet know. And if we don't, then one place I might suggest that research begin would be to ask of the plant the same question that was asked of the honeybee: "Can you be pessimistic?"

Given my commitment to a pragmatic philosophical approach in this work, I conclude my discussion of the "Possibility of Plant Feelings" here by relating my own trouble in taking seriously the possibility of plant emotions. Intellectually, I wholeheartedly embrace the possibility. But when initially encountering the work of Backster (1968, 2003), Dubrov and Pushkin (1982), and others I'll discuss below as being part of the "kook-fringe", I found myself worrying about offering their work as the backbone of an assertion that plants can feel. I say this because, at best, those authors have been categorically dismissed by most established researchers. Because of the exploration I've subsequently done into animal emotions and plant electrical and chemical processes, however, I'm now forced to ask an altogether different question: Given all of the surrounding literature on animal emotions and plant electrical communication, how is it possible that plant emotions have *not* already been studied? The only answer I am able to formulate has two parts. First, it is a result of what Brenner et al. (2006) said: stigmatization and its resultant self-censorship. Second, it is a result of what I've been arguing: that dualistic ontological commitments obviate the conceptual possibility of plant emotions. As a result, any effort to seek evidence to support them is eradicated before it can even begin. Ultimately, I've traced my own hesitation to embrace the possibility of plant emotions as my holding onto the very ontological commitments against which I've been making strenuous objections throughout this book. I will put them to the side of the path for now, and press on to consider the merits of kook-fringe arguments.

9.4.5 The "Kook-Fringe"

The extant research into plant emotions by the "kook-fringe" is relevant to my discussion for two reasons. First, though much of the research is riddled with problematic application of the scientific method, I find the response of plant emotion researchers to such critiques to be valuable in an ontological sense. Specifically, those researchers don't generally argue so much for the validity of their work within the bounds of the scientific method. Instead they question the application of the scientific method to what they see as a relational situation incompatible with such an approach. Second, despite the problems of method and design, some intriguing evidence still emerges even from within the confines of mainstream scientific approaches.

As an example of the first reason, I examine the response of Dubrov and Pushkin (1982) to critiques of their research into emotional communication between humans and plants. In response to critiques of their methodology, they say that they "learned that not only the state of the subject [a human trying to communicate with the plant], but also the state of the plant, is essential for a successful experiment" (p. 95). In addition, they suggest that "far from all the subjects were capable of establishing communication with plants. Apparently, these differences were due to individual differences in the psycho-energetic systems in our subjects" (p. 95). In both of these quotations, the authors reveal that, to them, the plants are not dualistically reduced, material objects but Selves in relationship with the human attempting to relate with them.

Backster's (1968, 2003) work on plant-human communication is perhaps best known from amongst those in this kook-fringe group. Because of this, it has probably received the worst treatment by the scientific establishment. Amongst his experiments, he explores the electrical response of plants to the killing of brine shrimp, and finds a correlation between plant electrical responses and the death of the shrimp (1968). Horowitz, Lewis, and Gasteiger (1975) attempted to replicate his results with what they saw as a more rigorous application of the scientific method and did not find the correlation that Backster had. When countering their and others' refutation of his work, Backster points out what he sees as flaws in *their* methodology. Like Dubrov and Pushkin (1982), what he points out about those flaws takes on a relational tone. For example, Backster (2003) says,

> Any kind of experimenter contact with the plants prior to the actual experimental usage can compromise the experiment by allowing prior attunement between the plant and the researcher rather than the more subtle stimulus provided by the death of the brine shrimp. (p. 70)

In general, Backster suggests that there is a "possible contamination of the experiment results by the conscious intent of the researcher being communicated to the biological material being tested" (p. 76). He notes that this may have occurred in Horowitz et al. (1975), and also Kmetz's (1977) research. In the latter experiment, plants were kept in a holding room for 7 days and their leaves bathed in distilled water by the researchers prior to the experiment. One can see that in Backster's approach, plants are included as relational partners able to influence, and be influenced by, thoughts and feelings.

This more relational approach is on display less controversially, since it is applied to laboratory animals, in recommendations by Poole (1997) regarding their treatment. He says,

> One of the most important aspects of the life of laboratory animals are their relations with human handlers and care givers...Good, kindly treatment and simple humane training are beneficial both in reducing stress and in producing animals which...will...be the best subjects for scientific investigation. (p. 122)

Poole suggests that because stress changes various physiological characteristics, that "these uncontrolled variables make [the animals subjected to the stress] unsuitable subjects for scientific studies" (p. 122). Though his argument is based in the material effects of what I take to be more-than-materially originated elements like stress, it is still relevant in the sense that, when one has a subject aware of the experimenter and affected by her, the experimenter must be accounted for as a variable. As Galston and Slayman (1979) quote Vogel saying about this relational approach, it "runs counter to the philosophy of many scientists who do not realize that...the experimenters must become part of their experiments" (p. 344). The kook-fringe researchers are suggesting the very same thing with regard to plants.

As sympathetic as Solfvin (2009) is to the concept of biocommunication in Backster's work, he appears, however, unable to grasp the undergirding relational ontology. For example, Backster suggests that spontaneity in experimenter-plant interaction is necessary for biocommunication, where measuring that communication is a function of allowing the interaction to occur, diligently recording the events contained therein, and then going back afterward to see if the plant responded electrically to them. Solfvin critiques this methodology by saying it is "logically equivalent to shooting your arrow at a blank wall and drawing the target afterward! The independent and dependent variables are confused!" (para. 7).

What Solfvin fails to see, however, is that for Backster, it's not possible to control or predefine what is relationally stimulating, as it's a result of *this particular relationship* at *this point in time*. If the target is predefined, and created unilaterally out of human intent for something other than genuine encounter and relationship, then of course the experiment will fail to direct the plant's responses toward the human's desired target. Put differently, in a truly relational approach where more-than-human beings such as plants are relational Selves, *it's not possible to draw the target ahead of time*. Instead, one must take the vastly different approach of drawing the target afterward, time and again, and only when a pattern emerges as to where those targets cluster will one see the repeatability that is the byproduct of genuine relationship. The target, then, is not the end-product of a relational interaction, it's the byproduct. Posthumanism is right to point out that in experimental intra-actions, the experimenter is as much a part of the experiment as the subject. They are also right to suggest that the dynamism and unpredictability of this interaction will necessarily influence whatever results are obtained. It's a constant, dynamic, moment-to-moment, feedback loop of generativity. But, as I've pointed out previously in this book, the posthumanist view of the "asymmetry" between plant and human experimenter, sponsored as it is by materialism, still allows for the plant to be treated as if it has very little, if any, substantive relational contribution. Thus, posthumanism won't get us far enough.

If the plant was another human, and the experimenter made no secret of her attempts to manipulate that other human and her thoughts and feelings about it, one can see more clearly how such elements would become variables not controlled for in the experimental interaction, ultimately confounding the results. On the kook-fringe, the plant is every bit as much a relational Self as a human is, responding not only to the manipulations the experimenter *thinks* she is applying, but to the experimenter's thoughts, feelings and other attitudes as well. Lastly, if one takes the plant as a relational Self, even if the desired end result of an experiment was closeness of relationship, without the goal being *mutually agreed upon* between human and plant, it still will not be achieved.

Backster, too, is critical of the scientific approach when discussing that method's insistence on things like repeatability. Of this, he says, "The...problem in this kind of [traditional scientific] research is that Mother Nature does not want to jump through the hoop ten times in a row, simply because someone wants her to" (in Galston & Slayman, 1979, p. 344). Thus, to understand if a plant is responding *in relationship*, one has to account for the relationship, not just the things that one participant believes she is controlling within it (and from outside of it, as if that's possible). The researcher must be a recognized relational partner whose influence on a living other can be neither reliably controlled nor systematized unilaterally.

Galston and Slayman, on the other hand, object to the Backster's approach by calling it generally a "corpus of fallacious or unprovable claims" (p. 344). In making this accusation, they do make the valid point that relational approaches can be dangerous if used as a means of dismissing any work that refutes them—where "Negative results [are] discounted because the experimenter is not 'in tune,' and [thus] only positive results are accepted" (p. 344). But, just as one cannot dismiss traditional scientific methods out of hand when they disagree with one's relationally generated results, neither can one wholly dismiss relationally generated results (and the methods by which they are generated) when they are at odds with the results gleaned through traditional scientific methods. Therefore, for Galston and Slayman and others to judge Backster's work from wholly within the ontological boundaries of traditional science is to ignore that he has intentionally stepped outside of traditional science and, in doing so, raised legitimate questions about its methods if plants are not assumed to be biotic machines.

If Galston and Slayman had instead taken Backster's objections seriously, modified their own methodologies to control for the possibility of plant awareness, conducted experiments where the relationship was allowed to vary both for the plant and the researcher, and *still* found no evidence of plant response to the interaction, their critique would be a stronger one. There would have been, however, a deeper issue if Galston and Slayman had considered altering their approach in this way. That is, within the confines of traditional scientific approaches, where largely objective approaches are seen as the only legitimate conduit to knowledge, it may simply have been impossible for them to have altered their methods sufficiently to establish the kind of relationship Backster and others on the kook-fringe suggest is necessary to do good relational research.

As mentioned at the beginning of this "Kook-Fringe" discussion, the second feature of interest in kook-fringe research is that even within the confines of scientific investigation, some intriguing results emerge. For example, researchers Dubrov and Pushkin (1982) suggested that Kaznacheev, Mikhailova, and Kartashov's (1980) work on electromagnetic communication between plant cells painted a portrait of a plant "empathy" (Dubrov & Pushkin, 1982, p. 93) In trying to find evidence of plant response to emotional experience in humans, Puthoff and Fontes (1975) found that "electrical activity of plants in close proximity to a human subject viewing slides of putative emotional content...did show in some cases (20%) statistically significant evidence of correlation with [human] subject GSR" (p. 32). From this, they concluded that "there is evidence for a degree of correlation [between human GSR in response to exposure to emotional content and the electrical fluctuation in electrically shielded plants that sat nearby] beyond that expected by chance" (p. 33). This would seem to correspond most with Pert's view of the electrical communicability of emotions. But, because the plants were electrically shielded, the communication had to have occurred by some non-electrical means. This would support my suggestion that there is the possibility of direct, more-than-material communication of feelings.

At one point, the television show Mythbusters (Rees, 2006) got involved in testing the electrical response of plants to human thoughts of harming them. In their initial experiment, they found that 35% of the time, the plant responded electrically to one of the hosts' thoughts of harming it. Of this result they say,

> The most surprising result was that the needle responded to [co-host] Tory's thoughts about harming the plant, but as the later experiments noted, these first tests didn't do a good job of isolating the variables. With [co-host] Grant and Tory standing right next to the sensitive polygraph, there were many ways in which their excitement and jumping around could have caused the instrument to respond.

By stating things this way, they seem to suggest that the next test they ran to control for the effect of standing too close to the polygraph showed no such effects, but this is not the case. In the first test both the hosts and the plant were inside a shipping container to control for external electrical interference. But, because of the concern just mentioned over host influence on the equipment, they moved the hosts outside of the shipping container, leaving the plant and polygraph equipment inside to shield them from the hosts. This time, when the host had thoughts of harming the plant, the "plant responded 28% of the time to Tory's stimuli."

The fact that there was a drop of 7% may indicate that the hosts were interfering with the equipment to some degree, but the fact that still, nearly 1/3 of the time, the plants responded to the hosts' thoughts of harming them is quite remarkable, and goes totally unacknowledged and unexplored in their writeup. Whether the percentage of 28%, 35% or both are statistically significant is not discussed by the show's hosts, but for them to see lay significance in the first set of results and to dismiss it in the second is curious, and could lead one to conclude that they simply didn't want to find significance (statistical or otherwise) in the second set of numbers. If they had, wouldn't they then, too, be accused of being part of the kook-fringe? Ultimately, much of this kind of plant research remains gimmicky and/or on the fringe, and yet it is strangely persistent. Critics have suggested that this is so because humans want to believe it (Galston & Slayman, 1979). But, in light of more recent research about plant neurobiology (Brenner et al., 2006), for example, the possibility of plant emotional capacity may be moving, ever so incrementally, closer to the center.

9.5 Feelings in "Inanimate Objects"

In this section. I will explore how, ontologically, feelings in "inanimate objects" might be possible. By "inanimate objects" I refer to those beings classified as "Not animated or alive; destitute of life, lifeless…that part of nature which is without sensation" ("Inanimate", 2018). Thus stones are "inanimate" while mold spores are "animate." To begin this discussion, I deconstruct the notion of "animism" in order to suggest that the differentiation between animate and inanimate is a largely dualist construct.

9.5.1 Bird-David's Animism

The OED defines animism as, "The attribution of life and personality (and sometimes a soul) to inanimate objects and natural phenomena" ("Animism", 2018). Since, to this point, I have contended that there is nothing ontologically problematic in nonhuman animals and plants being capable of experiencing and originating feelings, at first blush animistic theory may appear to align well with my position. Since my aim in this section is to argue for the ontological possibility of "inanimate objects" also being able to experience feelings, this would seem to strengthen such an alignment.

Yet, I must reject such a close coupling. After exploring several of the most accepted theories of animism, I've found that they suffer mostly from the same dualisms I've critiqued in other theories of human-nature relationships. In one respect, the actual experiences of human-nature relationships that these theorists are attempting to describe could, potentially, be the kinds of close human-nature relationships for which I argue in this book. But, because dualisms rear their heads in the theoretical interpretations of such experiences, animism ultimately falls short as a conceptual framework to adequately explain such experiences. It's unfortunate that this is the case, since my description of close human-nature relationship elements in large part owes its origins to the same experiences that these theorists are attempting to describe. I take their failure to recognize and reject the influence that their own dualistic ontologies have on their conceptual interpretation of those experiences as still more support for my contention that most modern ontologies of human-nature relationships require radical rehabilitation in order to allow for the possibility that what some humans are experiencing *are* close human-nature relationships.

To begin to articulate the reasons for my rejection, my point about inherent dualisms is illustrated by turning the tables on the very notion of animism. If we call mainstream modern thinking about animate and inanimate more-than-human beings "de-animism" and then define the term as, "The *post hoc* reduction and denial of life, individuality, *telos*, agency, and personality to all manner of more-than-human beings—animate, inanimate, thought of as phenomena, etc." then the point is made more clear.

The term "animism" was coined in the second half of the 19th century by Tylor (1871) and slowly evolved into the general notion of animism I quote in the OED definition above. A more recent proponent of animism, Bird-David (1999), seeks to rehabilitate the term by addressing what she identifies as dualisms (in her case, she calls them positivisms) at work in more traditional definitions. To Bird-David, there are two main positivisms that appear in these earlier construals. First, traditional theories generally take the approach of positioning animistic ideas as purely spiritual elements inside of a spiritual/material dichotomy (i.e., Cartesian dualism) where the material is seen as the only source of "'true' knowledge" (p. S68). Second, and as a consequence, traditional theories portray "'animists' [as humans who] understood the world childishly and erroneously" (p. S68).

Bird-David does not believe in the Cartesian dualistic interpretation of those said to have animist worldviews, nor does she believe that their understandings of the world are childish and erroneous. As an alternative, she suggests that animists such as the indigenous Nayaka in India (whom Bird-David studied) don't really see humans or more-than-human beings as individuals in the modern sense of the term. For example, to make room for why the Nayaka believe in the existence of "devaru," who to them are living, more-than-human, more-than-material "super-persons" (p. S69), Bird-David suggests that the Nayaka see "personhood" (p. S73) (attributed by the Nayaka to devaru as well as humans and other species), as a "composite of relationships" (p. S72). To the Nayaka, she explains, "kinship was primarily made and remade by recurring social actions of sharing and relating with, not by blood or by descent, not by biology or by myth or genealogy" (p. S73). For the Nayaka, a "person is sensed as 'one whom we share with.' It is sensed as a relative and is normally objectified as kin, using a kinship term" (p. S73). Bird-David employs the verb "to dividuate" (p. S72) as reflective of the Nayaka way to distinguish this "person." Of that verb's difference from the verb "to *in*dividuate," Bird-David says, "When I individuate a human being I am conscious of her 'in herself' (as a single separate entity); when I dividuate her I am conscious of how she relates with me" (p. S72). In some ways, her approach of downplaying the atomized individual is similar to posthumanist approaches. Bird-David applies the notion of dividuation and "the dividual" to the Nayaka's view of both human and devaru existence. And while this may at first seem like a move by Bird-David to acknowledge the existentiality of devaru as more-than-material, more-than-human relational partners for the human Nayaka, understood first as always being within a matrix of relations, it quickly becomes clear that she sees the devaru more as a concept than as a differentiable, living thing like a human being is a differentiable, if not wholly isolated, living thing.

To begin, she defines devaru in three ways: as composite "dividuals," as concepts of relational "events in-the-world" (p. S68), and as concepts representing human "social experiences" (p. S69). The latter two being purely conceptual, they clearly do not describe the devaru as relational Selves and thus do not warrant further consideration in the context of my attempt to suggest that all more-than-human beings are Selves. Regarding the first definition that is still worth exploring, Bird-David objects to thinking of devaru as individuals in a modernist sense by suggesting that "We cannot say...that Nayaka 'think with' this idea of personhood about their environment, to arrive by projection at the idea of devaru. The idea of 'person' as a 'mental representation'...is modernist" (p. S73). I believe she's right about this, and that this aspect of her critique mirrors my own in Chaps. 4 and 5 when I talk about the collapsing effect that dualist views of more-than-human beings as relationally inert, material objects have on theories of human-nature relationships. But, in terms of escaping such dualisms, Bird-David's alternative definition of devaru fares little better than that which she rejects. For example, she explains devaru in this way:

> [T]he devaru objectify sharing relationships between Nayaka and other beings. A hill devaru, say, objectifies Nayaka relationships with the hill; it makes known the relationships between Nayaka and that hill. Nayaka maintain social relationships with other beings not because, as Tylor holds, they *a priori* consider them persons. As and when and because they engage in and maintain relationships with other beings, they constitute them as kinds of person: they make them "relatives" by sharing with them and thus make them persons. (p. S73)

If such an explanation of devaru seems convoluted, I suggest that this is directly attributable to the fact that Bird-David is being forced, conceptually, to dance around the elephant in the room that is her own dualist ontology. That ontology requires that devaru not actually exist as living, differentiable Selves. Reading the part of the quotation above which says that "devaru objectify" or that a "hill devaru... objectifies Nayaka relationships with the hill" one realizes that the dualism of her ontology pushes her to employ a bit of grammatical sleight-of-hand in order to enable the coexistence of Nayaka meaning of devaru as actually existing and her own, which is that they are human-created concepts projected onto an external reality that is not actually delineated in this way. These statements about devaru objectifying things are not about devaru as actual, living Selves objectifying relationships between humans and more-than-human beings. Instead, the devaru are intermediary, conceptual objects created by actual, living *human* Selves as those humans objectify their relationships with more-than-human beings, at least according to what Bird-David is really saying.

It's interesting to note that here, the human as an individual who can conceive of a hill as a relation is unproblematic, yet the devaru as individuals conceiving of a human as a relation doesn't pass muster. This double-standard betrays the underlying ontology that allows for human Selves but not more-than-human Selves, and certainly not inanimate, more-than-material, more-than-human Selves. Stripping away the conceptual buildup around such dualisms shows the underlying belief that more-than-human beings as Selves is still just a projection by *in*dividual humans. In Bird-David's view, then, devaru are still at best a metaphor for the instantiation of human-nature relationships. Thus and perhaps unwittingly, Bird-David is in agreement with those earlier animists that she positions herself to critique (Harris, 1983; Hunter & Whitten, 1976; Tylor, 1871). In particular with their contention that human beliefs in spiritual more-than-human individuals is a mistake in thinking.

What's equally clear is that Bird-David does not want the Nayaka to be mistaken in this way and thus thought of as childish. The only way to "help" them then, given the ontological strictures within which she is operating, is to recast their notion of existential spiritual individuals as conceptual "persons" or "dividuals" that are relational "embodiments." These embodiments *are* bodies to the Nayaka, or so Bird-David argues. But, I disagree with what I see as a clearly dualistic interpretation of who they are describing when describing devaru. For example, she relates that "[The Nayaka man] Chathen (age 50)...said one morning that during the night he had seen an elephant devaru 'walking harmlessly' between our homes, and this is how he knew...that it was a devaru, not just an elephant" (p. S75). Because of Bird-David's ontology and her desire to "help" Chathen and his tribe avoid "being mistaken" that this was an actual spiritual being walking between their homes, she must interpret his statement to mean that there is a relation between him and the elephant that becomes the devaru *in Chathen's and other Nayakas' minds* when articulating such experiences. This is evidenced in her reference to another Nayaka man, Kungan, who

> once took me along on a gathering expedition, and on hearing an elephant and knowing by its sounds that it was alone and dangerous, he turned away and avoided it. He did not engage with this elephant and referred to it not as "elephant devaru" but simply as "elephant." The lack of mutual engagement prevented the kind of relatedness which would have *constituted* [emphasis added] this elephant (at this moment) as devaru while it might be perceived as devaru on other occasions. (p. S75)

Here, the devaru is "constituted" by the human engaging in an encounter-sparked relationship—that is, it is brought into being in the minds of the Nayaka during a material encounter with non-individual containing matter and then projected onto that external encounter as a distinguishable Self. The relationship is *not* between Kungan and a devaru in the form of an elephant, or between Kungan and the elephant that, in this case, is not also a devaru. That Bird-David believes that any particular elephant is or is not devaru based solely on engagement with the elephant by the Nayaka men betrays the dualistic, anthropocentric ontological filter she's laid over the particular experience of spiritual Selves by the Nayaka. Thus, the devaru are reduced from the more-than-material Selves to human-created concepts. Once more, we arrive at the *post hoc*, conceptually flattened landscape of a modernist, dualistic ontology.

If one is incredulous at the suggestion that a devaru can be more-than-material being occupying a non-devaru, material elephant for a time, or taking the form of a material elephant for a time, I suggest it's not because of any facts at one's disposal, but because of an ontology that *a priori* installs this notion: "This can't be." But, it can be, given a sufficiently accommodating ontology such as the one whose basis I am establishing in this book. In that ontology, this can be and *is*. What's more, it can be *known*, not believed, since the basis of knowledge in such an ontology is different from those requiring material proof of a material-only reality, as discussed in Part III and also below in Chap. 10. In a relational ontology, devaru are Selves (and perhaps more)-in-relation and can do these things and much more. This is a far more closely aligned reading of Nayaka ontology, and is perfectly legitimate when an epistemology (and its sponsoring ontology) that disallows experiences of actual, spiritual individuals is removed. As long as one's experience of such spiritual individuals is reliable and valid within a relational ontological matrix, this is not a problem.

Therefore, the basis of my rejection of even "progressive" forms of animistic theory like those offered by Bird-David is rooted in the failure of theorists like her to show that the rehabilitation of animism's epistemology lies in the rehabilitation of its ontology, and also in seeing that the Nayaka epistemology of existential spiritual individuals is wholly unproblematic when left within its own ontology. Figure 9.2 shows the types of epistemological/ontological alignments I see as possible when exploring animist worldviews. The first is that of Tylor and the other early theorists that Bird-David critiques. The second is Bird-David's own theory. The third is the close relational approach I espouse.

What the Nayaka view illustrates is that, prior to the dualistic reconceptualization of nature and the human relationship with it, more-than-material or spiritual beings were part of nature just like humans or stones. They were not *super*natural in the sense that they were "outside nature." In a relational ontology that admits irreducibly more-than-material and material elements alike, experience of devaru and

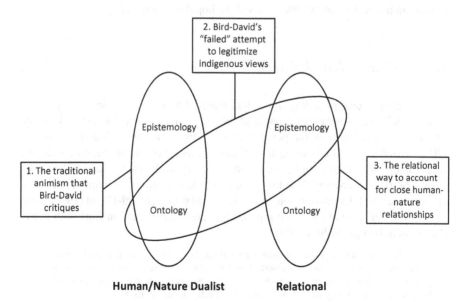

Fig. 9.2 Possible alignments of epistemology and ontology for theories of animism

other irreducibly spiritual entities is unproblematic. Of course, the experience of such beings by humans and the knowledge which emerges from such experiences ought to be subjected to the same rigorous vetting as any other knowledge based on experience. But, this does not equate with being subjected to the scientific method, with its intractably reductive, materialist constraints. This is not to say that such a view is inherently wrong, but neither is it a complete reflection of reality, nor without *its* particular limitations and biases.

If spiritual Selves do exist, then the material as origin of that which exists is short-circuited and the field of inquiry into who can and cannot have feelings is broken wide open. Human-like neurophysiology is no longer emotion's privileged bedfellow. Suddenly, human, elephant *and* devaru are potentially "in." So are stones like the one that the Nayaka woman Devi pointed out to Bird-David, where Devi said she'd "been digging deep down for roots in the forest when suddenly 'this devaru [stone] came towards her'" (p. S74). Again, my purpose here is not to establish as fact that the stone had a spirit that moved toward Devi, though I hold it to be no less plausible than the stone being inanimate matter. Instead, my point is that to view such a thing as *not possible* rests wholly on an unsubstantiated ontological position itself underwritten by seeing experience and thus reality as primarily material and largely without individualized more-than-human agency unless it's superadded by *post hoc* human conceptual processing and invention. Until modern humans begin to look at their experiences with all manner of more-than-human beings through an ontological lens liberated from such experience-dampening conceptual constraintes, it's impossible to determine just what exactly *is* happening in our relationships with more-than-human beings—"animate" or otherwise. Without such a shift, our perceptions, our knowledge, and as a result our understanding of human-nature relational experiences, will be hopelessly stunted.

9.5.2 Milton's Loved Nature

The problem of understanding more-than-material beings of all kinds—especially "inanimate" ones—as beings capable of feeling, is not restricted to animist theory, either. A good example of the pervasiveness of this view can be found in Milton's (2002) wonderful text exploring the effects of modernist beliefs on emotions and relationships with nature. In it, even though she explicitly acknowledges that some human beings experience close human-nature relationships with inanimate beings, at times she also falls into the dualistic ontological trap that Bird-David (1999) does. Describing one avenue by which humans feel engaged in relationships with more-than-human beings, Milton (2002) says,

> Given that we cannot avoid engaging with natural processes, given that such processes impact on our survival and livelihood, it is not surprising that people in many societies perceive personhood in the earth, the wind, the sun, the rain and nature as a whole. Relational epistemology is thus the foundation of religious ideas. (p. 50)

Here, Milton makes two key, mistaken assumptions. Like Bird-David (1999), she assumes that more-than-human beings cannot be persons in *the same way* that modern humans understand themselves to be persons. Posthumanists, for their efforts, do not resolve this issue. Their solution is to dispense with individuals entirely. But in so doing, their materialist ontology still dualistically supports the human-nature relational "asymmetry" that ultimately positions humans as relationally superior. Thus, the modernist issue of more-than-human beings not being persons in the same way that human beings are remains. The second mistake in Milton (2002) is that her assumption that spiritual or religions experiences are not the product of relationships between beings. Instead, to Milton, they are created by, and occur wholly inside of, humans operating in a material world with its evolutionarily driven desire for material survival. To her, they are based on the types of interchanges humans have within this drive. The chief features of these interchanges, according to her way of thinking, are material survival and, if I'm reading Milton correctly, occupational well-being (i.e., "livelihood"). That these are chiefly or initially material interchanges for Milton is clear. Thus, Milton falls into line with most materialists who believe that spiritual experience is not an interchange with the existential, more-than-material qualities of a relational, more-than-human Self, but instead are a *post hoc* human conceptual response to the individualistic meeting of human material "needs" with a passive, material, and, at times inanimate, other. Milton's examples of wind, sun and rain in the quote above are examples of such supposed "inanimate" beings.

To Milton, the religious is not based in a primary spiritual experience either. Her view is that "...we perform rituals of propitiation and conciliation in the hope of persuading [nature] to do what we want." (p. 51). Thus, to Milton we treat nature as a means to our own well-being. That these natural "elements" are not really persons as other humans are persons to Milton is betrayed by the fact that she subtly instantiates the ontological notion that, since the exchange *can only be* material, that all that we can and should seek with "inanimate" more-than-human beings is the kind of relationship that is evolution's *telos*—material survival for self and self's species. If rain, sun and wind as *animate*—feeling, thinking and acting—others existed instead, we might not propitiate or conciliate to get them to "do what we want," but instead to give them something based solely on their status as our close relational partners—like flowers for one's wife—where more-than-material Selves are relational ends in themselves. Milton reveals her deeply held belief that these more-than-human beings are inanimate and not Selves, when suggesting that the only way they get to be "persons" is through the same sort of affirmative action of expanded definitions of personhood that Bird-David (1999) employs to incorporate material exchange with a material object (and whatever trickle of relational feeling is created inside the human through such exchanges).

That such a view leaves intact the materialist ontology that prevents understanding of this exchange as actual, more-than-material, relational experience with living, more-than-human Selves seems utterly lost in this way of thinking even though this lies at the very heart of the Nayaka and others' views. Thus, though Bird-David (1999) and Milton (2002) are two examples of theorists who attempt to include all of nature within the notion of relations and personhood, by leaving intact the dualist ontologies that have historically negated the actuality of these more-than-human relational Selves, and thus truly close human-nature relational experiences, more-than-human beings are once more relegated to the status of unfeeling, potentially lifeless, passive material backdrops to the solely important and meaningful human endeavor. That the flimsy, material ties of evolution's *telos* of self-perpetuation are all that remains to such theorists in their attempts to *post hoc* conceptually explain why a human being could ever feel close with a more-than-human being is the height of reductionism, and strikes this author as nothing short of sad.

9.5.3 Feeling With?

In contrast, through experience rooted in the more-than-material (as well the material) even "inanimate" more-than-human beings can be fully capable of everything that makes a human a person, including feelings. Bose (in Shepherd, 2006) noted high degrees of similarity between the electrical behavior of "inanimate" metals and that of plants and living tissue such as muscle (p. 13). He found that all three responded with fatigue (a progressive reduction in electrical response to electrical stimulus) and, when the stimulation frequency was reduced, recovery of the normal electrical response. Noting that chemical agents can have either a stimulatory or "poisoning" effect on organisms—which can be seen when reflected in the amplifying or reducing cell electrical activity inside organisms—Bose also examined the effects of these chemical agents on metals. Remarkably, in the metals he studied, he found that the chemicals that worked as stimulants for living tissue did so with electrical behavior in metals as well, and those that worked as "poisons" (i.e., having the effect of reducing electrical sensitivity) on organic life also worked the same way on metals. If, as I've suggested above, electrical activity could be a material marker for feelings in all manner of beings, then here again there is a parallel in electrical activity between living beings and "inanimate" ones. Again, without the requisite levels of experience of the sharing of feelings with inanimate more-than-human beings, I cannot claim to have established the case for their existence. What I *can* safely conclude, however, is that when ontology is thoroughly cleansed of its dualisms, nothing remains of the definitive rejection of feeling in "inanimate" more-than-human beings, and the possibility that humans can feel with them. Loving a place, as the future poet did in Louv's anecdote mentioned in Chap. 1, is a perfect example of that possibility. What if that young girl was "feeling with" the rocks and stream in Noddings' sense of it? At this juncture, there is nothing in a non-dualist, relational ontology that stands in the way of such a possibility.

The key difference between my approach and that of Bird-David (1999) and Milton (2002), amongst others, is that in casting as ontological the critical features of close human-nature relationships, the definition of "person"—of close relational partner and what's necessary to be one—*need not change at all*. What must change are the ontological elements that falsely privilege humans as virtually the only ones who fit the definition. Since I've argued thus far against any appeal to the authority of material, human-like qualities for such privilege, one instead must turn, as I have, to an analysis of the origins of the elements of close relationships—feelings, thoughts and actions—and their more-than-material as well as material bases in all beings. This way, who is deemed as having feelings—who even counts as a "who" (instead of a "what") or a Self—is radically altered. Do stones have a soul or consciousness? I won't say, but I do know that those theorizing about human-nature relationships have rarely taken seriously such questions. They've instead allowed their dualistic ontologies to obliterate any reliable experience they themselves might have with stones that *would* suggest such a thing. In a close relational ontology, if one can feel what a stone feels through one's own experience or encounter with a stone, then the stone being a Self stands as distinctly possible and not even a little far-fetched. That it might seem absurd to modern minds speaks only to the hegemonic power of dualistically held materialist thinking, rather than any definitive experience confirming the absurdity.

Hallowell (2002), in considering Ojibwa ontology, addresses such a suggestion this way:

> [W]e cannot assume that objects, like the sun, are perceived as natural objects in [a modern] sense. If this were so, the anecdote about...old men [interacting with the sun, who is a person, and telling the sun to rise and set] could not be accepted as an actual event involving a 'social interaction' between human beings and an other-than-human person. Consequently, it would be an error to say that the Ojibwa 'personify' natural objects. This would imply that, at some point, the sun was first perceived as an inanimate, material thing. *There is, of course, no evidence for this* [emphasis added]. (p. 29)

Hallowell reinforces my contention that there is "no evidence" for material construal of experience. Thus, it must flow from a pre-existing ontology or worldview. Where I differ from Hallowell is in his attribution of the source of worldviews both in the Ojibwa and in modern humans. He begins with a perfect explanation of the difference between modern views of animistic ontologies but eventually succumbs to dualistic ontologies that deny the possible realities such ontologies sponsor. He does this when suggesting that all worldviews are "culturally constituted and symbolically mediated through language" (p. 20). I believe this view to be incorrect for the precise reason that he thinks it's incorrect to characterize Ojibwa ontological definitions of the sun-as-person as a "personification." That is, if he thinks it incorrect to assume that humans have control over their perceptions such that they can personify inanimate objects in a materially external world, he also ought to reject the notion that humans have control over culturally constituting the relational Selves they encounter in their experiences. To suggest that the modern view and Ojibwa view are equally valid and possible is to imply that the actual, external world is somehow an inert backdrop onto which these equally valid and real, culturally constituted views are projected by the human. The two views are incompatible. One is simply more accurately descriptive of reality than the other. The rub lies in how we weigh the accuracy of each view. By holding ontology the way he has, Hallowell still fails to move the locus of control of reality out of human minds. Thus, what is "real" in Hallowell's view must still begin in a purely physical world. The other possibility is that there is no external physical world at all, and then he is an idealist. Either way, he fails to establish the legitimacy of Ojibwa worldviews by unwittingly carrying his own dualist perspective into his interpretation of them.

I believe that the specific error he makes is in seeing culture as a product of human contrivance—that it is purely *social*, or cultural—instead of human-nature relational as well. By seeing it as only socially constructed, he leaves in place the ontological stance that humans and more-than-human beings are foundationally separate from each other, with human culture fully forming prior to being influenced relationally or otherwise by the more-than-human world. That this is materially inaccurate is plain. One would not have a single molecule of ATP, the basic unit of energy in cells, to burn in thinking about how to "constitute" a culture without the material and more-than-material interchange with more-than-human beings that makes food acquisition possible. I contend, then, that humans don't see the more-than-human world through a lens of socially constructed culture, they see it through a lens of human-nature relationships first, despite all attempts to conceptually eradicate the latter's influence and existence in modernist views.

By situating human culture within human-nature relations, my hope is to make it more clear that human thinking and collective culture is directly "constituted" by human-nature relationships. That's Preston's (2003) thesis in his volume *Grounding Knowledge*, though again, like many others, he takes us down a phenomenological road that is ultimately rooted in material interchange. But, as Martín Prechtel (personal communication) has pointed out to me and many others countless times, culture is the human response to more-than-human beings with which we live, be they "animate" or "inanimate," with them having material and more-than-material elements.

That a modern society living in New Jersey can have a radically different culture than, say, the Lenape Indians of the Northeastern shore that preceded them is not a function of cultures being constituted differently in the mysterious, more-than-human-free vacuum where human cultures are born and raised. If it were, we'd be right back to Lélé and Norgaard's "quicksand of values," with only human preference to guide us to the "right" relational point of view. Instead, these radically different cultures form because one group of humans holds an ontology that allows them to conceptually re-interpret their perceptions and experiences of their human-nature relationships to the point where it's as if they weren't having them—or they were occurring in partnership with a radically reduced object instead of a more-than-human, relational Self. The result is the view of culture as largely socially constructed. The result is also a constellation of modern cultures that now labor to "socially construct" more benign human-nature relationships out of a daily existence conceived to be largely human-nature relation-*less*. That this effort has failed to prevent impacts on the more-than-human world in materially negative ways is obvious. And though I take this material impact to be a byproduct of flawed relationships, it is usually taken to be the totality of the context for discussion of "environmental problems." That these impacts are an outgrowth of the failure to acknowledge, and allow to develop, deeper, close relations with more-than-human beings is clear when situating these issues within the relational ontology for which I argue in this volume.

If the development of close human-nature relationships were to happen for modern people living along the northeastern coast of the United States—if they were to discard the dualist ontologies that makes them almost wholly blind to the close relations they are in, and could be in, with more-than-human beings—I suggest that their culture would bear a striking resemblance to that of the Lenape. That's because it would be the product of a relationship with the *same actual more-than-human Selves* in that place, which would foundationally influence whatever human cultures alighted upon its living, teeming-with-Selves shores. History and social influences would certainly make that culture different from the Lenape in some ways, too, but the grounding in close human-nature relationships would make that difference one of degree, not kind as its currently envisioned by Hallowell and most other animist theorists.

Thus, what is and is not considered animate is the thing that is actually socially constructed, not the nature of the things falling into either bucket in our human-nature relationship theories. Even the existence of this dividing line is the product of modern societies that routinely assess the value, or lack, of what is to be put to human use next. Though not prone to considering conspiracy theories, such a way of dualistically reducing the inanimate to "raw material" strikes this author as conveniently self-serving. It arrives there by having taken the long way around through installation of dualistic ontologies, but such a path seems to have afforded little encounter with what more-than-human beings—even "inanimate" ones—are actually like.

9.5.4   Modern Human Perception and Feelings in the "Inanimate"

From the discussion of animist theory above, it should be clear that many indigenous cultures do, indeed, have knowledge that inanimate beings are alive and can have feelings. For example, Reinhard (2006) describes the views of South American Andean villagers who know that "traditional place spirits have feelings [and that m] ountains '...get hungry, have wives and children [the smaller mountains] and fight among themselves..." (p. 183). Or, to put it more succinctly, he quotes another villager as saying, "You must understand. For us the mountains are alive" (p. 184). Yet, as I acknowledged at the outset of this work, I eschewed reference to indigenous ontologies because to employ them risks framing the debate as a difference in *human created* concepts of ontology—a notion I have critiqued throughout the book. If, as is my belief, a close relational ontology is a *more accurate* reflection of the pragmatists' "primary experience," then even with dualistic conceptions in place for some humans, one ought to see evidence of the experience of sharing feelings, even with "inanimate" beings, "leak" through into perception. In this subsection, my purpose is both to explore this possibility through some examples, and in the process of doing so, explore how dualist thinking casts humans having feelings in the presence of inanimate beings to them *not* feeling with another Self. Unless otherwise noted in the following examples, the emphases in the quotations are mine.

A first area we might examine in societies with dualistic ontological commitment is poetry. As to whether poetry is relevant in an ontological discussion, Peirce (1960) refers to it in just such a context by saying that "nothing is truer than true poetry. And let me tell the scientific men that the artists are much finer and more accurate observers than they are..." (1.315, p. 358). If one presses the issue, suggesting that poets are engaging in the pathetic fallacy—that of attributing human attributes to all of nature—I suggest again that such a charge is almost wholly sponsored by dualistic assumptions that more-than-human beings cannot have certain attributes that humans have. Thus, I'll press forward with a first example from Mary Oliver's (1990) inquiry into the nature of the soul and who has it. She asks, "Is the soul solid, like iron?...Why should I have it, and not the anteater/who loves her children?...[and then] What about all the little stones, sitting alone in the moonlight?" (p. 1). Here, if one were materially or scientifically inclined, one might take hers to be a lyrical, anthropomorphic projection onto stones of being soul-bearing, or feeling alone or lonely. But, as I've discussed extensively, this interpretation is evidentially unfounded and ontologically circular. Thus, what if Mary Oliver was sitting there in the dark with those little stones while *perceiving* their feelings of aloneness? In a close relational ontology this is unproblematic. Her receiving their feeling and responding to it can be taken as more-than-material evidence for its occurrence because such communication of feelings is a direct and accurate way to assess the existence of such feelings.

Helen Keller (1904), for whom the senses were such an important doorway because of the difference of form they took in her, says, "What a joy it is...to clamber over a stone wall into green fields that tumble and roll and climb *in riotous gladness*" (p. 125). Diplomat and author Dag Hammarskjold (Lipsey, 2013) says,

> As a husband embraces his wife's body in faithful tenderness, so the bare ground and trees are embraced by the still, high light of morning. I feel an ache of longing to share in this embrace...*A longing like carnal desire...returned by the whispers of the trees, the fragrance of the soil, the caresses of the wind, the embrace of water and light* [emphasis added]. (p. 106)

Here, the "inanimate" soil, wind, water and light return his feeling of desire in their voices and actions. The artist Egon Schiele observed, "In late summer, when you see a tree showing signs of the coming fall, you experience it deep inside your entire heart and soul; I want to paint this wistfulness" (Museum of Fine Arts Boston, 2018).

To move on to an example of feelings being discussed, but denied as real by their authors, I consider the famed naturalist, Henry David Thoreau. Cronon (1996), arguing that human views of nature in the 19th century included fear and awe before falling out of fashion, quotes Thoreau's description of an encounter with Mt. Katahdin in Maine:

> Vast, Titanic, inhuman Nature...does not smile on [the human climber] as in the plains. She seems to say sternly, why came ye here before your time? This ground is not prepared for you. Is it not enough that I smile in the valleys? I have never made this soil for thy feet, this air for thy breathing, these rocks for thy neighbors. I cannot pity nor fondle thee here, but forever relentlessly drive thee hence to where I am kind. Why seek me where I have not called thee, and then complain because you find me but a stepmother? (p. 11)

While Thoreau's description reveals that he himself believes he is engaging in anthropomorphizing activity by describing "inhuman" Nature as *"seeming to say"* things in a human voice, I suggest that this could be because he is a naturalist of an albeit earlier form of modern society. Thus, he was already touched by the civilization against which he often railed, but of which he was at the same time a product. In a close relational ontology, Thoreau could easily have been perceiving exactly what was being felt and communicated *by the mountain*: rage at his presence. Many indigenous cultures proscribe entry into certain holy places. For example, people are no longer allowed to climb Uluru, or Ayers Rock, in Australia because it is sacred to the Anangu aboriginal people there. This could be exactly how the Abenaki understood Mt. Katahdin. The feeling of anger that Thoreau felt could be a response to the feelings the mountain was communicating. Therefore, while an indigenous view can be dualistically dismissed, through a close relational ontology the mountain *can be* angry and Thoreau *can know it directly* through communication of feelings.

In the same discussion by Cronon, in referencing Wordsworth the same ontology makes an appearance when he takes more-than-human persons to be *post hoc* human conceptualizations only. For example, Wordsworth opines:

Winds thwarting winds, bewildered and forlorn,
The torrents shooting from the clear blue sky,
The rocks that muttered close upon our ears,
Black drizzling crags that spake by the way-side
As if a voice were in them... [emphasis added] (p. 11)

Again through a close relational lens, Wordsworth hearing the voice of the place as Thoreau did is unproblematic. And he, like Thoreau, reduces that voice to a metaphor—"as if" a voice were in the winds, rocks and crags instead of the physical sound being their actual voices and the feeling conveyed being what was conveyed by the rocks.

Cronon goes on to suggest that the fear of nature that Wordsworth and Thoreau express in such passages is wholly dependent on social definitions of wild nature, as is the lack of such fear that comes later in the environmental movement. What goes unstated, but is an unavoidable complement to such a thesis, however, is that the more-than-human beings with whom we are in relations must in actuality be a kind of relational *tabula rasa*, forming only the backdrop onto which humans, by drawing from some isolated locale inside themselves, project their concepts of what more-than-human beings are like. By extension, the more-than-human beings themselves have no more-than-material nature of their own to be perceived. In this kind of thinking, because more-than-human beings have no inherent relational qualities or attributes, modern humans are then free to experience fear, joy, indifference or any other unilaterally sourced feeling when in the presence of more-than-human beings.

I believe it's clear at this juncture that I disagree with such a thesis, and offer instead the possibility that Cronon is observing one step in the dualistic, *post hoc conceptual movement away from close relational interchange with more-than-human beings*. At the later stages of this movement (in which we modern humans find ourselves presently), feelings of fear (save, perhaps, fear of the material power of storms), joy, and everything else that emanates from both animate and "inanimate" more-than-human beings and our close relations with them, is largely lost to modernity. From this point of view, the words of Wordsworth and Thoreau become a weigh station along this path to the present, descriptive of a time when modern humans still *felt* more immediately unpleasant things that more-than-human Selves were feeling and conveying to them. Back then, the step necessary to take those kinds of feeling experiences out of the realm of perception and into the realm of conception was to characterize them as internal to the human—as an act of anthropomorphism. 150 years later, modern humans are loathe even to discuss these as potentially real relational interchanges, so mothballed have these relational sense experiences become. Now, when modern humans hear from or feel their more-than-human relational partners, they simply don't—and *can't*—believe that this is what's occurring. They can't even perceive the anger—or any other feeling—choosing to dismiss these voices as anthropomorphic when in reality, by doing so, they are being anthropocentric.

In the present-day, nature writing still affords us a view into the possibility of perceiving feelings in "inanimate," more-than-human beings. For example, desert naturalist Krutch (2010) says,

> Of some spot of earth one may feel that one would like it if one could really see or really know it...It could give something to me and I, perhaps, something to it—if only some sort of *love and understanding*. The desire to stay, to enter in, is not a whim or a notion but a passion. *Verweile doch, du bist so schön!* [Stay awhile, *you are so beautiful*]. If I do not somehow possess this, if I never learn what it was that called out, what it was that was being offered, I shall feel all my life that I have missed something intended for me. (p. 5)

Here, Krutch speaks of exchanging love and understanding with a place, and of the place's recognition of his beauty. In one of his more well-known short essays, the "grandfather of the Wilderness movement" in the United States, Aldo Leopold (1949/1989), suggests that through a mountain's experience, he or she is *more* capable than a human being of feeling. Again, the materially minded might categorize his essay as quaint allegory—an affected tangle of anthropomorphic tripe—but in a close relational ontology, his experience is *perception* and a response to the mountain herself. In describing the howl of wolves and the response of others to it he says that "every living thing (and perhaps many a dead one as well) pays heed to that call" (p. 129) and relates the meaning of that howl to the view of the deer, pine tree, coyote, cowman and hunter. But, he says, "behind [the perception of] obvious hopes and fears [in the howl] there lies a deeper meaning, known only to the mountain itself" (p. 129).

As I argued in Part III, the meaning of things like a wolf's howl do not inhere in the listener, but instead exist as a more-than-material thing belonging, in part, to the wolf. Leopold echoes this suggestion when he says that "Those unable to decipher the hidden meaning know nevertheless that it is there...Only the ineducable tyro can fail to sense the presence or absence of wolves, or the fact that mountains have a secret opinion about them" (p. 129). Noting that without wolves, deer will eat away at the plants that make their homes on mountains, Leopold says that "a mountain live[s] in mortal fear of its deer" (p. 132). Thus, mountains not only fear deer, but they also love wolves for keeping deer at bay. To the mountain, then, the sound of their howls is one of well-being. For one, I think Leopold mistaken in his suggestions about the mountain's negative feelings about the deer. It strikes me as being colored, in my estimation, by the encroachment in his thinking of the materialism of scientific ecology that he practiced. But still, you can't fault the effort at interpretation of what, at some level, he takes to be an inanimate Self.

John Muir (1901), that great naturalist and nature wanderer, directs us to "Climb the mountains and get *their good tidings* [emphasis added]...[where t]he winds will blow *their own freshness* [emphasis added] into you..." (p. 56). He (1911) also says, of one journey with companions, "How deep our sleep last night in the mountain's heart" (p. 42). If a mountain has a heart, one could reasonably infer that he also has feelings. Other examples include environmental philosopher Jack Turner (1989), who says, "The mountains have many moods" (p. 81). There is Karen Warren (1990), the ecofeminist theorist who describes what has been interpreted by some to be a human-projected feeling of friendship between a climber and a cliff. Rachel Carson (1941) says in *Under the Sea Wind* that "The tundra was gay..." (p. 61) with its fall colors. Mildred Cable (1942), a missionary traveling the Gobi Desert in the 1930s, says, "The clearness and watchfulness of each planet suggests a personal and friendly interest toward the wayfarer" (p. 107). These sorts of literary references are numerous in modern societies, straying even into works of fiction that are apropos of my discussion. For example, that keen-eyed lamenter of the Industrial Revolution, Charles Dickens (1905), portrays a scene where "a melancholy wind sounded through the deserted fields...The sadness of the scene imparted a sombre tinge to the feelings of Mr. Winkle" (p. 28). Through a close relational lens, Dickens' depiction of Mr. Winkle is not that of a man alone, but of a man in a relational experience with the wind—a relationship imbued with existential sadness in the wind and Mr. Winkle alike.

To conclude, I'd like to note that I am not attempting to "hijack" any of these examples as "proof" of my thesis. I readily acknowledge that the authors themselves may not think my exploration of the meaning of their writing to be accurate—that they may very well not literally think the tundra gay or that the planets are taking a friendly interest in passing humans. But, I reiterate that in a society in which hegemonic, dualistic ontology has such a pervasive influence, it is reasonable to look to such literary descriptions as potential echoes of relational feelings coming through to a human writer from more-than-human Selves, regardless of whether the authors recognize it as such. If the reader is still not convinced, and one's response is still to see my extrapolation from these examples as unfounded, I point out one last thing: With a dualistic ontology positioned to deny the occurrence of feelings in any being that is not materially similar to humans, human experiences of feelings coincidental to experiences with "inanimate" more-than-human beings *must* be superadded by the human. There is simply no other avenue to explain their presence. Thus, an ascription of anthropomorphism *must* flow from that viewpoint. That such an unavoidable conclusion is based wholly in assumption, and not evidence, should raise serious concerns for those still cleaving to such a view.

9.6 A Re-Envisioned Environmental Conservation Example

To date, a response by the modern human world to the feelings of more-than-human beings has been almost entirely absent from serious, mainstream scientific and academic institutions. The response has been to literally imagine the possibility out of existence. But if we allow for understanding of our close human-nature relational experiences as rooted in part in feelings, what then does it mean for more-than-human beings to have feelings about us? If, for example, they have the capacity to be angry or annoyed like Mt. Katahdin might be with Thoreau, what does it mean for them to be angry or annoyed *at us*? What should our response to such anger or annoyance be? I suggest that, in the simplest and most initial stages, it would be to *listen*.

To see what effect such listening would have, let's look at an example of a conservation effort such as bird banding. The USGS's Bird Banding Laboratory (2015) describes bird banding data as contributing to

> an understanding of the life history and population dynamics of various species… [which], in turn, is critical to evaluations of management actions, including the establishment of appropriate exploitation rates for hunting; assessments for actions to improve the status of a species…or assessments of the effects of changes of habitats on all bird species. (para. 1)

The use of such ecological data is fairly "boilerplate" for modern academic and scientific endeavors of this sort. The understandings sought are dualistic, however, being material and largely anthropocentric. That's not to say that bird banding data doesn't also help birds, but it is almost exclusively to help them materially survive one or another pressure applied by modern humans, pressures which modern humans are disinclined to relieve, at least in full. Of course, if we were to relieve them entirely, we'd also obviate the need to determine via banding just which pressure is most disadvantageous to the birds in the first place. Such circular cause-and-effect is what I characterize as the "madness" of the modern conservation proposition: helping more-than-human beings survive a human-nature relationship imposed by humans that humans know is detrimental, yet generally do little to address at its source. To draw a parallel with the inefficiencies and logical sleight-of-hand that large bureaucracies such as governments employ and put it more strongly, bird banding is an example of humans "convening committees" to study the outcome of what we already *prima facie* know to be bad for birds. We cannot have our highway system, our smart phones, and even our wind power and let birds live how they will live. Not how we *think* they need to *materially* live—but how they will *choose* to live *for themselves*.

Further, by unilaterally imposing our will upon birds through traps to band them, we do absolutely nothing to address the fact that such a relationship between human and more-than-human is no different than it is in something such as large scale deforestation that would spur us to band and then study them in the first place. The only difference is in what *we modern humans* find acceptable to do to the birds inside of a largely dualistic and anthropocentric frame of reference. By carrying on in our relationships with *this* mist net and *that* bird caught in it without a thought as to how *that bird feels* to be treated thusly by us, is to simply continue imposing the same old dualistic human-nature relationships. To the birds, I imagine, any good intentions behind these acts are both invisible and irrelevant. I remember a colleague some years back describing with glee her trip to band passerines. With regard to trying to band one species, she said with a wide grin, "And the Black-capped chickadees fought the hardest of all!" Given the level of consternation dualistically influenced scientific thinkers such as her display at objections one might raise to such "innocuous" efforts to help birds I said nothing, but was inwardly saddened. I remember thinking, "Good for *them!* They're fighting because they're angry at being accosted." Of course someone could object to my characterization of the chickadee this way, but when dualisms are eliminated, I don't think the characterization at all anthropomorphic. Hold down a human being and forcibly encircle her ankle with an aluminum ring and she will likely become very angry. If a bird has a capacity for anger, then he, too, will feel angry. Given that honeybees feel pessimistic, it's not a stretch to allow for this possibility. If one were to suggest that the feelings of anger don't arise in response to the act itself but in the metacognitive human-only understanding of what the act means, I suggest first that the feelings are at least in part instantaneous and instinctual both in humans and birds, and second that, as I'll address in Chap. 10, various cognitive abilities in birds, that could include anger or thoughts about it, cannot be uncritically dismissed.

Ultimately, I'm not trying to characterize bird banding as a sin, as it does have beneficial effects for birds. When there's a war on and someone is shot and bleeding, you have to staunch the flow of blood first and foremost. But if conservation efforts are only ever this kind of triage, and no ceasefire is either sought or even thought necessary or possible, then the war will continue to produce those shot and bleeding, and everyone will eventually lose. Ultimately, within a human-nature relational ontology inclusive of more-than-human feelings, the sin lay not in the banding, but in the lack of consideration of the feelings of those banded. If, in whatever way works, banding were suggested to the birds, and they agreed to be monitored, then such an activity ceases to be detrimental to that particular human-nature relationship, and ceases to add to the larger-scale relational ignorance of more-than-human feelings. How could we know how the birds feel about being monitored? We'd have to employ the kinds of knowing and feeling that I discussed in Part III. We might also take their very presence or absence, or their level of resistance, as a way of them saying whether they agree or not. We might also take our own feelings in response to the encounter with them as reflective of their sharing them with us via Peircian feelings or some other means. In other words, humans would have to work hard to *listen* to the response the birds have—to perceive it unfiltered by dualisms—and honor whatever comes through. If the birds did not agree, and expressed an unhappiness with the possibility, then the researcher would have to abandon his efforts. Such a suggestion is anathema, I suppose, to the modern researcher, but this is what it means to have feelings and their communication between human and more-than-human become an essential part of human-nature relationships. If one were to conclude that chickadees would always fight and thus would never agree, I think that's possible. Perhaps chickadees just don't want that kind of contact with humans. The point is, if such preferences on the part of birds were *always* honored in our modern societies, we'd quickly arrive at the place I suggested above—we'd eliminate at the source the need to band them at all.

References

Aftanas, L., & Golocheikine, S. (2001). Human anterior and frontal midline theta and lower alpha reflect emotionally positive state and internalized attention: High-resolution EEG investigation of meditation. *Neuroscience Letters, 310*(1), 57–60.

Ahern, G. L., & Schwartz, G. E. (1985). Differential lateralization for positive and negative emotion in the human brain: EEG spectral analysis. *Neuropsychologia, 23*(6), 745–755.

Allen, C., & Bekoff, M. (1999). *Species of mind: The philosophy and biology of cognitive ethology.* Cambridge, MA: MIT Press.

Animism. (2018, June). *Oxford English dictionary online.* Retrieved from http://www.oed.com

Backster, C. (1968). Evidence of a primary perception in plant life. *International Journal of Parapsychology, 10*(4), 329–348.

Backster, C. (2003). *Primary perception: Biocommunication with plants, living foods, and human cells.* Anza, CA: White Rose Millennium Press.

Barad, K. (2007). *Meeting the universe halfway: Quantum physics and the entanglement of matter and meaning.* Durham, NC: Duke University Press.

Bartels, A., & Zeki, S. (2000). The neural basis of romantic love. *Neuroreport, 11*(17), 3829–3834.

Bates, L. A., Lee, P. C., Njiraini, N., Poole, J. H., Sayialel, K., Sayialel, S., et al. (2008). Do elephants show empathy? *Journal of Consciousness Studies, 15*(10–11), 204–225.

Bateson, M., Desire, S., Gartside, S. E., & Wright, G. A. (2011). Agitated honeybees exhibit pessimistic cognitive biases. *Current Biology, 21*(12), 1070–1073.

Bekoff, M. (2000). Animal emotions: Exploring passionate natures. *Bioscience, 50*(10), 861–870.

Bekoff, M. (2007). *The emotional lives of animals: A leading scientist explores animal joy, sorrow, and empathy – and why they matter.* Novaro, CA: New World Library.

Bekoff, M., & Pierce, J. (2009). *Wild justice: The moral lives of animals.* Chicago: University of Chicago Press.

Berns, G. S., Brooks, A. M., & Spivak, M. (2012). Functional MRI in awake unrestrained dogs. *PLoS One, 7*(5), e38027.

Bird-David, N. (1999). "Animism" revisited: Personhood, environment, and relational epistemology 1. *Current Anthropology, 40*(S1), S67–S91.

Brenner, E. D., Stahlberg, R., Mancuso, S., Vivanco, J., Baluška, F., & Van Volkenburgh, E. (2006). Plant neurobiology: An integrated view of plant signaling. *Trends in Plant Science, 11*(8), 413–419.

Buber, M. (1970). *I and thou.* (W. Kaufmann, Trans.). New York: Charles Scribner's Sons. (Original work published 1923).

Bucci, W. (2001). Pathways of emotional communication. *Psychoanalytic Inquiry, 21*(1), 40–70.

Budagovskii, A., Turovtseva, N., & Budagovskii, I. (2001). Coherent electromagnetic fields and remote cell interaction. *Biophysics-Pergamon then Maik Nauka-c/c of Biofizika, 46*(5), 860–866.

Byrne, S. E., & Rothschild, A. J. (1998). Loss of antidepressant efficacy during maintenance therapy: Possible mechanisms and treatments. *The Journal of Clinical Psychiatry, 59*, 279–288.

Cable, M. (1942). *The Gobi desert.* London: Hodder and Stoughton.

Carson, R. L. (1941). *Under the sea-wind: A naturalist's picture of ocean life.* New York: Signet.

Celizic, M. (2007, November 8). Dolphins save surfer from becoming shark's bait. *Today.*

Li, M., Chai, Q., Kaixiang, T., Wahab, A., & Abut, H. (2009). EEG emotion recognition system. In *In-vehicle corpus and signal processing for driver behavior* (pp. 125–135). Boston: Springer.

Cifra, M., Fields, J. Z., & Farhadi, A. (2011). Electromagnetic cellular interactions. *Progress in Biophysics and Molecular Biology, 105*(3), 223–246.

Cronon, W. (1996). The trouble with wilderness: Or, getting back to the wrong nature. *Environmental History, 1*(1), 7–28.

Custance, D., & Mayer, J. (2012). Empathic-like responding by domestic dogs (canis familiaris) to distress in humans: An exploratory study. *Animal Cognition, 15*(5), 851–859.

Davies, E. (2004). New functions for electrical signals in plants. *New Phytologist, 161*(3), 607–610.

Dickens, C. (1905). *The Pickwick papers*. London: The Amalgamated Press, Ltd.

Dubrov, A. P., & Pushkin, V. N. (1982). *Parapsychology and contemporary science*. New York: Consultants Bureau.

Esslen, M., Pascual-Marqui, R., Hell, D., Kochi, K., & Lehmann, D. (2004). Brain areas and time course of emotional processing. *NeuroImage, 21*(4), 1189–1203.

Falik, O., Reides, P., Gersani, M., & Novoplansky, A. (2003). Self/non-self discrimination in roots. *Journal of Ecology, 91*(4), 525–531.

Fisher, H. (2004). *Why we love: The nature and chemistry of romantic love*. New York: Henry Holt.

Foster, J. B. (2000). *Marx's ecology: Materialism and nature*. New York: Monthly Review Press.

Freedman, J. (2007, January 26). *The physics of emotion: Candace pert on feeling go(o)d*. Retrieved from http://www.6seconds.org/2007/01/26/the-physics-of-emotion-candace-pert-on-feeling-good

Fromm, J., & Lautner, S. (2007). Electrical signals and their physiological significance in plants. *Plant, Cell & Environment, 30*(3), 249–257.

Galston, A. W., & Slayman, C. L. (1979). The not-so-secret life of plants. *American Scientist, 67*(3), 337–344.

Griffin, D. R., & Speck, G. B. (2004). New evidence of animal consciousness. *Animal Cognition, 7*(1), 5–18.

Hallowell, A. I. (2002). Ojibwa ontology, behavior, and world view. In G. Harvey (Ed.), *Readings in indigenous religion* (pp. 18–49). New York: Continuum.

Harris, M. (1983). *Cultural anthropology*. New York: Harper & Row.

Hazan, C., & Shaver, P. R. (1994). Attachment as an organizational framework for research on close relationships. *Psychological Inquiry, 5*(1), 1–22.

Hernandez, S. (2012, February 19). Sturgeon bay cat saves owner's life. *Green Bay Press Gazette*.

Hirata, S., Matsuda, G., Ueno, A., Fukushima, H., Fuwa, K., Sugama, K., et al. (2013). Brain response to affective pictures in the chimpanzee. *Scientific Reports, 3*, 1342.

Horowitz, K. A., Lewis, D. C., & Gasteiger, E. L. (1975). Plant "primary perception": Electrophysiological unresponsiveness to brine shrimp killing. *Science, 189*(4201), 478–480.

Hunter, D. E., & Whitten, P. (Eds.). (1976). *Encyclopedia of anthropology*. New York: Harper and Row.

Inanimate. (2018, June). *Oxford English dictionary online*. Retrieved from http://www.oed.com

Kaznacheev, V., Mikhailova, L., & Kartashov, N. (1980). Distant intercellular electromagnetic interaction between two tissue cultures. *Bulletin of Experimental Biology and Medicine, 89*(3), 345–348.

Keller, E. F. (1983). *A feeling for the organism, 10th anniversary edition: The life and work of Barbara McClintock*. San Francisco: W. H. Freeman and Company.

Keller, H. (1904). *The story of my life*. New York: Doubleday.

Kelley, H. H., Berscheid, E., Christensen, A., Harvey, J. H., Huston, T. L., Levinger, G., et al. (1983). *Close relationships*. New York: W. H. Freeman and Company.

Kesey, K. (1987). *Demon box*. New York: Penguin Group USA.

Kessler, N. (2012). Chaos or relationalism? A Pragmatist Metaphysical Foundation for human-nature relationships. *The Trumpeter, 28*(1), 43–75.

Kmetz, J. M. (1977). A study of primary perception in plants and animal life. *Journal of the American Society for Psychical Research, 71*(2), 157–170.

Kreibig, S. D., Wilhelm, F. H., Roth, W. T., & Gross, J. J. (2007). Cardiovascular, electrodermal, and respiratory response patterns to fear-and sadness-inducing films. *Psychophysiology, 44*(5), 787–806.

Krutch, J. W. (2010). *The desert year*. Iowa City, IA: University of Iowa Press.

Kuzin, A. M., Surkenova, G. N., Budagovskii, A. V., & Gudi, G. A. (1997). Secondary biogenic radiation of gamma-irradiated human blood. [Vtorichnoe biogennoe izluchenie gamma-obluchennoi krovi cheloveka]. *Radiatsionnaia Biologiia, Radioecologiia/Rossiiskaia Akademiia Nauk, 37*(4), 577–580.

Laing, R. D. (2001). *The divided self: An existential study in sanity and madness*. London: Routledge (Original work published in 1971).

Leopold, A. (1989). *A sand county almanac, and sketches here and there*. New York: Oxford University Press (Original work published 1949).

Lipsey, R. (2013). *Hammarskjöld: A life*. Ann Arbor, MI: University of Michigan Press.

Low, P., Panksepp, J., Reiss, D., Edelman, D., & Van Swinderen, B. (2012). *The Cambridge declaration on consciousness*. Retrieved from http://fcmconference.org/img/ CambridgeDeclarationOnConsciousness.pdf

Merchant, C. (1980). *The death of nature: Women, ecology, and the scientific revolution*. San Francisco: Harper & Row.

Midgley, M. (2004). *The myths we live by*. London: Routledge.

Milton, K. (2002). *Loving nature: Towards an ecology of emotion*. London: Routledge.

Muir, J. (1901). *Our national parks*. Boston: Houghton Mifflin.

Muir, J. (1911). *My first summer in the sierra*. Boston: Houghton Mifflin.

Museum of Fine Arts, Boston (2018). *Museum label for artist, Egon Schiele*. Klimt and Schiele: Drawn Exhibition, February 25, 2018 – May 28, 2018.

Musumeci, F., Scordino, A., Triglia, A., Blandino, G., & Milazzo, I. (1999). Intercellular communication during yeast cell growth. *EPL (Europhysics Letters), 47*(6), 736–742.

Nikolaev, Y. A. (2000). Distant interactions in bacteria. *Microbiology, 69*(5), 497–503.

Noddings, N. (1984). *Caring: A feminine approach to ethics & moral education*. Berkeley: University of California Press.

Nollman, J. (1987). *Animal dreaming:(the art and science of interspecies communication)*. New York: Bantam.

Oliver, M. (1990). *House of light*. Boston: Beacon Press.

Pampallona, S., Bollini, P., Tibaldi, G., Kupelnick, B., & Munizza, C. (2004). Combined pharmacotherapy and psychological treatment for depression: A systematic review. *Archives of General Psychiatry, 61*(7), 714–719.

Panksepp, J. (1998). *Affective neuroscience: The foundations of human and animal emotions*. Oxford: Oxford University Press.

Pecchinenda, A., & Smith, C. A. (1996). The affective significance of skin conductance activity during a difficult problem-solving task. *Cognition & Emotion, 10*(5), 481–504.

Peirce, C. S. (1960). *Collected papers of Charles Sanders Peirce* (C. Hartshorne, P. Weiss, & A. W. Burks, Eds.). Cambridge, MA: Harvard University Press.

Pert, C. B. (2006). *Everything you need to know to feel go(o)d*. Carlsbad, CA: Hay House.

Pierce, J. (2008). Mice in the sink. *Environmental Philosophy, 5*(1), 75–96.

Plumwood, V. (1993). *Feminism and the mastery of nature*. London: Routledge.

Poole, T. (1997). Happy animals make good science. *Laboratory Animals, 31*(2), 116–124.

Preston, C. J. (2003). *Grounding knowledge: Environmental philosophy, epistemology, and place*. Athens, GA: University of Georgia Press.

Preston, S. D., & de Waal, F. B. M. (2002). The communication of emotions and the possibility of empathy in animals. In S. G. Post, L. G. Underwood, J. P. Schloss, & W. B. Hurlbut (Eds.), *Altruism and altruistic love: Science, philosophy, and religion in dialogue* (pp. 284–308). Oxford, UK: Oxford University Press.

Puthoff, H., & Fontes, R. G. (1975). *Organic biofield sensor*. Menlo Park, CA: Stanford Research Institute.

Radner, D., & Radner, M. (1989). *Animal consciousness*. Amherst, NY: Prometheus Books.

Rees, P. (2006). *Deadly straw. Mythbusters*. Sydney, Australia: Beyond Television Productions.

Reinhard, J. (2006). *The ice maiden: Inca mummies, mountain gods, and sacred sites in the Andes*. Washington, DC: National Geographic Books.

Rollin, B. E. (1990). How the animals lost their minds: Animal mentation and scientific ideology. In M. Bekoff & D. Jamieson (Eds.), *Interpretation and explanation in the study of animal behavior* (pp. 375–393). Boulder, CO: Westview Press.

Schmidt, L. A., & Trainor, L. J. (2001). Frontal brain electrical activity (EEG) distinguishes valence and intensity of musical emotions. *Cognition & Emotion, 15*(4), 487–500.

Servan-Schreiber, D. (2004). *The instinct to heal.* Emmaus, PA: Rodale.

Sheets-Johnstone, M. (2011). *The primacy of movement.* Amsterdam: John Benjamins Publishing.

Shepherd, V. (2006). At the roots of plant neurobiology. In A. G. Volkov (Ed.), *Plant electrophysiology* (pp. 3–44). Heidelberg: Springer.

Shepherd, V. (2012). At the roots of plant neurobiology: A brief history of the biophysical research of JC Bose. *Science and Culture, 78*, 196–210.

Simons, P. (1992). *The action plant: Movement and nervous behaviour in plants.* Oxford: Blackwell.

Sinha, R., & Parsons, O. A. (1996). Multivariate response patterning of fear and anger. *Cognition & Emotion, 10*(2), 173–198.

Skutch, A. F. (1996). *The minds of birds.* College Station, TX: Texas A&M University Press.

Solfvin, J. (2009). [Review of the book: Primary perception: Biocommunication with plants, living foods, and human cells, C. Backster]. *Journal of Parapsychology, 73*, June 1, 2015. Retrieved from http://www.thefreelibrary.com/Primary+Perception%3A+Biocommunication+With+Plants,+Living+Foods,+and...-a0219588969

Trewavas, A. (2003). Aspects of plant intelligence. *Annals of Botany, 92*(1), 1–20.

Turner, J. (1989). The abstract wild. *Witness, 3*(4), 81–95.

Tylor, E. B. (1871). *Primitive culture: Researches into the development of mythology, philosophy, religion, art, and custom.* London: Murray.

USGS Bird Banding Laboratory. (2015). Home page. Retrieved from https://www.pwrc.usgs.gov/BBl/homepage/gswhy.cfm

Van West, P., Morris, B. M., Reid, B., Appiah, A. A., Osborne, M. C., Campbell, T. A., et al. (2002). Oomycete plant pathogens use electric fields to target roots. *Molecular Plant-Microbe Interactions, 15*(8), 790–798.

Warren, K. J. (1990). The power and the promise of ecological feminism. *Environmental Ethics, 12*(2), 125–146.

Weisberg, Z. (2009). The broken promises of monsters: Haraway, animals and the humanist legacy. *Journal for Critical Animal Studies, 7*(2), 22–62.

Yoon, C. K. (2008, June 10). Loyal to its roots. *New York Times.*

Zeki, S., & Romaya, J. P. (2010). The brain reaction to viewing faces of opposite- and same-sex romantic partners. *PLoS One, 5*(12), e15802.

Chapter 10
Thoughts

Missives such as "a penny for your thoughts" have long been fodder for human intimacy. The surprise of what the other thinks, the beauty of what or *how* the other thinks, the depth, kindness, similarity and also intriguing dissimilarity of what the other thinks—these are the lifeblood of close relationships. Unexchanged or unexpressed thoughts about the other and the relationship can also work for or against closeness. For example, an avoidant attachment style (Collin & Feeney, 2000) by one person can work against the development of close bonds with another. Indeed, it could be said that thoughts of almost every type are deeply influential in both the development and maintenance of close relationships.

In this chapter, I explore the possibility of more-than-human beings having thoughts in addition to feelings, and if they have them, the possibility that they are the kind that can contribute to close human-nature relationships. The overriding sentiment in the literature on human-nature relationships is that most don't believe more-than-human beings to have this capacity in any substantial sense. I believe this position can be traced to two interlocking assumptions. First, that minds (as the originators of thoughts) are exclusively the product of a material structures and processes (neurophysiology) and second, that only a human-like neurophysiology is capable of producing the kind of mind that contributes the kinds of thoughts needed to engage in close relationships.

An example of these assumptions at work can be seen in Thomas' (1971) description of an ant. He says, "[a] solitary ant, afield, cannot be considered to have much of anything on his mind; indeed, with only a few neurons strung together by fibers, he can't be imagined to have a mind at all, much less a thought" (p. 12). Though he was writing on another topic, it doesn't seem a stretch to suggest here that Thomas would take the ant's lack of mind and thoughts as prohibitive of her capacity to have a close relationship either with those of her own kind or with humans. Would that assumption have to be altered, however, if the ant did not need a human-like neurophysiology to have a mind? Constraining the problem in this way, the answer ought

© Springer Nature Switzerland AG 2019

N. H. Kessler, *Ontology and Closeness in Human-Nature Relationships*,
AESS Interdisciplinary Environmental Studies and Sciences Series,
https://doi.org/10.1007/978-3-319-99274-7_10

to be that it is at least possible. It is my intention in this chapter to demonstrate this very possibility for the ant and many others, and thus the possibility that more-than-human beings of all kinds can have thoughts that contribute to close relationships.

10.1   The Mind-Brain Problem

I begin by discussing the first of the two interlocking assumptions I articulate above, that the mind is sourced purely in the material. To explore the validity of this assumption, I examine the latest literature on the relationship between minds and their supposed material originators, the brain and its attendant neurophysiological structures and processes.

The first area I'll explore in the literature on the relationship between minds and material neurophysiology is what has been termed the "mind-brain problem" (e.g., Uttal, 2005). Beliefs about the relationship between the mind and the brain has its roots in the long-running discourse called Theory of Mind. This discourse centers around the nature of the mind and its relationship to the body and material world in general. The "mind-brain problem" is a more specific casting of this larger Theory of Mind discourse, and has arisen in response to scientific discoveries of the last 50–100 years that have given theorists more confidence in positioning neurophysiological structures and processes as the sole source of the mind. The mind-brain discourse focuses on the human mind and brain largely because (a) they are taken to be the most complex and (b) knowledge of their existence is considered the most obtainable. While I've thus far avoided reliance on heavily anthropocentric theories because of their inbuilt dualisms, here I choose to focus on them because, if my postulation that more-than-human beings also have minds (and minds capable of producing the kinds of thoughts needed for close relationships) is true, then exploring the most accepted explanations for how the most accepted minds originate can show one of two things: First, that more-than-human beings might also meet these conditions, or second, how any corrected flaws in such explanations must admit other potential sources for human and more-than-human minds equally.

Returning to Theory of Mind more generally, one of the most influential theories is offered by Descartes (Kim, 2000). For him, mind and body are to be taken as metaphysically distinct—they are made of foundationally different "substances." For Descartes, "physical things...have a single defining feature [of] extension... [meaning that] things *occupy space*" (Rowlands, 2010, p. 11). The mind, on the other hand, because it is not physical does not occupy space. Hasker (2001) summarizes the Cartesian dualist view of the mind-body relationship thusly: "Cartesian dualism...accepts [that there is a] chasm [between the mind and body], postulating [that] the soul[/mind]...must be added to the body *ab extra* by a special divine act of creation" (p. 189).

Today, Descartes' dualistic view has largely been rejected. One element of that rejection is rooted in appeals to scientific discoveries showing (or so it is argued) that "brains are the *de facto* causal basis of consciousness" (McGinn, 2005, p. 438). Hasker (2001), an emergence theorist (see "Emergence" discussion in this section below), echoes this sentiment when stating that his own mind-brain hypothesis is rooted in "the well-confirmed results of natural science, including research on neurophysiology" (p. 188). What he believes these results show unequivocally is that *"the human mind is produced by the human brain and is not a separate element"* (p. 189). Griffin (2013), in arguing that some nonhuman animals have consciousness, also accepts the material origin of minds when stating that

> questions about animal mentality can best be approached from the viewpoint of a materialist who assumes that mental experiences result from physiological processes occurring in the central nervous system...[Thus in explaining those mental experiences] there is no need to call on immaterial factors, vitalism or divine intervention. (p. 17)

Such rejections of Cartesian dualism seem, however, to rely predominantly on materialist ontology. Kim (1997), for example, traces the predominance of materialist ontology back to the formulation of the *brain-state theory* or *central state identity theory* back in the mid-1950s. This theory holds that "specific types of mental states...[are] identical with particular types of brain states" (Cunningham, 2000, p. 24) or in a more liberal formulation, "every mental state is identical with *some* physical state, but instances of the same type of mental state...might occur as different types of physical states" (p. 25). Kim (1997) suggests that, though this theory quickly fell out of favor, it "helped set the basic parameters for the debates that were to follow—the broadly physicalist assumptions and aspirations that still guide and constrain our thinking today" (p. 185). Thus, the mind-brain problem became one of "finding a place for the mind in a world that is fundamentally and essentially physical" (p. 186). This is a powerful, *a priori* ontological imprint upon theories designed to explain the origins and nature of the mind. By extension, discussions such as mine around what beings have minds, and thus thoughts, feels this imprint very forcefully. In the rest of this section, I'll explore the central elements of modern mind-brain theories and consider what kind of support they ultimately provide for the physicalist/materialist presumption to being the origin of the mind. Kim separates materialist mind-brain discourse into three families of theory. I'll briefly articulate each family, then assess its ontological merits.

10.1.1   Supervenience

The first family of theory that Kim articulates is that of *supervenience*. The main tenet of supervenience is that the mental must "supervene on" the physical. That is, the mental is wholly dependent on the physical. But, elements of supervenience theory make it a problematic explanatory framework for the dependence of the mind on the brain. For example, supervenience theory allows for *multiple realizability*, which is the notion that "a single mental kind (property, state, event) can be realized by many distinct physical kinds" (Bickle, 2013, para. 1). As an example, Casacuberta, Ayala, and Vallverdú (2010) describe the occurrence of pain, where "philosophers have asserted that a wide variety of physical properties, states, or events, sharing no features in common at that level of description, can all realize the same pain" (p. 365). Thus, while a dependency may exist between particular physical properties and the mental ones that supposedly supervene upon them, there is no ontological impediment to a diversity of physical properties underlying certain mental properties. Thus, it remains a mystery what utility supervenience theories have in helping mind-brain theorists better understand just which physical properties create the conditions for certain mental ones.

A second weakness of supervenience theory is in explaining just how certain physical properties cause certain mental ones. That is, it does little or nothing to explain the *causal* link of the physical with the mental. As Kim (1997) aptly explains,

> mind-body supervenience itself is not an explanatory theory; it merely states a pattern of property covariation between the mental and the physical, and points to the existence of a dependency relation between the two. Yet it is wholly silent on the nature of the dependence relation that might explain why the mental supervenes on the physical. (p. 190)

I believe that in trying to explain the material origin of the mind in the brain, answering the question of *how* the brain *causes* the mind is the *primary* ontological question. As I'll show in considering the next two families of theories as well, very little explanation of what this link is and how it operates has been forthcoming. Thus, in supervenience theory as well as what follows, the Cartesian dualist "chasm" has yet to be spanned.

10.1.2 Physical Realizationism

The second family of mind-brain theories is *physical realizationism*. This family holds that mental properties are "'realized'...by (or in) physical properties, though not identical with, or reducible to, or definable in terms of, them" (Kim, 1997, p. 186). The implication here is that "the mental is exclusively physically realized— that is, there are no nonphysical realizations of mental properties" (p. 189). This rules out, obviously, any sort of ontologically more-than-material elements in the foundation of mind. Instead, it positions "mental properties [as] 'second-order' properties defined over first-order physical properties" (p. 190). Physical realization theories thus position mental properties as *functional* properties, which means they are defined only in terms of their roles as causal intermediaries between material, sensory inputs and materially observable behavioral outputs. As such, "physical states and properties are the only occupants, or realizers, of these causal roles definitive of mental properties" (p. 194).

The implication of this notion, according to Kim, is that mental properties can only be "defined in terms of causal...relations among first-order [physical] properties" (p. 195), or as Kim puts it, "To be in a mental state is to be in a state with such-and-such as its typical [material] causes and such-and-such as its typical [material] effects" (p. 195). Kim goes on to explain that in physical realizationism, mental states are best defined by the way in which they connect the input and output conditions which give rise to them. They may have distinctive intrinsic properties, but their only relevance is in "their capacity to get causally or nomologically hooked up with other [physical] properties" (p. 196).

One might note that the implication of such a concept is that the mind and mental states are approached in an almost purely utilitarian manner. Their existential relevance becomes their ability to be "seen" in a connective capacity between physical realities. Thus, conceptually, their role as irreducible ends in themselves is reduced or eliminated. Kim describes the effect of functionalizing the mental in this way:

> It follows then that if mental properties are functional properties, they are not tied to the compositional/structural details of their realizers, since these are intrinsic features; any base properties with the right causal/nomological relations to other properties can serve as their realizers. And any mechanism that gets activated by the right input and that, when activated, triggers the right response serves as a realizer (in an extended sense) of a psychological capacity or function. (p. 196)

Such a stance puts mental phenomena on particularly slippery existential footing. They are not things *per se*, but are to be understood by those physical complexes that create the conditions for their instantiation and those physical realms in which the effects of their instantiation are expressed. This is very much the stuff of subjective idealism and utilitarianism, and strikes me intuitively as missing something essential about consciousness and/or "Self-ness" that is the *source* of so much impact on the material and more-than-material. That such a notion is a *reduction* of the mental to the physical is echoed by Kim when he says,

> To reduce a property, or phenomenon, we first construe it-or reconstrue it functionally, in terms of its causal/nomic relations to other properties and phenomena. To reduce temperature, for example, we must first construe it, not as an intrinsic property, but as an extrinsic property characterized relationally, in terms of causal/nomic relations. (p. 197)

I suggest that by making the mental subservient to the physical, the concept of physical realization obscures the nature and origin of the mind (or at least its relevance) while not committing to rejecting its existence outright. Though convenient, since it straddles two potentially contradictory ontological stances, such an approach contradicts the notion of consciousness that rose up in the face of behaviorist reductionism (Sperry, 1987) and that is now recognized and widely accepted as an irreducible reality—at least when considering human minds. By neglecting the mental as a primary element in mind-brain relationships, physical realizationism does not go far enough in defining the mind such that we can have a firm grasp of its nature and role in the mind-brain relationship, or even in what essential role it plays in the relation of physical elements with each other.

Physical realizationism also carries the notion of multiple realizability referenced in my discussion of supervenience above, thus the door again is opened for myriad physiological beings to have mental states so long as the precipitating and resultant material conditions are met. As to the implications of this, it's worth quoting Kim's (1997) suprising exposition on this at length:

> It has long been a platitude in philosophy of mind/psychology that mental properties can have diverse and variable realizers in different species and systems, and that the formal/abstract character of mental properties, standardly taken to be a consequence of this fact, is just what makes cognitive science possible—a scientific investigation of cognitive properties as such, across the diverse biological species and perhaps *nonbiological cognitive systems* [emphasis added], independently of the particulars of their physical implementations. In fact, some have even speculated about the possibility of *nonphysical realizations of psychologies* [emphasis

added]; it is a seductive thought that there may be contingent empirical laws of cognition, or psychology, that are valid for cognizers as such, whether they are protein-based biological organisms like us and other earthly creatures, electromechanical robots, noncarbon-based intelligent extraterrestrials, immaterial Cartesian souls, heavenly angels, and even the omniscient one itself!...Even when we bring in the materialist constraint of physical realizationism, the idea of universal laws of cognition and psychology, contingent and empirical, and applicable to all nomologically possible physical systems with mentality, is heady stuff, indeed. (p. 196)

Though Kim fans out the cards in his ontological deck quite widely here, he ultimately fails to pursue any of the more-than-material implications that he suggests follow at least logically from physical realizationism's ontological position. Instead, he wedges them somewhere between unserious (at least to him) notions of extraterrestrials and angels, and in so doing is able to categorically dismiss them as legitimate subjects of scholarly inquiry. I feel confident that his stating this is without conscious understanding of the ontological flimsiness of the materialist position that affords him the opportunity to do so.

Kim suggests that physical realizationism is a stronger notion than supervenience because it is an explanatory theory while supervenience theory is not. What makes physical realizationism an explanatory theory, according to Kim, is that it specifies a reason that the mental supervenes on the physical: "because every mental property is a second-order functional property with physical realizers" (p. 197). But, just because this is offered as an explanation does not necessitate that it is a sufficient one. It is an explanation *of sorts*, but one can see from it that the explanation is based purely on a conceptual arrangement of what is already pre-supposed about the mental-physical relationship—that the physical is primary. That is not actually an explanation, as we must require of our explanations that they have some fundamental chain of argumentation and empirical support in order to be taken as legitimate. This take on physical realizationism has neither. It presumes the primacy of the physical and then uses that presumption to craft a statement that essentially supports the presumption. That is circular reasoning. What it doesn't do is justify *how* or *why particular* mental properties are second order functional properties of *particular* physical realizers, much less why it is, ontologically, that the mental is second order to the physical at all. Thus, my initial and primary question above, of *how* the physical determines the mental, remains unanswered. By ontologically reducing the mental to "'nothing over and above' having one of its physical realizers" (p. 197) one is no closer to understanding whether the mental exists in its own right, or its nature if it does. As with supervenience, the Cartesian chasm between mind and brain yet again remains.

10.1.3 Emergence

Emergence theory, or emergentism, encapsulates a family of theories that arose in response to the discomfort some theorists had with the possibility, in supervenience and physical realizationism, that the mental can be reduced without remainder to the physical. In both supervenience and physical realizationism, by the laws of transitivity (if A produces B and B produces C, then A produces C) the physical alone is deterministic of the mental. Put another way, in these theories there is nothing the mind can do that cannot be explained by the workings of the physical which produced it. Thus, it is the physical, not the mental, that has causal power.

Through intuition and primary experience, emergence theorists see the mental as irreducible to the physical in this way. They see the consciousness that is at the core of the mental as greater than the sum of its physical underpinnings. To explain how the mental can be both irreducible to the physical yet originate there, the notion of "emergence" was developed. In it, consciousness is seen as a novel property emerging from the physical brain and its supporting material structures while differing from them in qualitative (i.e., irreducible) ways.

While there are many theories of emergentism, I'll base my discussion on two that offer a robust treatment of the concept. One is offered by Hasker (2001) and the other by O'Connor (1994). In both, two key features distinguish emergence from simple supervenience and physical realizationism. The first is that the mind "exerts a causal influence over the behavior of its possessor" (p. 95)—that is, it has a causal influence over the physical from which it emerges. As a result, as Hasker (2001) explains, "if...consciousness is emergent...the behavior of the physical components of the brain (neurons, and substructures within neurons) will be *different*, in virtue of the causal influence of consciousness, than it would be without this property" (p. 174). The second feature is that this influence of the mind cannot be traced back, deterministically, to the physical from which it emerges. This is because, according to these theorists, the laws governing the physical and mental are qualitatively different. This qualitative difference is attributed to the mind emerging as a "higher order" level of existence, and is "governed" by "laws" that are fundamentally "different because of the influence of the new property that emerges in consequence of the higher-level organization" (p. 174). As O'Connor (1994) puts it, an emergent property such as consciousness, "could not be deduced from [even] a complete knowledge of the properties of its components [even if one were one in possession of such]" (p. 96).

O'Connor says that the main work of emergence theory is to explain

> (i) the nature of the dependency of the emergent property [in this case, consciousness or the mind] upon the lower level properties of the object [in this case, the body or the brain] and
> (ii) the general nature of an emergent's causal influence. (p. 95)

Figure 10.1 depicts both elements in operation as the mind and brain interact. I think O'Connor is right to distill the main work of emergence theory to these two elements, but I differ from him in that I believe the first area of work to be of primary importance, and the second to be either of lesser importance, or at least dependent upon the first for its coherence. As I've stated so far in this discussion, *how* the mind emerges from the physical is essential to understanding the nature of any influence that the mind can have upon the physical or vice versa. The most damning criticism of Descartes' substance dualism centered on the relative weakness of his explanation for how the more-than-material soul or mind, lacking spatial extension, could causatively interact with the physical (Robb & Heil, 2013). With an emergentism of the mind, we have an inverse of this problem and so an inverse

Fig. 10.1 Theorized
process of "emergence" of
the human mind

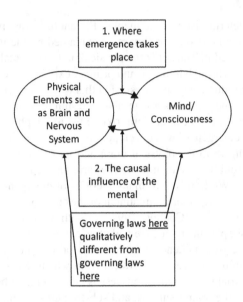

of the criticism. In emergentism, an explanation of how the physical, being qualitatively different from the mind, can still produce that mind, must be offered. Yet, the explanations offered tend to be deeply inadequate. Since emergentism stands as one of the more widely proffered present-day mechanisms for the ontological relationship between the physical and other, seemingly non-physical elements such as the mind, consciousness, intuition, etc., the failure to articulate an adequate answer to the question about the mind-brain relationship is extremely problematic. I'll explore the three main explanations given for the emergence phenomenon, and note the ways in which they fail to sufficiently explain it.

The first way that emergence theorists have grappled with explaining how the mind emerges from the brain is simply *to admit ignorance*. For example, in explaining how "conscious life and experience" can exist "in the nature of matter itself" Hasker (2001) admits that *"we have no insight whatever* [emphasis added] into how this is the case" (p. 194), and adds that

> [i]n saying that 'the existence of consciousness can be explained by the causal interactions between elements of the brain at the micro [physical] level' [John Searle, 1992] clearly doesn't mean that we can now give such explanations; it's obvious we can't do this, since we don't yet know even the Humean-type correlations that govern the emergence of consciousness. (p. 173)

An earlier proponent of emergentism, Sperry (1969), says, "At present even the general principles by which cerebral circuits produce conscious effects remain obscure" (p. 535). Kim (1997) says that even if we could trace mental properties back to physical ones, it still would not "automatically promise us an intelligible account of why the supervenience [of emergence] obtains. (p. 194)

That this is insufficient support for the claim that the mind emerges from the physical is obvious. It is sponsored first by confidence in a material reality, which in turn underwrites the assumption that while the explanation of emergence is not known, it is ultimately or eventually discoverable as a materially sourced thing. I'll explore this—what I call the "fallacy of discoverable materiality"—in more detail below, so suffice to define it here as the unbending faith that if a material explanation for some phenomenon does not present itself, because the world is assumed to be material, eventually a material-only explanation can (and will) present itself.

The second way that theorists attempt to justify the position that the mind emerges from the physical is *by process of elimination*. That is, they begin with the assumption of a material reality and use it to eliminate any other explanation. For example, an early emergentist referred to in both Kim (1997) and O'Connor (1994) and whose basic construction of emergence is still relevant today is Alexander (1920). He suggests that the

> existence of emergent qualities...is something to be noted, as some would say, under the compulsion of brute empirical fact, or, as I should prefer to say in less harsh terms, to be accepted with the 'natural piety' of the investigator. It admits no explanation. (pp.46–47)

To insert that which is unspoken into his statement, what Alexander is really suggesting is that because mind and consciousness exist, and *because the world is assumed to be radically material*, then it is an "unavoidable" fact that the mind emerges from the material. Kim (1997) also injects this assumption when he says that "there appears to be no alternative but to accept the higher-level properties involved as emergent properties—intrinsic properties in their own right whose supervenience on their base microproperties must be taken as brute, unexplainable correlations" (p. 204). O'Connor (1994), too, adopts this stance when saying that

> it is true...that...even the quirky phenomenon [those said to have emerged] could be described in terms of functions from the base-level [physical] properties alone. But this does not motivate the repudiation of the presence of emergent properties. For the laws adequate to describe the quirky phenomenon will themselves have a very odd complexity, involving tacked-on disjuncts to cover the special cases. And this, surely, demands explanation in terms of the properties of the object exhibiting the strange behavior, an explanation that the postulation of an emergent property seems to provide. (p. 98)

Once the assumption of a material reality is exposed for what it is: largely dualist assumption, the kind of reasoning these authors use appears much weaker. It's one thing to say that emergence is a better *conceptual* or explanatory "black box" within which to consider these phenomena. It's quite another to say that because of this, one ought to conclude, without any explanation of the actual mechanism of emergence itself, that it *exists*.

To push beyond the "only available alternative" argument, however, O'Connor acknowledges that there "will always be an alternative explanatory possibility available to the theorist determined to...posit the presence of further (hitherto undetected) [material] micro-properties" (p. 98). But he rejects such a move as "implausible" by asking:

> Why does such a micro-property make its presence known only in highly complex systems of a certain sort? How is it that such a fundamental property can be so causally isolated from other microproperties so as to be discernible only in circumstances that are otherwise noteworthy only for the complex macro-properties which are instantiated? The presence of an emergent property is by far the more natural assumption to make... (pp. 98–99)

In essence, by his having eliminated (1) more-than-material explanations through ontological assumption and (2) other material explanations because the unique-to-the-macro-property appears to be qualitatively different from the microproperties that are its supposed origin, then emergence is the only explanation. O'Connor also suggests that emergence is a "far more *natural* assumption." But, the naturalness of the assumption is sponsored wholly by defining that which is "natural" as that which is accepted to occur in the way that it does in a materialist ontology. In the relational ontology I espouse, for example, where more-than-material elements are also present, when intuition tells one that something is qualitatively different from the material with which it has been observed to co-occur, and when no material *explanation* consistent with other, accepted explanations of known material processes can be found, then a *more-than-material* explanation is the more "natural" assumption.

At this juncture it is critical to point out that the *process of emergence* is yet to be articulated in any substantive way. "Yes," we can say for the sake of the emergentist's argument, "we have a material world." "Yes," we can say again for the sake of this argument, "the macro is qualitatively different from the micro." Allowing this, we are still not offered a single statement on *how* the mind emerges from the physical under such conditions. In O'Connor's explanation, then, emergence becomes a word or concept wall-papered over the existential void left when the two ontological poles of mind-brain theory—more-than-materialism and pure physicalism—are eliminated. It has no substance of its own. Thus, emergence as a thing unto itself *does not exist*.

One way that emergence theorists try to sidestep this ineluctable fact is to avoid discussing what emergence is, and instead discuss the "conditions" or circumstances" under which it occurs. O'Connor talks about "uncovering [the] precise causal conditions under which emergence occurs" (p. 97). Kim (1997) speaks of the lack of understanding of why certain experiences, such as pain, "emerge just when... [certain] physiological conditions obtain...[and] why, and how, mentality makes its appearance when certain propitious configurations of biological conditions occur" (p. 200). Though Kim goes on to make a complex logical argument for it having to happen, he offers no explanation as to how it does. Thus, again, we are left with a void filled by the word "emergence" and nothing more.

Hasker's (2001) theory suffers the same fate, predicating the nature of emergence on the acceptance of "well-confirmed results" of science coupled with materialist ontology. "In rejecting [Cartesian] dualisms," he writes, "we implicitly affirm that the human mind is produced by the human brain and is not a separate element 'added to' the brain from the outside. This leads to the further conclusion that mental properties are emergent" (p. 189). This is yet another argument for emergentism gleaned from the pure assumption of materialist ontology. It does not offer any evidentiary support for such a concept of ontology. At least Cartesian dualists and reductive materialists have *attempted* the explanation. Emergentists, finding neither acceptable, seek a middle ground. In doing so, however, they fail to see the potentially unresolvable ontological tension that exists there. I'll explore the effect this conceptual fence-straddling has on the very viability of emergence as a theory next.

The third way that emergence is justified is by attempting to "have one's onto-logical cake and eat it." To explain what I mean by this, amongst emergence theory's basic claims, two are of relevance to my critique that emergentism has attempted to straddle an ontological fence and failed. The first is that the mind, or consciousness, is material in origin, emerging wholly from some configuration of its material microstructures. The second is that this same mind cannot be reduced to, or explained by, the laws that govern those microstructures. Putting the latter another way, the mind is governed by laws qualitatively different from those governing its material microstructures and thus itself is qualitatively different from them.

The main problem with these claims is that the support for each of them is rooted in incompatible ontological, or at least epistemological, bases. Emergentists have unwittingly created this situation, I believe, in response to their general dissatisfaction with the reductive nature of strict materialist explanations of the mind and consciousness while clinging to the same materialist ontology that underwrites those theories about the relationship of the mind to the material. Hasker says of "eliminative" materialism that it fails to account for things like "consciousness... mental content and intentionality" and that at times it makes "highly counterintuitive claims" (p. 23). Echoing this, Sperry (1980) says that before emergence theory, "[traditional] materialist-[pure] behaviorist reasoning effectively outweighed all the [valid] intuitive, natural and omnipresent subjectivist pressures and arguments [for consciousness]" (p. 198). As one can see from these two quotations, intuition plays a significant role in undergirding emergence theory's second claim that the mind and consciousness are both existential and irreducible to any of the material micro-properties from which they emerge. Given the reliance of emergentists on their intu-ition, one can also see that the primary experience of mind and consciousness that produces it is being held *prima facie* as a source of truth. To wit: Griffin (1981) says of consciousness that it is really the only thing of which we *do* have direct experi-ence, and therefore, knowledge. The implication here is that emergence theory's second claim rests upon truth known *not from material evidence*.

As for emergentism's first claim that the mind is materially sourced, it does rely wholly on material evidence. For example, Hasker (2001) rests his form of emer-gence theory on the assumption that "the well-confirmed results of natural science, including research on neurophysiology...[are] in the main true...[and]...informa-tive about the real nature of things" (p. 188). Again, I quote Griffin (2013), who accepts the material as the origin of the mind when unequivocally stating that

> questions about animal mentality can best be approached from the viewpoint of a material-ist who assumes that mental experiences result from physiological processes occurring in the central nervous system...[Thus in explaining those mental experiences] there is no need to call on immaterial factors, vitalism or divine intervention. (p. 17)

But, as I've argued in several places in this book, such scientific approaches mistak-enly equate correlation and causation. In addition, these scientific investigations take as relevant and valid only those data which are externally observable (thus of a material nature) and repeatable. One may notice that this is the hallmark of the sci-entific method, with its roots in "systematic observation...[and quantitative] measurement" ("Scientific Method", 2018). Granted, the scientific method can be applied to the phenomena of mind by investigating the materially observable *effects* of such mental activity (as Sperry, 1980, has suggested), but I find it striking that the functioning and nature of the activity itself remains virtually impenetrable to current methods for gathering knowledge of the material attributes of the phenomenon itself.

Indeed, this should be troubling for *any* theory that the mind is, and has its origins purely within, the material. Certainly, if so many other material structures and processes have yielded material results to material inquiry, one would think that by now we'd have *some* information about these particular material elements. And yet, as I point out when discussing emergence theory's appeal to ignorance, it is admitted that there is virtually no knowledge of it whatever. If the materialist responds to this void in knowledge by saying that it is only a matter of time until the material structure and processes are discovered, then she is in danger of appealing to what I brand the "fallacy of discoverable materialism."

I define this fallacy as the belief that all things are (eventually) explained by a material ontology. This, in and of itself, is not a fallacy since a fallacy is the act of reaching a conclusion based on poor logical premises, and this is only a belief without (at least stated) premises. But it becomes a fallacy when one investigates the ways in which this belief is most often reached. The first way is by arguing that since the world *is* material, then it *must be* that all things will eventually reveal themselves as material. Popper (1977) calls this "promissory materialism" (p. 96). Put another way, the logic is such that since materialism is *assumed to be* true, it must be *actually be* true. This is circular and fallacious.

The second way the fallacy is constructed is as follows: since all investigations of phenomena have yielded some material attributes (and more and more with time), then eventually all phenomena will yield material, *and only material,* results. This statement has two problematic elements. First, there is an unwarranted assumption that because there is material evidence, that this is the only kind possible. Such an assumption is supposed to be lent even greater weight by a line of reasoning contending that as more material evidence accumulates it becomes harder and harder to cleave to more-than-material alternatives. One immediately identifiable weakness of this belief is the fact that no researcher has looked for evidence of any other kind. More importantly, however, this belief is only possible if material evidence is taken as support only for a material ontology. As I state in Part III above, this cannot be uncritically accepted. In a complementary material and more-than-material ontology, material evidence can be taken as support for the presence of more-than-material elements.

The second problematic element in this fallacy is that in a material and more-than-material ontology, the accumulation of material evidence is not an explanation as to *why* a thing is what it is, it is only a description of its effects—it is only the "what" with the more-than-material as the "why." In such a view, the accumulation of the material "what" through experimentation and other means can be seen not to be a growing body of evidence, but instead an ever finer splitting of the same, original "hair" of "what." That is, it's not really cumulative. It is descriptive but not explanatory. *Why* is a mind produced by the material structures and processes of the brain instead of, say, a songbird or a quark? A more-than-material answer might be that that there is a soul whose nature it is to be a mated with the material in the mind. The material is only the "why" when one subscribes to an ontology where there are no more-than-material causes, which means the material phenomena either has no cause—that is, it's the result of the random nature of the universe (see Kessler, 2012)—or the cause will again, eventually, reveal itself as material. Ultimately, the fallacy of discoverable materialism originates within assumed materialist ontologies, thus is generally self-fulfilling and invalid.

To return to the incompatibility between intuitive-experiential ways of knowing (i.e., potentially *poetic* ways) and material-scientific ways, I suggest that if one takes truth to be arrived at through the latter (as emergentists do in the first claim above), then the mind in the second claim above is threatened. A materialist world demands material answers, and at the moment the actual material *process* of emergence has none. Likewise, neither do the workings of the mind, except in neurophysiological research which again either *correlate* the brain's activity with the mind's and offers it as causation, or overlays the mind's workings with an assumed materialist ontology that disallows non-material explanations for that which is observed.

Ultimately, application of the scientific method to the truth of the mind gleaned from intuition and experience may still reveal the mind to be material in origin, but this is not *de facto* true, and is not currently *known as true* in a material sense. To support this point, in Part III where I discuss poetic knowledge, I quoted Nobel Prize winning geneticist Barbara McClintock as saying that "you work with so-called scientific methods to put it into [the scientific method's way of knowing] *after* you know" (Keller, 1983, p. 389). The source of truth for her in this example is intuitive and experiential, and may be a stronger support for the knowledge of something than a subsequent attempt at accumulating material confirmation of that knowledge. If truth can be arrived at or discerned via intuitive-experiential means, then all manner of things not easily compartmentalized into a materialist ontology or epistemology (not just the emergentist's human mind) can be taken as real or potentially real. If that's the case, then we may be in a better position to make the claim that there are entities quite unlike human beings that possess minds.

As one can now see, by trying to step outside the strictures of traditional materialist views, emergence theorists have embraced an explanation for the nature of the mind that is at odds with a strict materialist ontological adherence. As a result, trying to fit the mind back into a materialist world without any material support for it becomes profoundly awkward. For example, to make his claims coherent, Hasker (2001) suggests that consciousness occupies physical space. Specifically, he argues that due to the "close *natural* [emphasis added] connection...between mind and brain...it is *natural* to conclude that the emergent consciousness is itself a spatial entity" (p. 192). Here we have an attempt to dissolve the dualism of Descartes' nonspatial mind and spatial body, but it runs into trouble right away. First, it's a bit deceiving for Hasker to piggyback the naturalness of his conclusion on the natural-ness of the connection between mind and brain. The latter is generally accepted, the former generally rejected—that is, most people *don't* believe that the mind, or con-sciousness, occupies space. Second and again, it only becomes "natural" to con-clude that consciousness occupies space when holding a pre-ordained materialist ontology. Since assumption is generally weak support for any claim, there is little else of it here.

By moving aside these weak materialist ontological claims, two strange things result. First, *the process*—the "how"—by which the mind springs from the material microstructures within which it is supposedly rooted becomes no less mysterious or magical than the more-than-material soul offered in Cartesian dualist ontology. Emergence becomes at most a post-dated, blank check to be filled out and drawn on a materialist bank account sometime in a future where more material funds are pres-ent. Second, with the elimination of self-fulfilling materialism, emergence as an existential process becomes wholly empty, a ghost left behind in the exact shape of the Cartesian soul. Emptied of substance or explanation, the origins of the mind "wander" emergence theories without definition. What's more, they become really *undefinable* since they are positioned as being different from all other known mate-rial things, and thus can gather no material evidence or support for themselves—at least not yet.

In my complementary material and more-than-material ontological view, such an untenable outcome is, ultimately, support for a very plausible alternative: *that the mind is more-than-material in origin*. This is not to say, as Hasker does, that it is "a separate element 'added to' the brain from the outside" (p. 189). If the brain and mind are material and more-than-material in origin, respectively, the mind isn't "in" the brain exactly. Such an orientation of the two to each other already suffers from a materialist dualism that positions the brain as the house of the mind. Perhaps the brain exists in the mind, as much as the spatial notion of "in" can cohere around a thing that does not easily obey spatial laws. Maybe it would be more accurate to say that the mind and brain co-occur—neither existing within, nor independent of, the other. This would be a true interdependence within an ontologically radical *rela-tional* dualism in the Cartesian sense of irreducibly distinct ontological elements. In emergentism, given the reliance on relational knowledge and the primacy of experi-ence, it is the more-than-material which far more comfortably or "naturally" accom-modates the definition of the mind as emergence theory itself has postulated it.

Ultimately, then, I believe it's reasonable to conclude that emergence theory's description of the origin and nature of the mind fails to address these ontological and epistemological fissures, devolving into an attempt to have one's ontological cake and eat it, too. If that's true, then I suggest we can decouple minds, conscious-ness and their thoughts from any narrow material bases. In doing so, the monopoly control over the capacity to possess thoughts that contribute to close relationships, heretofore granted by humans only to human and human-like physiologies, can be comfortably rejected.

10.2 "If I Only Had a Brain," the Plant Thinks...

In the previous section, I explored what I consider to be the best materialist onto-logical explanations offered for the origin of minds from certain material neuro-physiological structures and processes. From that discussion, I concluded that explanations such as emergence are riddled with ontological and epistemological contradictions such that a potentially more-than-material sourcing of the mind is worth consideration. In the rest of this chapter, I'll take up discussion of a specific type of being previously not thought capable of having a mind as a means to explore this ontological possibility. The broad category of being I'll consider is a plant.

I begin my inquiry in the field of plant neurobiology, specifically, in plant intel-ligence theories, because in this family of theories is noted a capacity in plants that is usually attributed to mental activity in humans and those possessing human-like neurophysiology. It must be noted before I begin that most plant intelligence theo-rists don't suggest that plants have minds. In fact, most are strictly materialist in their ontological commitments. But, I believe that such assertions are based on a presupposition that plants *can't* have mental activity, therefore they don't. In other words, they use the same self-fulfilling materialist argument I've critiqued through-out this work. I will first explore the evidence given for plant intelligence, then cri-tique materialist explanations for that intelligence in order to bring forward the possibility that plant intelligence is a product of mental activity that is not based in human-like neurophysiology.

10.2.1 Plant Intelligence

There is a growing belief that plants display intelligence (Carello, Vaz, Blau & Petrusz, 2012; Marder, 2012; Trewavas, 2003, 2005). Some of this belief is based in experimental evidence while other portions are based in an attempt to modify the definition of intelligence to rid it of self-serving, anthropocentric elements. Garzón (2007) comments on the latter effort when saying, "Intelligence is usually cashed out in animal or anthropocentric terms, in such a way that plants plainly fail to meet the conditions for animal or human-like intelligence..." (p. 209). To move away from such restrictions, Trewavas (2003) defines intelligence more generally as "adaptively variable growth and development during the lifetime of the individual" (p. 1). Carello et al. (2012) do likewise, characterizing intelligence as "end-directed behavior marked by the making of meaningful distinctions made possible by perception-action cycles...[where the end-directed behavior shows] evidence of three aspects of...agency...namely, *prospectivity, retrospectivity, and flexibility*" (p. 241). As Carello et al. define them, *prospectivity* is the "ability to be forward-looking, changing...behavior in anticipation of what will be" (p. 248), *retrospectiv-ity* is defined as behavior that is "reflective of what has been" while *flexibility* is the possession of "options as to how [one] can behave to accomplish [ones] end" (p. 248).

Experimental evidence shows that plants act in ways that meet these definitions. For example, Carello et al. describe the behavior of the parasitic plant, Dodder, saying,

> It will coil around a good host or bend away from a poor host, and it will do so without first taking up any food from either host and without being in the presence of other potential hosts for comparison. (p. 243)

Marder (2012) notes that plants can "detect and react to different sounds, bending root tips toward [a] sound source" (p. 1368). Trewavas (2003) observes that in response to competition, the stilt palm will relocate itself to full sunlight by putting down new stilt roots and letting the old ones die off in order to slowly walk to the new location (p. 15). Note that in the previous sentence I don't put quotation marks around the word "walk." To do so is anthropocentric, where the non-quoted "walking" is that which is achieved by animals with legs, starting with humans and extending outward to include those with four or more legs and that don't grow roots down into the ground with each step. It is not that plants "walking" needs qualification as walking, it's that the definition of walking that necessitates the quotation marks when applied to plant locomotion is too narrow and anthropocentric to accommodate the fact that walking is simply the act of ground-based, self-propelled locomotion across some spatial expanse.

Returning to the examples of intelligent plant behavior, they are just two examples of plants' capacities for the memory, variation in response, foresight, choice and intention that Trewavas argues are the basis for intelligence. Because plants display these behaviors, it might be tempting to conclude that they think. In fact, were the examples I just cited offered as instances of human behavior, that's exactly what one would likely conclude. A human faces competition, he thinks about going somewhere where that competition is reduced and then acts on that thought. A human likes a particular food she tried at a party, thinks about ways she can find it in a store, and goes to that store to get it. In truth, the coupling of human intelligence with thought is so tight that to ponder one without the other seems nonsensical. Obviously, at this juncture, this same coupling is not commonly attributed to plants.

10.2.2 Explanations for Plant Intelligence

Having examined how plant intelligence theorists reorient definitions of intelligence to expunge them of anthropocentric biases, the work now lies in assessing the efficacy of that effort. One might expect that such a reorientation would put humans and plants on equal footing, as that is, to some degree, the intent. But there still appears to be a qualitative difference between human-as-intelligent and plant-as-intelligent. Further, this appears to be the case mainly because plant intelligence theorists tend to discount the role of cognition and consciousness in intelligence.

They do this, in turn, because they want to include plants in definitions of intelligence while at the same time holding a pre-existing, dualist assumption that cognition and consciousness in plants is not possible. Thus, the attempt to define intelligence to include animal and plant still results in a bifurcated notion. I'll trace the effect of this bifurcation by exploring three ways that plant intelligence is characterized in the literature, and comparing it with human-like intelligence.

The first characterization of intelligence under this broadened definition is that of intelligence as *microstructural*. Trewavas' (2002) explanation of how plant behavior constitutes intelligence is focused at the cellular level. He suggests that certain cell behavior in plants, taken collectively, produces their intelligence. "Internally, plant cells and tissues communicate with each other...[thus a] huge reservoir of individual cell behaviours can be coordinated to produce many varieties of organism behaviour" (p. 841). But, in Trewavas' estimation, such coordinated cellular activity does not produce a mind. The title of the essay: "Mindless Mastery" is but one indication of his belief, another being his question: "How is such intelligent behaviour computed without a brain [and, one must suppose, a mind]?" (p. 841). To answer this question. Trewavas notes that "Cellular calcium mediates most plant signals, and calcium waves inside cells offer computational possibilities" (p. 841). He also notes that "If the calcium signaling system has a formal equivalence to a neural network, it should be able to compute, remember, and learn even though it is confined to single cells" (1999, p. 4). But if calcium wave effects are intracellular and intelligence is attributed to *inter*cellular activity, this is not a plausible explanation.

Like Marder (2012), Carello et al. (2012) and others, Trewavas (1999) offers such explanations because of an overarching belief that plant intelligence is based in the microstructurally material. That is, it doesn't ever coalesce or "emerge" into a "higher"-order mind. Such a dualist belief is echoed in my discussion of posthumanist theorists Cudworth and Hobden (2015) in Chap. 3. There, it is agency that is disembodied in "lower order" more-than-human beings such as plants when they ascribe the "*intrinsic capacity*" for agency to humans and some animals while granting only a disembodied agential capacity to "*situated relations*" that are made up of the more-than-human world (p. 10). In the context of intelligence, Marder (2012) describes this "mindless" and "lower" plant intelligence by equating it with autonomic "intelligence" in humans. He says that "human subjects are attentive to numerous environmental factors, such as the temperature, without taking cognizance of the fact. This somatic attention perhaps comes closest to that of plants" (p. 1368). But, I note that somatic "attention" is a far cry from "intelligence" as we think of it. I'd also argue that it is a far too limiting an equivalence to make given the more advanced memory, learning, intention and other behaviors that plants have been shown to exhibit (see a far more detailed exploration of this below). Further, while we may think of human "attention" to temperature as a culminating form of the random genius of evolution, we don't think of it as *intelligence*. By virtue of its chaotic and mindless origins, we tend to think of it as decoupled from the kind of conscious mental activity that produces the intelligence we most highly prize. If, on the other hand, this attentiveness is thought of as a product of some being's unconscious mind that is "above" categorization as autonomic, then the inextricable relationship between mind and intelligence reappears, and must then include plants.

To explain how plant intelligence comes to be produced at the microstructural level, though, plant intelligence theorists once more bring us back to the "black box" explanation of ontologically material emergence. For example, to explain how intelligence can be the result of cellular activity in plants, Trewavas (2003) says that intelligence "is indeed an emergent property...just as it is in the brains of animals" (p. 17). But, by taking this tack, Trewavas has painted himself into a conceptual corner. If what he asserts is true, then the plant must also have the possibility of a mind regardless of whether that mind is a wholly determined sum of microstructural parts or is what emergence theorists contend, that it is a qualitatively different and irreducible thing emerging when certain material microstructural conditions come to pass. If the cellular networks of both plants and animals are similar enough to produce intelligence, then it becomes extraordinarily difficult to argue that they wouldn't produce what is thought of as such an inextricable part of that intelligence, at least in humans—a mind. The second possibility is that, if the plant is a "mindless master" as Trewavas believes, the logic that supports that conclusion must also be applied to the conclusion that humans don't have a mind, or their intelligence is not sourced there either.

The "safe harbor" that plant intelligence theorists holding to a materially microstructural explanation of intelligence might seek from such conceptually troubling waters—and the one forming a third possible explanation of microstructural intelligence—is the possibility that neural networks of a *certain level of sophistication* produce minds and the kinds of intelligence we believe to be products of those minds. But if that's the case, then one is confronted with two different *kinds* of intelligence—the mental and the microstructural. According to plant theorists, though, there is a continuity of intelligence between plant and animal, and so this explanation must also be rejected. I suppose a fourth possibility is that mental intelligence— intelligence based in thought—*is* the more sophisticated form of intelligence. But, the plausibility of this—that a mind simply comes on the scene to produce a certain type of intelligence or work with intelligence rooted elsewhere (how they would interact being a mystery as well) seems less than wholly credible and more than a little self-servingly anthropocentric.

The next explanation given for this mindless plant intelligence is that of *decentralization*. As Marder (2012) points out, in plant signaling and behavior theory, there has been a movement away from the notion that intelligence is a product of individuals in the sense that it does not require an individual as a gathering or coordinating entity. This would seem to align well with posthumanist notions of intraaction and disembodied agency advanced by Barad (2007), Bennett (2010) and others. To Marder's (2012) point, he asks, "Is it necessary to postulate the existence of a complete cognitive map or a coherent global representation of foraging space in plants to account for their spatial orientation?" (p. 1369). What he is asking is, does there need to be a cognitive "entity" or element to coordinate the information gathered by the plant and the intelligence displayed in response to it? According to Marder and others, there does not. Marder instead suggests that the concept of "decentralized memory" may be applicable to plant intelligence. When applied, Marder likens decentralization to a computing model, with "the plant's non-totalized intelligence...explicable, in cognitive terms, as a parallel processing model, with every organ of intentionality playing the role of a parallel processor...[where] modular memory and modular intelligence...do not correspond to the organismic logic" (p. 1370). This notion seems tenuous given that there still exists no explanation for how such parallel processing comes to be *coordinated*, as it must in a computer, and who exerts the coordinating influence. This is the stuff of ontology, and so may be slightly out of place in plant intelligence theory, but the viability of such theories depends on ontological as well as logical coherence and must be addressed.

Garzón (2007) is another theorist who suggests that elements of intelligence can operate in a decentralized way. His description is worth quoting at length:

> As animals grow and develop, attention is paid to many different processes at layers of organization that range from the subcellular level to the level of tissues and organs... In this way, no individual as such shows up in the cognitive equation...[Thus, c]ausal efficacy operates at levels below the level of the individual in such a way that no self is subject of scientific research unless broken down into its microconstituents and their interactions...In short, it is the interactions that take place among processing units where the emergence of intelligence or cognition resides. (p. 209)

Here, we again see emergence as an explanatory framework for how uncoordinated constituent elements of intelligence come together to actually be intelligence vs., say, random actions constituting maladaptation or stupidity. Tracing Marder's (2012) notion of decentralized intelligence back to one of his theoretical sources leads us again to emergence, where Cruse and Wehner's (2011) notion of decentralized memory in honey bees relies upon it. They suggest that the ability of honey bees to map novel paths through territory is not, as had been previously argued for other organisms, the product of accessing a cognitive map of the territory. Since those authors hold that honey bees are incapable of cognition, they suggest that "map-like behaviour as observed in honey bees arises as an emergent property from a decentralized system" (p. 1).

Ultimately, in order for a decentralized notion of intelligence that sponsors the denial of minds to plants to be considered valid, an explanation for just *how* these disparate, material, mechanical elements *do* come to be coordinated in what we recognize as intelligence is required. In humans, the explanation is easy: It occurs in some level of our consciousness whether we're aware of it or not. But, for those not believing that plants have consciousness, the explanation becomes infinitely more difficult. Because intelligence is a coordination, then this coordination *must occur somewhere* and *be performed by some 'one.'* Where? And by whom? For those not like humans, decentralized intelligence theories postulate that it simply "emerges" from thin air. As in my discussion of consciousness as an emergent phenomenon above, on ontological grounds this assertion cannot be uncritically offered nor accepted.

Embodied cognition is the third and last explanatory framework for plant intelligence that I'll consider. Garzón (2007), the chief proponent of its applicability in plants (it is traditionally applied to human and nonhuman animals), explains the notion by saying that

> embodied cognitive science rejects the metaphor of cognition as a centralized process. Cognition is rather an emergent and extended self-organizing phenomenon whose explanation requires the simultaneous scientific understanding of neural, body and environmental factors as they interact with each other in real time. (p. 209)

For Garzón, both plants and animals are "open systems coupled with their environments" where there is a "continuous interplay of [individuals] in relation to the environmental contingencies that impinge upon them" (p. 209).

In this new definition, "cognition is...a biological phenomenon...that...exhibits itself as a capability to manipulate the environment in ways that systematically benefit a living organism" (Garzón & Keijzer, 2011, pp. 161–162). As one can see in this definition, thought is eliminated as part of cognition. As Garzón and Keijzer point out, it's possible to include plants in such notions of cognition if three constraints on embodied cognition, when previously applied to certain animals, are modified. The first is that a cognitive agent must be able to freely move about. Garzón suggests that this constraint has less to do with a requirement for cognition and more to do with observing evidence for its effects in what has traditionally been only animals. The second constraint is offline processing, which "allows an organism to dissociate its behavior from the immediately impinging stimuli and to act in ways that are guided by forms of knowledge" (p. 163). Garzón and Keijzer say that evidence for this ability abounds in plants. The third, and most interesting in relation to my discussion, is that the organism's cognitive organization must be a "globally organized cohering unit, not a collection of individual stimulus–response relations" (p. 168). Given that Garzón and Keijzer's materialist ontological commitments carry the suggestion that plants don't have minds, I predict that this constraint would give them the most trouble, and it does. In response, the authors are only able to offer the example of it being possible that plants have a sense of the geometric layout of the environment "without us knowing it" (p. 163). This is intriguing because they are essentially saying that there is nothing *ontological* preventing plant cognition from being a globally organized cohering unit. All that is missing is the gathering of empirical evidence for it.

10.2.3   Critique of Elements of Explanatory Theories

I'll now go on to critique four particularly troublesome elements from the three explanations of plant intelligence just discussed. The first area I'll critique is the element of coordination/coherence. The requirement for global coherence in embodied cognition theory parallels Trewavas' need for the "coordination" of cellular activity. In both, however, the influence of the materialist ontological commitments poses certain problems for the requirement when applied to plants. These problems exist for two reasons. First, if there is to be coordination or coherence, there must be a "place" (material or otherwise) where it occurs. In materialist ontologies currently, plants don't have such a place. In humans and some other animals, there is the brain and/or the mind and its thoughts—conscious or otherwise—that make sense of these individual stimulus–response relations. Plants are currently denied this apparatus.

The second reason that materialist ontology poses a problem for plants is that in order for coherence to be coherence, it must have a governing pattern, or "holism" to use Trewavas' (2006) term. I think it particularly relevant that, for Trewavas to be able speak of the coordination required by intelligence, he must invoke an ontological term, "holism," that is completely out of keeping with the generally chaotic metaphysics (Kessler, 2012) of his materialist approach. What and/or who is responsible for the coherent pattern of intelligence in plants? This essential ontological question has yet to be answered, or even acknowledged. The commitment to a materialist ontology in these theories currently only allows for human-like neural networks to be the source of the consciousness and particular form of cognition that is argued to serve this coordinating purpose in humans and some nonhuman animals. Therefore, if plants have coordination, they will either have to be "granted" some form of mind or they will have to be thought to have something so similar to it that the notion of a truly decentralized or "mindless" intelligence becomes destabilized. What is the alternative to consciousness as a coordinating influence for materialists? I suppose there would be the resort to emergence as an explanatory phenomenon even though its very existence can only be inferred, and then only when a materialist ontology is, through assumption, pre-installed. In my estimation, emergence makes an appearance whenever materialism bumps up against its own limitations in satisfactorily explaining the origin of things so intuitively different from its material portrayal of the nature of reality in general. Consciousness and intelligence certainly fit this description. Given this, and as discussed earlier in the chapter, I suggest that emergence is more of the materialist's *deus ex machina* than any kind of plausible explanatory framework.

The next area worth critiquing is the explanation of the relationship between thinking and intelligence. While both embodied cognition and decentralization "open the door" for plant intelligence, one may also notice that they do so not by suggesting that plants can think, but by severing the connection between thinking and intelligence. As I stated earlier in this chapter, thinking is the lifeblood of notions of human intelligence. To eradicate it in order to broaden definitions of intelligence—or to functionalize them so radically differently in order to enable the inclusion of a being that, ontologically, one can't justify as having thoughts because of one's limiting ontology—is intuitively unsatisfying. By undertaking this modification of the definition of intelligence, one notices also that theorists get to leave intact the main supposition that only those with human or human-like neurophysiology can have mental experiences. This is the case even though the materialist understanding that allows inclusion of plants as intelligent is still wielded arbitrarily to then deny them thoughts while granting them to humans and human-like animals under similar circumstances.

The distilled set of claims undergirding the denial of thought to plants is as follows:

1. Both plants and humans have learning, memory, etc. that help comprise intelligence.
2. The capacity for these things in humans is rooted in their capacity to think.
3. The "source" of the thoughts that undergird these attributes in humans are causally connected with the brain and its attendant neurophysiology.
4. 3 is true even though, as some emergentists admit, the mind is qualitatively different from the brain.
5. 3 is true even though there is no material evidence to explain how the mind emerges from the brain.
6. Therefore, because plants don't have a human-like brain or other neurophysiology, they are not capable of thoughts.

This argument is rife with problems, yet constitutes the only undergirding of the denial of thought to plants. I can say this because there have been virtually no empirical studies of plant thinking, therefore no direct empirical evidence for its denial. In the rare, fringe instances that this topic has been studied, such studies have been excoriated by mainstream scholarship (see the "Kook-Fringe" discussion in Chap. 9). On the other hand, studies of human thinking are countless, and correlations between minds and brains, sponsored by an ontological assumption that brains are the sources of minds, have undergirded the argument that human-like neurophysiology sponsors mental experience. Yet this is still not support for (a) such neurophysiology being the *only* kind to support thought and (b) human thought being the only *kind* of thought. Thus and again, the difference in attribution of mental experience to human-like animals and not to plants can mostly be ascribed to the unsubstantiated assumption that human or human-like physiology is the only kind that produces it. That's why emergence produces conscious intelligence in humans and human-like animals, while in plants it at best only produces something like decentralized intelligence. If plants are presupposed to not have minds, then nothing mental need "emerge" for them from the microstructures responsible for the remarkably similar types of end-product-of-intelligence that we see in both humans and plants.

What I find particularly astonishing in the context of this discussion is that a process that is admittedly so thoroughly opaque—emergence—can be so consistently and unflinchingly fallen back upon to sponsor mental experience in one type of being and to deny it in another. This, despite both responding similarly to similar stimuli. This is a particularly disconcerting conceptual turn given that virtually the entire bifurcation of the explanation rests upon an assumption, not evidence, that plant neurophysiology cannot produce a mind. That the assumption is also offered as *evidence* can be plainly seen since, when we pull the thread of that assumption-*cum*-wellspring, we find it dangling out the other end of the discussion in the form of a side conclusion. It gets one's back up, such self-fulfilling conjecture, and makes one inclined to think that what lies in the deepest parts of the modern, human psyche is that *we don't want plants to think*, therefore we construct ever more conceptually circular and obfuscating material arguments to support this desire. It's embarrassing and intellectually dishonest, really. It is in need of earnest, critical examination and reassessment.

The third area to critique is the denial of individuality. According to Marder (2012), a plant's "decentralized structure means that, besides being non-hierarchical, it does not fall under the category of organismic life" (p. 1370). He contrasts this kind of life with a description of organismic life that is

> succinctly expressed by French philosopher and physician, Canguilhem: "To live is to radiate; it is to organize the milieu from and around a center of reference, which cannot itself be referred to without losing its original meaning." The plant does not organize its milieu; it is comprised of a series of internal communicative networks (e.g., biochemical and hormonal channels, or synaptic cell-cell communication) and external communication pathways that connect it to its environment. It is, thus, an open system, coupled with its environment...[A plants'] capacity for kin recognition is perhaps best understood not in the categories of "self" and "other"... (p. 1370)

Here, in one conceptual stroke, Marder has denied both thought and individuality to plants. And even though he is willing to apply the same framework to humans, because a human *can* "organize its milieu," the author affords individuality and thought-based intelligence to the human. But, don't the Red Alder trees and other "pioneer species" fix nitrogen in the soil for the benefit of themselves and other species that follow (Chapin, Walker, Fastie & Sharman, 1994)? Lichen break down stone into minerals (Lisci, Monte & Pacini, 2003). The "invasive" Tamarisk exudes salt from its leaves to inhibit competition (Adler, 2007, p. 88). Are these not examples of organization of one's milieu? To suggest otherwise is to take an artificially narrow and anthropocentric view of the act of organizing. Perhaps those acts of organization are seen as "lesser" because we either value what plants value less than what we value, or because we believe it *not to be a product of thought* or the action of a Self. But again, to deny thought and individuality based upon the presupposition of their absence is circular reasoning.

Maybe our denial of the organization of one's milieu to plants is rooted in the argument that we humans are aware of what we're doing at times when we organize our milieu, and that this self-awareness makes us unique. But there is clear evidence that plants can distinguish both between self and other (Gruntman & Novoplansky, 2004) and between those of their own species and those of other, potentially competing species (Chen, During & Anten, 2012). Thus the denial of thought and individuality to plants must become even more narrowly justified, the denial becoming almost purely presupposed instead of demonstrated. Ultimately, to face the possibility of plant thought and intelligence—an individual organizing its milieu—means facing the problem of locating just where, *within each individual plant*, the basis of such faculties resides. With a dualist ontology that conceptually reduces plants to "open systems" or "mindless masters" the problem is solved, but when such a solution is wholly sponsored by dualist assumption, this cannot be accepted.

Posthumanism might suggest that since there are no metaphysical individuals, that we need not worry about individual plants. But since humans, even if they are admitted to be situated within intra-actions, and form an intra-action as an individual, are rarely denied thought-based intelligence, then the only justification for a posthumanist denial of it to plants is the same dualist, materialist reductionism that I've pointed out, in Chap. 3, as persisting within posthumanism's materialist commitments. Any posthumanist argument for "asymmetry" in the origin of intelligence is rooted in this dualist ontology and must also be rejected. The other posthumanist option to explain a non-Self-ed form of intelligence for plants would follow the line of argumentation that Barad (2007) offers regarding the Brittlestar, where that being's thoughts are directly "*materially enacted*" (p. 375). I discuss the insufficiencies of such an argument in more depth in Chap. 3, but note here that such a notion fails on its face to describe how such direct material enactment comes to be intelligence when a mind, or any other coordinating influence, is absent.

The last element I'll critique is whether theorists are offering a qualitative or quantitative difference between human or other animal intelligence and plant intelligence. As I've been discussing up to this point, it appears that theorists (perhaps unwittingly) hold plant intelligence to be markedly different from human or human-like animal intelligence in its origins. This is despite the fact that plant intelligence theorists believe that the difference between the two kinds of intelligence is one of degree and thus is not qualitative. In arguing for his notion of a continuity of intelligence, Trewavas (2003) contrasts his view with those who see plants as "automatons" (p. 2). But, I believe that the qualifications that plant intelligence theorists impose on plant vs. human-like intelligence don't actually free plants of categorization as *automata*. In the sense that Trewavas uses it, the OED defines an automaton as: "An organism that functions purely involuntarily or mechanically; an animal, insect, etc., not motivated by higher consciousness or intelligence" ("Automaton", 2018). And while it may *appear* that defining plants as intelligent frees them from this category, it only does so by artificially restricting the definition of automata to a certain coarseness of granularity. To wit: if plant intelligence is not a product of consciousness, then regardless of the seeming variety of choices plants make (and the voluntarism with which they appear to make them) they are neither voluntary nor are they selected from any real variety. The variety only appears as such to the outside observer. They are all determined for the plant by biology and physical states, which means they are *wholly* determined.

Whatever mysterious origins one assigns to the variety and novelty of plant response, such as a non-cognitive emergence of intelligence, they are still materially determined and therefore not chosen by a conscious, choosing individual. In other words, were one able to know all material determiners of material or "physically intelligent" response, by employing a LaPlacian calculator we *can* know the response of the individual plant according to these theories. Thus, plants *are* automata, regardless of how complex the programming. Of course the same LaPlacian calculation can be performed for human beings if one's ontology is wholly or reductively material. This brings back the claim that emergence theorists are making—that there is a qualitative difference between consciousness and the microstructural material which is its supposed source. In that theory, the emergence of consciousness saves the day for humans, but no such rescue awaits the "intelligent" but ultimately "mindless" plant.

I feel comfortable using quotes around the word "intelligence" now because I no longer see plant intelligence theorists' definition of intelligence as entirely useful in the context of assessing a plant's behavior and whether the plant has the kind of faculties most people think of as intelligence—that is, based at least in part on thought. In these theories, the plant as individual is either only housing the acts of intelligence happening at a more microstructural level or the intelligence "emerges" from the larger matrix of environmental relations within which the individual plant is embedded. In the latter, the intelligence is floating out there in the ether some-where. A disembodied thing that is only differentiable *as* a thing because we think-ing humans identify it and its effects as such. As I quoted Marder (2012) earlier about the plant, he says its

> decentralized structure means that, besides being non-hierarchical, it does not fall under the category of organismic life....[and its] capacity for kin recognition [for example] is perhaps best understood not in the categories of "self" and "other" but in terms of the construction of a world in common and a clash between various life-worlds. (p. 1368)

Because of the ontological issues I've pointed out, any beachhead theorists have tried to establish for plants as intelligent beings—by making the differences between intelligence in plants, animals and humans a quantitative one—is illusory. Humans and human-like animals are still alone as intelligent *individuals*. If there's a doubt, Trewavas' (2003) words fairly glisten with the anthropocentrism that results from such dualistic reductions when he says,

> ...although as a species we are *clearly more intelligent* than other animals, it is unlikely that intelligence as a biological property originated only with *Homo sapiens*. There should therefore be aspects of intelligent behaviour in *lower* organisms from which our *superlative capabilities* are but the latest evolutionary expression [emphases added]. (p. 1)

Carello et al. (2012) also, eventually, succumb to the force of this dualistic, anthropocentric ontology, allowing it to demarcate a qualitative boundary between human and more-than-human. They ask, "So what are we to do with our gut feelings that physical and bacteria and plant intelligence are all different from animal intel-ligence, which is itself different from human intelligence?" (p. 260). To answer their own question, they advise us to maintain "an eye [as] to how these different embodi-ments might be ordered" (p. 260) with humans at the top, of course. Because of the material nervous systems that produce consciousness, despite the authors' belief that the differences are "orders of magnitude of...perception-action couplings" (p. 260), we humans and those like us are physiologically alone. We are intelligent in a qualitatively different way. The anthropocentrically motivated, qualitative dif-ferences in definitions of intelligence that plant intelligent theorists are working to leave behind remain with all their hegemonic force—perhaps even more forcefully since they go so unrecognized. To say that a *mindless* system will ultimately pro-duce intelligence via emergence or some other material mechanism is, ultimately, to equate self-perpetuating chaos with intelligence (again see Kessler, 2012)—to ren-der intelligence as a mindless artefact of the machinations of evolution. Logically, while all materialists must eventually be forced to this conclusion, I suggest that, intuitively, this does not equate with our sense of intelligence as something in which the mind figures so heavily. What we value so highly is missing from the equation, therefore it cannot stand as the basis of a definition of intelligence for human *or* for plant.

My conclusion is that plant intelligence theorists have failed to appreciably reorient the definition of intelligence or those who might have it. They have simply downgraded intelligence in the same materialist ontological landscape that positions humans as "highest." To adapt an Orwellian turn of phrase, Carello et al.'s argument states that all living things equally have intelligence, but some have it more equally than others. This will not do.

10.2.4 The Possibility of Mental Experiences in Plants

Putting aside any *a priori* rejection of the notion that plants can have minds, the difficulty of making the case for plants potentially having them lies primarily in the lack of any proof of a mind at work in an organism so different from humans—humans being the standard of comparison to permit the possibility of minds in more-than-human beings. But, as Adams (1928) notes, feeling confident that even another human is having certain kinds of mental experiences is not achieved by direct observation of those experiences, since a human being "can observe directly the processes of only one mind, his own" (p. 235). Noting this, Adams also notes that humans routinely assume a similarity of mental experience in other humans anyway, and arrive at this "'knowledge'... through comparison—more or less conscious and formal—of other behavior with our own, and by inference from analogy" (p. 235). Adams notes that, when concluding that a nonhuman animal has mental experiences similar to humans, this is achieved "by analogy with one's own [mental experiences]." Adams teases apart and formalizes this idea in the following postulate:

> Any experience or mental process in another organism can be inferred from structure, situation, history, and behavior only when a similar experience or mental process is or has been invariably associated with similar structure, situation, history and behavior in oneself; and the probability of the inference will be proportional to the degree of the similarity. (p. 252)

Here, Adams has succeeded in bypassing the materialist argument against minds in others, but in so doing, fails to account for his own anthropocentrism. Upon whose authority is it determined that minds are only possible in the ways humans have them? Adams has also failed to see that his argument conflates the degree of epistemological certainty a human individual can have about the mind of another with the probability of the other actually having that mind—the latter being a thing which has nothing to do with another's ability to perceive or measure it. To say that my confidence that another has a mind is the determiner of whether they should be thought to actually have that mind is a logical misstep and puts humans in the position of an unlimited perceiver of reality, which they certainly are not. Just as when Carruthers (2004) casts doubt on the argument that "the real existence of...thoughts is contingent upon the capacity of someone else to co-think them" here I suggest that the existence of a mind, or even the probability of its existence, cannot be contingent on any other individual's ability to infer it through analogy, observation, or any other means.

But, Adams' (1928) framework is not without its uses. His suggestion is right to the extent that, when we perceive what we take to be the product of mental activity, we *should* assume that it *can be* the product of mental activity. It is not through analogy to our own reaction alone that we do this, nor should it be through similarity alone. Some other criteria, such as those established by Garzón (2007), may be useful.

Garzón, for his part, suggests that for *cognition* to exist in any living system (including plants), it must be seen in the ability of that system to manipulate *representations* instead of simply reacting mechanically to external stimuli. He goes on to say that for a thing to be a representation it must meet two standards. First, the representation must be able to stand for "things or events that are temporarily unavailable" (p. 209). A moose's memory of a wolf attack that occurred at a lakeside the week before would be an example of a representation of an external stimuli that is not present. Second, Garzón says that the representation must have a detectable effect, and that effect must be traceable to the representation specifically and the external stimuli that originally created it. In the previous example, if the moose avoids that lakeside at a later time, one must be able to reliably trace that avoidance behavior to the previous incident with the wolf and its current influence in the form of an internal representation. Garzón says that defining cognition this way can help "to assess the cognitive capacities of any information-processing system whatsoever. Notice that it does not rely upon the existence of any specific brain tissue to perform computations" (p. 210). I note here that his qualification about brain tissue aligns with my suggestion that one cannot *assume* that thinking is dependent upon any human-like neurophysiology.

To apply his criteria, Garzón goes on to note that manipulations on representations are not only achieved by humans and those with similar neurophysiology, but also by bees and plants. He gives the example of how the "Leaf laminas of *Lavatera cretica* can, not only anticipate the direction of the sunrise, but also allow for this anticipatory behavior to be retained for a number of days in the absence of solar tracking" (p. 210). While he suggests that such behavior meets the first of his two standards for cognitive manipulation of representations, he says it remains to be seen whether it meets the second: that the actions are indeed caused by computations in response to the representational states. But, in referencing Carruthers, Garzón goes on to argue that if the non-reactive, non-associative, belief/desire mental architecture that Carruthers articulates as cognitive can be attributed to animals such as bees, then there is no reason not to apply it to plants if the causes of their actions meet the same criteria as those of bees. In fact, he concludes, "[n]othing...prevents other information-processing systems [such as plants] from possessing minds" (p. 210).

Coming as the conclusion before his summary remarks, this last sentence is no trivial statement on Garzón's part, and stands in direct contrast with his suggestions, in the same and other of his essays, that plant cognition could be the product of embodied cognition and decentralization. Thus, we see exposed in this essay—and perhaps even in Garzón's own beliefs—an ontological tension. On one hand, he aligns himself with the materialist-based explanation of plant cognition as mindlessly embodied or decentralized. On the other hand, he describes representational states which, if they exist, must exist at least in part inside an individual plant's mind. Therefore, the decentralized or embodied theories that eradicate the possibility of the individual plants being the source of cognition *cannot* exclusively apply. If Garzón believes individual plants meet his two criteria for cognition, then he *must* conclude that they have minds. There is no other place for those sources of cognition to exist.

To reinforce this conclusion, I point to an example from Nagel (1997), who states,

> When one of the first two leaves—cotyledons—of [the plant species *Bidens pilosus*], which basically grows symmetrical, is punctured and shortly afterwards both cotyledons and top are cut off, the plantlet continues to grow asymmetrically; it somehow 'stored the information' that it was punctured, or 'remembered' the injury. How such information is stored is still not well understood. (p. 218)

Nagel's use of quotes around storage and remembrance is illustrative since, at least in Garzón's notion of cognition, no quotations are necessary. It is real. The quotes instead betray Nagel's dualist ontological commitments. The individual plant *does* remember, and with Garzón as potential guide, we can suggest that the plant may remember it in his or her mind. How far, then, have we traveled from human-like neurophysiology as the exclusive sponsor of mental experience? When a plant encounters a root, she remembers what it was like to encounter a similar one previously and get around it. Does my use of a gender-specific appellation for the plant strike an odd chord? That's only because we like to refer to plants as "it" not because they don't have gender (most do, or have both genders simultaneously) but because we have *post hoc* conceptually negated their individuality, capacity for thought, Selfness, etc. But *she* remembers. And *she* thinks about what to do to get around the root and sets to work getting around it. To return to Adams' (1928) inference of mind through similarity to humans, with this rehabilitation of our ontology can we not now, in this moment of plant thought and decision, *see our human selves*?

I'd suggest that such *simpatico* in response to observing the behavior of another human being is a potential embarkation point for close relationship. It appears that there is nothing now, ontologically speaking, standing in the way of it serving a similar function in the context of human-plant relationships. In such a context, Backster's (1968) assertion that plants respond to a humans' intent to harm them seems far less fantastic—and leads to a host of other questions relevant to human-plant relationships. If plants can respond to humans cognitively, are these the roots of relational reciprocity? If plants have thoughts and feelings about what humans are doing, how does that change our response and responsibility to them? Can I clearcut a forest tract under such circumstances, or even do experiments like those carried out routinely in places like Hubbard Brook Experimental Forest in northern

New Hampshire? If nothing else, such possibilities radically alter the context and tenor of the attempt to answer to such questions. Natural "resources" become living others, and Hubbard Brook research such as experimental tree cutting and herbicide application (Likens, Bormann, Johnson, Fisher & Pierce, 1970) could be seen by some to be little different from vivisection.

10.2.5 Beyond Basic Mental Experience – Can Plants Be Conscious?

Pushing beyond the basic mental experience of activities like manipulating representations in the absence of precipitating stimuli, plants have many abilities that, if found in humans, would be attributed to more advanced mental activity. I'll examine a few of these before discussing the possibility of awareness and consciousness in plants. The first ability I'll examine is that of *self/other recognition*.

Awareness or recognition of self is often a criterion by which a being is judged to have mental activity commensurate with consciousness. In experiments on root growth and competition in plants, there appears to be an "overall coordinating mechanism of root distribution... including self-recognition" (Trewavas, 2005, p. 406). Trewavas goes on to say,

> Growing plant roots preferentially occupy vacant soil and deliberately avoid the root systems of competing, alien individuals. If roots of different individuals of the same species are forced to contact each other, decisions are then made to rapidly cease growth of the touching roots. (p. 406)

Thus, not only is there a self-other differentiation being made by plants, differentiation is made between others of the same species and those of different species.

When describing this behavior in the preceding paragraph, Trewavas (2003) uses terms like "prefer," "recognize," "avoid," and "decide." In the context of his discussion of consciousness in nonhuman animals, Griffin (1981) suggests that the belief in the applicability of terms like these to more-than-human beings is often dismissed by scientists as "childish sentimentality" (p. 116). But, Griffin notes, for the lay person there is an "intuitive impression" of nonhuman animals as having "sensations, feelings and intentions," and this is "based on our experience with patterns of animal behavior that appear sufficiently analogous to some of our own behavior" (p. 116). In my discussion of Adams (1928) above, I note the pitfalls of judging the presence of a mind in another based on any similarity to humans. Putting those pitfalls aside temporarily, though, I ask whether or not plant self-recognition and the behavioral response of not competing with one's own roots can be seen as analogous to human decisions not to engage in behavior destructive to self, kin or self's own species. What would be particularly interesting is an experiment to see if plants avoid competing with roots of plants genetically similar—that is, parent and offspring plants—more than plants of the same species, etc. Physiological differences aside, the self-other awareness behaviors are highly similar in human and plant.

While I admit that recognition of such similarities does not necessitate a conclusion that plants are conscious, if one strips away the assumption that plants *cannot be* conscious, then admitting to such parallels can be a first step toward legitimate inquiry into the possibility of consciousness.

Those adhering to an explanation of plant behavior as mindless might counter the drawing of such parallels by suggesting that for the plant, such decisions are the result of chemical and electrical phenomena (Gruntman & Novoplansky, 2004). In such an explanation, things like preference, recognition and other thought-based elements would not apply. But, such a claim would fail to account for the fact that such chemical and electrical phenomena also accompany human behavior and yet very few but the most reductive materialist or behavioralist thinkers would suggest that humans don't have minds or consciousness. Thus, in certain ontologies, and in certain lines of thinking such as Garzón's (2007), if the criteria for mental activity is met, then plant or no plant, mental activity exists. If that's the case here, then self-other recognition may be evidence of mental activity that moves beyond the basically cognitive toward consciousness.

Those seeking a material foothold in such a suggestion might ask *where* this more advanced cognitive activity takes place in plants. I think the plain answer is: Wherever their minds exist. In a material and more-than-material ontology, that does not mean in the material world only. Even in the material, if we transcend the fallacy that human brains produce minds thus all minds must be the product of human-like brains, I suggest that it's not conclusive what sorts of material structures and processes correlate with minds. Darwin theorized that plants had a type of "brain" at the tip of each root (Baluška, Mancuso, & Volkmann, 2005). Trewavas (1999) draws many parallels between animal neural functioning and plant calcium waves. Could a plant's mind correlate with such material structures and processes? Once the impediment of an ontology that *a priori* denies minds to plants is removed, there is nothing ontological preventing this possibility. As of right now, we have plain evidence of self-other recognition in plants that may very well, with more research, move potential plant mental activity beyond the basically cognitive and into consciousness.

Next I consider learning and memory, traits that plants exhibit and that are commonly associated with more-than-basic cognitive functioning when observed in animals. As an example of learning and memory in plants, Callaway, Pennings, and Richards (2003) say that "Morphological traits differed and water use efficiencies were higher for seedlings grown with annuals than for those grown with perennials" (p. 1119). This was due to more rapid water depletion in communities where annuals were present. In recognizing that water was more scarce, the plants in the study learned how to behave differently in different environment. Callaway et al. also speak about research in which clones of the species *T. repens* responded differently to different grass species in the greenhouse when the clone had previously been associated with that species in the field (p. 1119). Here we see that not only learning, but memory of what was learned, applies in a future scenario.

In Trewavas' (2003) theory of plant intelligence, he states that "Plants must...have access to an internal memory" (p. 2) that allows them to optimize ecological fitness.

They use this memory to store learned behaviors. That these behaviors cannot be reduced to automatic, reflexive and/or invariant biological responses is evidenced for Trewavas by things like "trial-and-error" learning where young plants will tend to initially overshoot, undershoot, or grow in the wrong direction before correcting themselves when turned at some angle from having being upright in relation to gravity (p. 4). Trewavas suggests that plant learning is based on the plant having a goal and then having an error-correcting mechanism that feeds back to the plant on the efficacy of the behavior in reaching the intended goal. Another example of learning and memory from Trewavas is when "Resistance to drought or cold can be enhanced by prior treatment to milder conditions of water stress or low temperatures" (p. 4).

While these are powerful examples of learning and memory, I think the scholarly reflex is still to dismiss them as something other than evidence of potential mental activity. But again, this is based largely in the assumption that plants can't have minds, and not in any evidence that they don't. Ultimately, whether this learning or memory is evidence of consciousness in plants is an open question. I'll explore the possibility that it does constitute such evidence in the "Might Plants Be Consciously Aware?" discussion just below.

The last ability I'll consider is that of *intent*, which is commonly believed to be associated with thought and consciousness. Intent can be defined in numerous ways, but Carruthers (2004) suggests that intent is the possession of a belief/desire mental architecture where one's beliefs influence one's desires, and the setting about of fulfilling those desires is based, in part, on those beliefs. Griffin (1981) defines intentions as mental experiences where "the intender pictures himself as an active participant in future events and makes choices as to which sort of image he will try to bring to reality" (p. 13). Garzón (2007) suggests that plants have this capacity, calling what they form "intentional systems" (p. 211). Marder (2012) calls intentionality a "crucial phenomenological concept, operative in plant life" (p. 1367), and Trewavas (2003) says that the walking behavior of stilt palm plants is one of many examples of "very clear" (p. 15) intentionality.

As to what intent signifies about mental experience and consciousness, Griffin (1981) quotes Longuet-Higgins as saying, "An organism which can have intentions I think is one which could be said to possess a mind [provided it has] ... the ability to form a plan, and make a decision-to adopt the plan." (p. 18). If plants can do this, then it is still to be seen whether such intentionality meets the more stringent standards of minded or consciousness-based intent laid out by Carruthers (2004). He defines minded intention as the possession of "distinct information states and goal states" that "interact with one another in ways that are sensitive to their contents in determining behavior" (p. 216). Carruthers concludes that honeybee behavior satisfies these conditions, and so it must be concluded that honeybees have minds.

While it is beyond the scope of this book to argue that plants also satisfy Carruthers' conditions, I note that no such experiments into plant behavior to examine this possibility exist, but there are studies that are in intriguing keeping with the possibility. For example, Carruthers speaks about experiments that show that

animals on a delayed reinforcement schedule in which the rewards only become available once the conditioned stimulus (e.g., an illuminated panel) has been present for a certain

amount of time, will only respond on each occasion after a fixed proportion of the interval
has elapsed. (p. 210)

In other words, they aren't "merely building an association between the illuminated
panel and the reward. It seems to require, in fact, that they should construct a representation of the reinforcement intervals, and act accordingly" (p. 210). A parallel
example for plants can be found in Trewavas (2005), who says, "When young trees
were provided with water only once a year, over the next several years they learned
to predict when the water would be supplied and synchronised their growth with its
appearance" (p. 408). Here it appears that the trees are constructing "a representation of the reinforcement intervals, and act[ing] accordingly."

An interesting side-effect of Carruthers' constraints arises when one examines
the ontology that underpins defining information or goal states. For example,
Carruthers (2004) notes that Australian Digger Wasp behavior does not qualify as
true belief/desire architecture because,

The wasp appears to have no conception of the overall goal of the sequence [of building a
tower-and-bell structure over her egg burrow whose purpose, according to Carruthers, is to
defend her eggs], nor any beliefs about the respective contributions made by the different
elements. (p. 212)

But, the way in which he sets the wasp's actions outside the belief/desire architecture
is based almost wholly on *his* definition of what the *wasp's* purpose is—a definition
which appears to be heavily influenced by the author's dualist ontology. He assumes
that her purpose in building this structure is to defend her eggs because the structure
can serve that purpose, and assesses the presence of beliefs and desires based on her
response to the disruption of this purpose. But we may ask if this is really true.

One example from Carruthers is illustrative. While a wasp is building her tower-
and-bell structure, an experimenter drills "a small hole...in the neck of the tower"
(p. 212) to gauge her reaction. Her reaction was to build "*another* tower-and-bell
structure...on top of the hole" (p. 212). Carruthers takes this as proof that the wasp
"seems to lack the resources to cope with a minor repair" (p. 212). Since the purpose
of protecting her eggs is not served, at least not directly, Carruthers concludes that
she must be ignorant of it. But, if there is a purpose toward which any being works
and, in being disrupted, the being responds with some behavior that seems out of
keeping with the purpose identified, *two* conclusions are possible. The first is the
one that Carruthers reaches: that the being is ignorant of the purpose. The second is
that the purpose identified by the observer is not the correct purpose—thus the being
is still behaving according to her belief/desire architecture, it just has not been identified correctly by the observer. In the case of this wasp, what if the wasp's purpose
is to build a structure that has no flaws, and when the experimenter pokes a hole in
it, she prefers to build an entirely new one so that there exists a tower-and-bell structure without flaws, scars or repairs? Who knows why she wants to do it this way, but
perhaps in her *thinking*, the notion of "repair" simply doesn't exist. There is the
tower-and-bell structure, and when there is a hole put in it, it ceases to be what she
conceives of as the right structure. Put differently, the structure ceases to meet her
mental representation of it and the desire to bring that representation into reality.
Thus, she begins again. In such a scenario, Carruthers is simply wrong about the

purpose he divines for her building of this second structure, constraining his think-ing only to purely material-evolutionary and utilitarian purposes for the wasp.

To bolster my suggestion, I note that many humans labor toward purposes that are not obviously, if at all, material-evolutionary ones. Humans can be loyal and chivalrous. They make art. These are purposes that do not easily translate into evolu-tion's supposed materialist *telos*. Some may still position them as end-products of evolution's work on our species over hundreds of thousands of years, but even stipu-lating this for the sake of argument, the human may not *know* the material-evolutionary purpose and yet still have a belief/desire architecture that drives them onward toward another purpose, one that bespeaks conscious thought. None but the most reductive materialists and perhaps pure behaviorists would argue otherwise about humans. Why, then, is the wasp not afforded the same courtesy? I suggest that the cause of this—and this is again crucial—is that there is a *pre-existing assump-tion* that the wasp can have no purpose other than an obviously material one. If she is already believed to have no mind, then certainly she can have no conception of what her tower-and-bell structure ought to be like except for one that, in brute physi-cal terms, "does the job" of protecting her egg burrow. She cannot *be* the artist (and thus respond as one when her construction is interrupted). Therefore, when she *does* respond as the artist, instead of it being recognized as such, she is stripped of *all* purpose that is not directly material.

Here again, we see the dualist sleight-of-hand at work. Beneath the surface the conclusion of mindlessness is drawn *a priori,* then the "facts" fitted so that it can be reached *a posteriori.* The Digger Wasp is operating in the arena of an ontologically rigged game. I suggest that what we see in wasp or plant behavior may sit well out-side of what dualistically constrained ontologies permit, yet still be purpose that humans know well and exercise routinely themselves. When one strips away the pre-existing assumption that the wasp is incapable of a belief/desire architecture, one finds it staring one in the face. There is nothing in the Digger Wasp's own actions that work against this possibility.

Yes, one may respond, but are the Wasp's eggs protected when she responds as she does to the hole being poked in her tower-and-bell structure? Maybe. But if it doesn't, this doesn't mean a purposelessness, it means that in striving toward the goal she has in her mind for, in my example, a perfect tower-and-bell structure, she fails to see that building a new structure may work against the protection of her egg burrow. When humans fall in love, they may do many seemingly unreasonable things even though, as many materialists argue, love is species-perpetuation in action. Does the human have no other, less tangible purpose, then, in pursuing the one he loves? Of course he does. Sometimes, though (and sadly) he still may fail.

10.2.6 Might Plants Be Consciously Aware?

Having explored the self-other recognition, learning, memory and intent that plants exhibit, the next logical step is to consider whether these are examples of conscious awareness. Griffin (1981) says that awareness can be defined as "the experiencing of interrelated mental images...[and i]f...events [that occur in these images] include participation by oneself, we say the organism is self-aware" (p. 12). For such aware-ness/self-awareness to be considered a form of consciousness, Griffin says that these images will be used "by that organism to regulate its behavior" (p. 15). I note that the latter qualification for consciousness closely resembles the one given by Garzón (2007) for cognition, where representational states are used by a being to regulate his behavior. The key difference is whether Garzón's representational states are mental images. If they are, then his definition *is* Griffin's definition for one type of conscious awareness.

As I state above, Garzón concludes that it's at least possible that plants have minds if the second of his two criteria can be shown to have been met—that the representation inside the plant has a detectable effect on her behavior, and that effect is traceable to the representation specifically and the stimuli that created it. While there is little in the way of evidence showing this at present in plants, neither is there any against it. We simply don't know.

At this juncture, if we are to entertain the possibility that plants are consciously aware, it's fair to ask whether that awareness is like human conscious awareness. To try to answer this, I'll begin by saying that rooting the mind in more-than-material as well as material ontology, more-than-human beings can unproblematically have conscious awareness in the way humans do. Mental images would not exclusively grow from the human-like material elements upon which the present human-nature relationship literature hangs its theoretical hat. Neither is there evidence that differ-ent physiologies, if they did produce conscious awareness, couldn't produce the same kind. My discussion of Kim's (1997) take on multiple realizability above pro-vides more detail on such a possibility.

But, if we believe that more-than-human beings have a *different* type of con-scious awareness, we must delve to the root of plant conscious awareness to see if it is facilitative or prohibitive of the kind of thinking that contributes to close relation-ships. I will say that by defining the type of conscious awareness I describe above as "basic," Griffin (2013) implies that the "better" kind of consciousness is not this kind, but the more "complex" (and not coincidentally, human), kind. But, we must ask what evidence exists to support such a claim. Beyond the irremediably anthro-pocentric argument "because it's our [human] kind," we must ask what makes human consciousness, if it is unique, more conducive to close relationship? Its level of complexity is regularly pointed out in the literature (e.g., Griffin, 2013; Panksepp, 1998), but even the way in which human consciousness is defined as more complex is itself hopelessly confounded by anthropocentrism. *If* plant awareness is less com-plex at an individual level, it may be far more complex at a relational level. In Buber's (1923/1970) thought, the immediacy of a less detached self in I-Thou rela-tion is far superior to the self in an I-It relation—that is, interactive awareness is superior to objective awareness. The former is a conflation of self and other (without loss of the individuality of either) along the lines of Peirce's (1960) shared feeling or the sudden intimacy of "feeling with" that Noddings (1984) notes and that I referred to in Chaps. 6, 7, and 9. In such a context, plants may have more capacity to be relationally aware, and because of this, have a *far more* complex awareness than a typical human. If plant awareness is not less complex at an individual level, but just differently complex, then again, plant consciousness is unproblematic.

I conclude by relating an experience my sister (personal communication) had with two plants. With the possibility of consciousness in mind, it is highly provocative. Many years ago, my sister had two beans she had planted in a window box in her apartment. Both sprouted and began to grow. Soon, one started to look unhealthy and, as the days went on, wither and begin to tilt over. Upon coming home from work one day she found that the healthy bean plant had bent itself over in an unnatural position, away from the sun, and was literally holding the other plant up as a human would an ailing companion. The day after that, when she returned from work again, the unhealthy sprout was laying on the soil dead while the healthy one had returned to its previous, upright position. In the context of my discussion above, no longer must we be forced by such observations to seek a mindless, material explanation for such a sequence of events. We could, but we could also begin to see it for what it, in my opinion, could also so easily be: one being caring for another in continuity with how a human would. If true, that this is the basis for common ground—for closeness—cannot be reasonably denied.

10.3 Thoughts in the "Inanimate"?

In the previous section on plants, I mostly explored the ways in which an adjustment to anthropocentric, materialist and individualist ontology may clear the way for realizing that plants may be thinking, conscious beings. I have not undertaken an exploration of "inanimate" beings partly because it's very difficult, given the dearth of any research into such a possibility, to even approach the subject. I will say, however, that I find the work of Bose to which I refer above intriguing on this front. Specifically, his position that the property of "life," which he saw reflected in material electrical activity, crosses the boundary usually drawn between animate and inanimate beings is intriguing. Since I've argued in the previous chapter that electrical activity may be associated with feelings (and here suggest it could be associated with thoughts), then if electrical activity is similar under similar circumstances for a Western Hemlock and a river stone, then the possibility of finding thoughts in that stone must also exist.

10.4 Exchange of Thoughts

In arguing that humans have close relationships with companion animals in ways similar to interhuman closeness, Sanders (2003) says that there is a "rich body of literature [focused on]... how people come to define their animal companions as unique individuals, *comprehend their mental experience* [emphasis added], and organize everyday exchanges based on these understandings" (p. 408). For Sanders, then, thoughts and their interaction between human and more-than-human relational partners leads to closeness.

A particularly interesting feature of Sanders' argument for the interaction of minds is his suggestion that, instead of minds being distinct and thus ultimately unknowable by others, they are "constructed and shared through interaction" (p. 419). Because they are knowable through relationship, Sanders believes that it is reasonable to draw conclusions about another's thoughts even if that other is not human. This is yet another perspective that helps us understand what I discuss about Adams (1928) above. When applied to interhuman relationships, the interpretation and exchange of thoughts constitutes the thought vector in my close human-nature relational model. By "[c]asting off the linguicentric and anthropomorphic restraints of conventional views of mind" (pp. 418–419), this can be achieved between humans and more-than-human beings as well.

For example, Sanders speaks of a dog owner who trains her dogs to sit through a meal humans are having and to get some table food only at the end. One of the dogs, Toby, is fine to wait except when butter is on the table. Normally when he spies butter, he squirms around and makes other gestures of impatience, but one night

> Toby thought through the problem of how to get the bread and butter. He...left the room. We heard him rummaging around in the house, up and down the stairs, until he finally appeared with a treasure: a roasted pig ear...He had obviously hidden this one. He laid it at my feet... Toby reasoned and came up with the idea of trade. (p. 419)

Sanders suggests that it's perfectly reasonable for the dog owner to have drawn the conclusion she did about Toby's thoughts of trading for butter. And if this is the case, then here is an example of the exchange of thoughts, one that is achieved through unscientific means. This more relational knowledge is based on intuition, familiarity and experience instead of repetition and quantifiable observation. Not only does Toby's owner know Toby's thoughts, but Toby knows hers, and knows that, if presented with an alternative means of persuasion, out of appreciation for it she may give him the bread with lots of butter that she does as an exception to her mealtime rules. This is just one event contributing to their closeness.

The same closeness of relationship has the potential to form between humans and wild animals as well. For example, Houston (personal communication) related a story to me about a friend of hers who'd gone camping with a partner and who, at one point at their campsite, crouched down,

> pointing to a wolf who had come close to their campsite: both stared at each other for some time—i.e. the wolf and the two people. Then the wolf moved off into the bushes. The people crept forward, crouching down to see if they could see the wolf only to see the wolf crouching down to see them!

In this scenario, there appears to be a mutual and instantaneous recognition of the others' thought processes and a sharing of those processes as the campers recognized their own curiosity in the wolf. What kind of closeness might develop between these humans and this wolf were they to permanently reside in the wolf's environs? We don't know, but the mutual understanding of thoughts is clearly possible here, and would form one pillar of that potential closeness.

Another example of the potential for closeness via the exchange of thoughts comes from Nollman (1987), whose work on communicating with animals via music I've referenced elsewhere in this volume. In his writing he offers an anecdote from an interaction with a buffalo herd in Yellowstone National Park. Aware that buffalo have a reputation of being aggressive toward humans that come too close to the herd, Nollman was interested in seeing if, by playing music and taking a slow approach while projecting congeniality, he might garner a non-aggressive reaction from the individuals in the herd. Over a period of more than an hour, by playing music and projecting calm he slowly walked toward them. At some point during this walk, he noted entering into a heightened sense of awareness of everything and, when he was close to them, beginning to interact with the herd leader across a "ring" that he could actually see, and that seemed to be a "signal [from the herd] that defined the group territory" (p. 125). Each time Nollman touched the ring, it drew a reaction from the herd leader of pawing the ground as indication that the ring was there, and that Nollman was now as close as the bull wanted him to be. Nollman says that at some point,

> The herd leader looked at me directly...I stepped on the ring and he pawed the ground. I leaned to the left and looked at him, then leaned to the right. Intuitively, I felt that he was slowly becoming aware that we were in this dance together, and that it was based on harmony and not on threat...I felt that the alpha bull had come to the realization that, indeed, this two-legged had actually seen the ring, and recognized its purpose. The two of us had invented a dance, a communication. It told the entire herd that I meant no harm. (pp. 125–126)

Here there is a perception on the part of both human and buffalo of thoughts on the part of the other, and what's more, a coming to a realization for both human and buffalo of what the other was thinking. Nollman realizes that the buffalo is showing him the boundary of the herd and what is and is not acceptable for Nollman in relation to that boundary. For his part, the herd leader realizes that Nollman understands this and thus he harbors no aggressive thoughts about the herd. Thus, Nollman's presence is accepted without a response of aggression. In this example, I'd again suggest that if Nollman lived in Yellowstone, a frequent exchange of these and other kinds of congenial thoughts between him and this herd and its leader would contribute to the development of a profound closeness between them.

If the above examples are instances of closeness or its beginning, then I must ask why this isn't possible between humans and plants. As I've argued above, it's quite possible that plants think. If we see them remembering, learning and recognizing themselves and others, can we not also exchange these thoughts as we would with a nonhuman animal, and, in wonder at such different yet similar and amazing beings, begin to develop closeness? I referred elsewhere in the book to corn geneticist Barbara McClintock, and so reiterate here her description of interactions she's had with corn plants when she says of the plants, "These were my friends...you look at these things, they become part of you" (Keller, 1983, p. 165). I suggest that her words were carefully chosen, especially given her awareness of the scientific world and its general prohibition against such characterizations.

To say "friend" is to say something beyond the propositional knowledge I discuss in Chap. 7. It is relational knowledge—knowledge by personal acquaintance. Friendship is a type of close relationship, and just as Sanders (2003) suggests that humans have actual friendships with companion animals, McClintock may very well mean the very same thing when she describes her relationship with corn plants. This is reinforced in her statement that, "'Every time I walk on grass I feel sorry because I know the grass is screaming at me'" (Keller, 1983, p. 388). Now, perhaps it's only because she's studied the material changes in plants under stress and knows they "scream" when being trod upon. But what if, even if unbeknownst to her, she was getting a sense of their thoughts, either through shared, Peircian feelings or maybe, like Sanders suggests, by reasonably inferring their thoughts about being stepped upon based on her intimacy with them—based on her *knowing* them as friends. Of course we can't know if this is why McClintock came to characterize their response as a "scream," vs. some other descriptor, but the possibility that it was due to her friendship with the corn is, in both my estimation and my ontology, plausible.

As one last example, I discuss Buber's (1947/2003) relations with more-than-human beings. Here is one relevant passage describing Buber's encounter with a horse he refers to as "my darling" (p. 26) in a stable on his grandparents' estate:

> I must say that what I experienced in touch with the animal was the Other...which, however, did not remain strange...but rather let me draw near and touch it. When I stroked the mighty mane...it was as though the element of vitality itself bordered on my skin, something that was not I, was certainly not akin to me, palpably the other...and yet it let me approach, confided itself to me, placed itself elementally in the relation of Thou and Thou with me. The horse, even when I had not begun by pouring oats for him into the manger, very gently raised his massive head, ears flicking, then snorted quietly, as a conspirator given a signal meant to be recognizable only by his fellow-conspirator; and I was approved. But once—I do not know what came over the child, at any rate it was childlike enough—it struck me about the stroking, what fun it gave me, and suddenly I became conscious of my hand. The game went on as before, but something had changed, it was no longer the same thing. And the next day, after giving him a rich feed, when I stroked my friend's head he did not raise his head. A few years later, when I thought back to the incident, I no longer supposed that the animal had noticed my defection. But at the time I considered myself judged. (pp. 26–27)

I first note here the strangeness of Buber's move, in the second-to-last sentence, to *re*-interpret his experience with the horse as one that did not contain the intimacy he originally felt. He reinterprets the experience from being an example of I-Thou to being one of I-It, where only *he* felt a judgment that the horse did not convey. In other words, Buber is suggesting that he felt judged but had not, in fact, *been* judged. Given that it is the yearning for, and real possibility of, the I-Thou relation that is virtually the entirety of Buber's philosophical project, it strikes me as odd that Buber would *post hoc* transform the experience as he did. The encounter is meant by Buber as a description of what he means by I-Thou, and the original encounter *was* an I-Thou *experience*. If it wasn't, then why offer it to the reader at all? To *post hoc* re-interpret it as I-It tugs at the very fabric of the I-Thou project. I don't think this was his intention, nor a slip-up in presenting his argument. But if it wasn't, then why do it?

Given the growing hegemonic force of the materialist conceptualizations of more-than-human beings during Buber's lifetime, I begin to wonder whether he sensed the risk he would be taking by not qualifying this example of I-Thou experience with a more-than-human being, and so moved to obviate any backlash. While saying this, I also acknowledge that Buber himself did not necessarily see the potential for close human-nature relationships that I do, though he hints at it in several places (Berry, 2012). But, whatever the reason for his *post hoc* interpretation of the experience related, I take his original description of it *as experienced* to be the more accurate one. Thus, I suggest that between him and the horse there was an exchange of thoughts that, while not uttered in English and really, in the end, being inextricable from the kinds of feelings that intertwine with such thoughts in any actual closeness-producing encounter between relational partners, leads the boy to call the horse by the relationally intimate term, "darling." Buber (1947/2003) describes the horse as a fellow "conspirator" giving Buber, through communication of thoughts, a signal meant only to be "recognizable" by him. Thus Buber sensed the horse's thoughts of approval. And even though the horse as a "Thou" as opposed to an "It" rests on a knife's edge in Buber's relation of the story, when employing the close relational ontology I suggest, the failure on the horse's part to raise his head that second day *is* the horse's existential expression of the thought of judgment, one that stands out as the most powerful and central element in the story. It is the precise point where the transformation from I-Thou to I-It takes place. In a close relational ontology, with dualisms as a distorting lens discarded, such possibilities need no censure. Thoughts and their exchange, leading as they do both toward and away from closeness under varying circumstances, are entirely possible.

References

Adams, D. K. (1928). The inference of mind. *Psychological Review, 35*(3), 235–252.

Adler, R. W. (2007). *Restoring Colorado river ecosystems: A troubled sense of immensity.* Washington, DC: Island Press.

Alexander, S. (1920). *Space, time, and deity: The Gifford lectures at Glasgow, 1916–1918.* London: Macmillan.

Automaton. (2018, June). *Oxford English Dictionary Online.* Retrieved from http://www.oed.com.

Backster, C. (1968). Evidence of a primary perception in plant life. *International Journal of Parapsychology, 10*(4), 329–348.

Baluška, F., Mancuso, S., & Volkmann, D. (2005). *Communication in plants.* Berlin: Springer.

Barad, K. (2007). *Meeting the universe halfway: Quantum physics and the entanglement of matter and meaning.* Durham, NC: Duke University Press.

Bennett, J. (2010). A vitalist stopover on the way to a new materialism. In D. Coole & S. Frost (Eds.), *New materialisms: Ontology, agency, and politics* (pp. 47–69). Durham, NC: Duke University Press.

Berry, D. L. (2012). *Mutuality: The vision of Martin Buber.* Albany, NY: State University of New York Press.

Bickle, J. (2013). Multiple realizability. *The Stanford Encyclopedia of Philosophy* (Fall 2013 Edition). Retrieved from http://plato.stanford.edu/entries/multiple-realizability.

Buber, M. (1970). *I and thou*. (W. Kaufmann, Trans.). New York: Charles Scribner's Sons. (Original work published 1923).

Buber, M. (2003). *Between man and man*. (R. Gregor-Smith Trans.). London: Routledge. (Original work published 1947).

Callaway, R. M., Pennings, S. C., & Richards, C. L. (2003). Phenotypic plasticity and interactions among plants. *Ecology, 84*(5), 1115–1128.

Calvo Garzón, F. (2007). The quest for cognition in plant neurobiology. *Plant Signaling & Behavior, 2*(4), 208–211.

Calvo Garzón, P., & Keijzer, F. (2011). Plants: Adaptive behavior, root-brains, and minimal cognition. *Adaptive Behavior, 19*(3), 155–171.

Carello, C., Vaz, D., Blau, J. J., & Petrusz, S. (2012). Unnerving intelligence. *Ecological Psychology, 24*(3), 241–264.

Carruthers, P. (2004). On being simple minded. *American Philosophical Quarterly, 41*(3), 205–220.

Casacuberta, D., Ayala, S., & Vallverdú, J. (2010). Embodying cognition: A morphological perspective. In J. Vallverdú (Ed.), *Thinking machines and the philosophy of computer science: Concepts and principles* (pp. 344–366). Hershey, PA: Information Science Reference.

Chapin, F. S., Walker, L. R., Fastie, C. L., & Sharman, L. C. (1994). Mechanisms of primary succession following deglaciation at Glacier Bay, Alaska. *Ecological Monographs, 64*(2), 149–175.

Chen, B. J., During, H. J., & Anten, N. P. (2012). Detect thy neighbor: Identity recognition at the root level in plants. *Plant Science, 195*, 157–167.

Collins, N. L., & Feeney, B. C. (2000). A safe haven: An attachment theory perspective on support seeking and caregiving in intimate relationships. *Journal of Personality and Social Psychology, 78*(6), 1053–1073.

Cruse, H., & Wehner, R. (2011). No need for a cognitive map: Decentralized memory for insect navigation. *PLoS Computational Biology, 7*(3), e1002009.

Cudworth, E., & Hobden, S. (2015). Liberation for Straw Dogs? Old materialism, new materialism, and the challenge of an emancipatory posthumanism. *Globalizations, 12*(1), 134–148.

Cunningham, S. (2000). *What is a mind?: An integrative introduction to the philosophy of mind*. Cambridge, MA: Hackett Publishing.

Griffin, D. R. (1981). *The question of animal awareness: Evolutionary continuity of mental experience*. New York: Rockefeller University Press.

Griffin, D. R. (2013). *Animal minds: Beyond cognition to consciousness*. Chicago: University of Chicago Press.

Gruntman, M., & Novoplansky, A. (2004). Physiologically mediated self/non-self discrimination in roots. *Proceedings of the National Academy of Sciences of the United States of America, 101*(11), 3863–3867.

Hasker, W. (2001). *The emergent self*. Ithaca, NY: Cornell University Press.

Keller, E. F. (1983). *A feeling for the organism, 10th anniversary edition: The life and work of Barbara McClintock*. San Francisco: W. H. Freeman and Company.

Kessler, N. (2012). Chaos or Relationalism? A Pragmatist Metaphysical Foundation for Human-Nature Relationships. *The Trumpeter, 28*(1), 43–75.

Kim, J. (1997). The mind-body problem: Taking stock after forty years. *Noûs, 31*(s11), 185–207.

Kim, J. (2000). *Mind in a physical world: An essay on the mind-body problem and mental causation*. Cambridge, MA: MIT press.

Likens, G. E., Bormann, F. H., Johnson, N. M., Fisher, D., & Pierce, R. S. (1970). Effects of forest cutting and herbicide treatment on nutrient budgets in the Hubbard Brook watershed-ecosystem. *Ecological Monographs, 40*(1), 23–47.

Lisci, M., Monte, M., & Pacini, E. (2003). Lichens and higher plants on stone: A review. *International Biodeterioration & Biodegradation, 51*(1), 1–17.

Marder, M. (2012). Plant intentionality and the phenomenological framework of plant intelligence. *Plant Signaling & Behavior, 7*(11), 1365–1372.

McGinn, C. (2005). Can we solve the mind-body problem? In T. O'Connor & D. Robb (Eds.), *Philosophy of mind: Contemporary readings* (pp. 438–457). London: Routledge.

Nagel, A. H. (1997). Are plants conscious? *Journal of Consciousness Studies, 4*(3), 215–230.

Noddings, N. (1984). *Caring: A feminine approach to ethics & moral education.* Berkeley: University of California Press.

Nollman, J. (1987). *Animal dreaming: (The art and science of interspecies communication).* New York: Bantam.

O'Connor, T. (1994). Emergent properties. *American Philosophical Quarterly, 31*(2), 91–104.

Panksepp, J. (1998). *Affective neuroscience: The foundations of human and animal emotions.* Oxford: Oxford University Press.

Peirce, C. S. (1960). *Collected papers of Charles Sanders Peirce* (C. Hartshorne, P. Weiss, & A. W. Burks, Eds.). Cambridge, MA: Harvard University Press.

Popper, K. R. (1977). Part I. In K. R. Popper & J. C. Eccles (Eds.), *The self and its brain* (pp. 3–224). Berlin: Springer International.

Robb, D., & Heil, J. (2013). Mental causation. *The Stanford Encyclopedia of Philosophy* (Fall 2013 Edition). Retrieved from http://plato.stanford.edu/entries/mental-causation.

Rowlands, M. (2010). *The new science of the mind: From extended mind to embodied phenomenology.* Cambridge, MA: MIT Press.

Sanders, C. R. (2003). Actions speak louder than words: Close relationships between humans and nonhuman animals. *Symbolic Interaction, 26*(3), 405–426.

Scientific Method. (2018, June). *Oxford English Dictionary Online.* Retrieved from http://www.oed.com.

Sperry, R. W. (1969). A modified concept of consciousness. *Psychological Review, 76*(6), 532–536.

Sperry, R. W. (1980). Mind-brain interaction: Mentalism, yes; dualism, no. *Neuroscience, 5*(2), 195–206.

Sperry, R. W. (1987). Structure and significance of the consciousness revolution. *The Journal of Mind and Behavior, 8*(1), 37–65.

Thomas, L. (1971). *The lives of a cell: Notes of a biology watcher.* New York: Viking Press.

Trewavas, A. (1999). Le calcium, C'est la vie: Calcium makes waves. *Plant Physiology, 120*(1), 1–6.

Trewavas, A. (2002). Plant intelligence: Mindless mastery. *Nature, 415*(6874), 841–841.

Trewavas, A. (2003). Aspects of plant intelligence. *Annals of Botany, 92*(1), 1–20.

Trewavas, A. (2005). Plant intelligence. *Naturwissenschaften, 92*(9), 401–413.

Uttal, W. R. (2005). *Neural theories of mind: Why the mind-brain problem may never be solved.* Mahwah, NJ: Lawrence Erlbaum Associates.

Part V
Conclusion

Chapter 11
An Example of Modern Closeness

I've argued in this volume that some humans have failed to establish a substantive basis for their claims that humans or human-like animals alone possess the elements necessary for close relationships. While I accept that more-than-human beings have a great diversity of feelings and thoughts, some that can be quite dissimilar from their human relational partners, I believe there is no basis to allow such differences to undergird the conceptual eradication of any means of communication and mutual understanding in a close relational sense. I have argued that the confidence with which modern humans make their claims to near-exclusivity in close relationships is based not in evidence and experience, but in human/nature dualisms that underwrite the interpretation of experience—of thoughts and feelings—as a near-exclusive human domain. Wielded thusly, thoughts and feelings become the currency not of relationships wherever and between whomever they occur, but of human separation and supposed superiority. This is dualist through and through. Posthumanists, for their part, seek to eradicate this superiority, but by failing to expunge their largely materialist ontological commitments of lurking dualisms, they perpetuate, rather than alleviate, the entrenched relational "asymmetry" that leaves the human as relationally superior. By working in this book to correct flawed interpretations of experience rooted in such dualistic ontological stances, my hope is that I've opened a door upon vistas where human-nature relationships are full of the potential for closeness in an interhuman sense.

My position has been that closeness in human-nature relationships is occurring for us all the time, regardless of the cultural or social background or situatedness of the humans involved. Thus, the love and care that human-nature relationship theorists espouse as ameliorative of environmental problems, for example, is not something we must dauntingly manufacture inside ourselves and project onto either an inert and passive, or lively, material other. It is a love and care we innately feel *with* the other, and that indeed we *fight against* each time our dualistically influenced professions, social interactions and institutions tell us to do so. Such a notion reminds me of a lunch seminar at which I presented a broad take on environmental

© Springer Nature Switzerland AG 2019

N. H. Kessler, *Ontology and Closeness in Human-Nature Relationships*,
AESS Interdisciplinary Environmental Studies and Sciences Series,
https://doi.org/10.1007/978-3-319-99274-7_11

philosophy incorporating some of these ideas. At one point, a student raised his hand to relate an experience of wanting to hug some great old trees while doing field work. His teachers told him not to do so because, they said, if everyone did that it would damage the trees' root systems. I have heard many stories like this, where a human being was moved out of passion and care (and potentially, by detecting some relational response from the more-than-human other as well) to make some gesture toward a more-than-human being only to have that impulse squelched by either the person himself or by the that person's peers or superiors. In the case of this student, the superficial reason given to him for avoiding such behavior was damage to the roots. But, the hard truth is that in a modern society saturated with dualisms, we often simply bridle in the presence of such unfiltered feelings. So deeply are such dualist responses embedded that in both thought and feeling, no room remains for hugging even though hugging may be the best and most accurate relational response one can have.

To conclude, I offer one last example of how this need not be the case. Before exploring it, I'll pause one last time to note that while some might suggest that the encounter described is really only one of infatuation and not reciprocating love, I'd respond only by saying that yes, it *could* only be that. But, the only way it can be *concluded to be that* is via the dualist reduction of close relational possibilities against which I've argued strenuously throughout this volume. In other words, while it *could be that*, it could *also be something far deeper* and well beyond the bounds of comfortable modernist thought. Thus, without further ado I offer an example from a podcast of the show *Radiolab* (WNYC Studios, 2010). The subject of this particular episode is the possibility of thinking and feeling in animals and whether humans can know what these thoughts and feelings are. For the final segment of the podcast, the hosts interview National Geographic photographer, Paul Nicklen, about an experience he had in the Arctic with a female Leopard Seal. He begins by describing how in his first encounter with her, she began trying to feed him. First, she'd bring him live penguins held in her mouth, then when he didn't appear able to eat those, she brought him ones she'd already killed for him. With Nicklen again failing to eat what she brought, she graduated to showing him *how* to eat the penguins, by splitting the skin and stripping it off, etc. She'd shred them in front of him, plop dead penguins on top of his camera, and so on. He describes all of this happening not over minutes or hours, but over a four-day period with this one female seal.

At one point, one of the hosts asks him, "And when you're in the water, you know, day after day, what's happening for you at this point? Are you still just a guy with a camera or..." Nicklen stops him and says, "I was starting to fall in love with this seal. It's just, uh, this animal that's just so intelligent, so powerful, that could kill you in an instant yet you're, I mean..." The other host interrupts and asks, "When you say you were in love, were you in love with the *idea* of this or did you really like *her*?" Nicklen replies quickly,

> I really liked her. She was beautiful. She was big. She had this beautiful face. Beautiful silver color to her. She kinda glowed underwater. I'm just so in love with this seal at this point. I'm not sleeping at night. I have a hard time eating. I just can't wait to see her. I can't...the first thing in the morning, the first sign of light I'm in that Zodiak...

Then Nicklen relates how on the fourth day, thinking perhaps that he was annoying her instead of there being a mutually felt, developing intimacy between them, he moved off into the colony to photograph other seals. But, at some point she came and found him again and began to do an underwater dance, making ballet-like moves and issuing a deep guttural sound that shook his whole body. At that moment, Nicklen in his self-doubt was *still* wondering whether she had grown tired of him and so was now attacking him to drive him out of her feeding grounds. But, just at the moment he was thinking this, another seal that had snuck up on them shot out from behind his back and she chased the other seal away. In addition, this other seal had a penguin in his mouth and she went and took the penguin from the other seal and brought it back to Nicklen, dropping it off in front of him so he could have it. Then Nicklen pauses in the story and says to the two hosts, "I was getting emotional reliving that…It's very powerful."

After this, one host asks, "Have you ever been in love with an animal quite this way before?" And right away Nicklen says, "Never." The host then follows by asking, "Have you ever had an experience with another human that rivals this?" To which Nicklen responds, "Perhaps when I was a kid with my mom. Someone taking care of you and feeling safe and nurtured or protected. But I've never had that in my life as an adult."

As I first pondered how to approach writing this concluding chapter, I thought I'd write about how more-than-human beings are trying to share thoughts and feelings with us, but that we modern humans aren't listening like we should. And then, just to take a break and seek out some examples that I might sprinkle into the rest of the book, I chanced upon this episode of RadioLab. How auspicious to find within it an example of the kind of closeness for which I've been arguing throughout. Here was a modern man taking photographs of, and falling in love with, a seal who was clearly enamored of him, too. Imagine if Nicklen lived in the Arctic and knew that leopard seal her whole life. What *is* a marriage if not that? What is closeness if their experience together is not closeness—or at least its roots? The only difference between that and interhuman closeness is that when it occurs between humans it's *human* closeness. And in the end, that's no difference at all.

Reference

WYNC Studios (Producer). (2010, April 2). *Animal minds*. RadioLab.

Index

A
Actor-network theory (ANT), 169
Affective agency, 84
Affective contagion, 38
Agency, 49, 50, 52, 53
  definition, 80
  without agents, 80–86
Agential realism, 50, 77, 81, 85
Agents
  agency without, 80–86
Altruism, 91, 247
American pragmatism, 15, 185
Amphibious landing, 66
Animal emotions, 259, 260
Animal empathy
  altruistic nature, 240
  definition, 247
  domestic cats, 248
  elephants, 247
  monkeys, 241
  morality, 240
  music, 241
Animal feelings, 242–246, 249
  denial of feelings
    materialists, 245
    ontological shift, 243
    positivism, 242
    posthumanism, 244
    religious totalitarianism, 243
    self-fulfilling, 244
  empathy, 247
  grief
    bull, 246
    chimpanzees, 245
    elephants, 245

    magpie, 246
    Scarlet-rumped Tanager, 246
  insect (*see* Insect feelings)
  romantic love, 248, 249
  shared feeling, 242
Animality, 95
Animism
  definition, 265
  devaru, 267, 268
  dividuation, 266
  dualisms, 265, 266
  epistemology and ontology, 269
  human Selves, 267
  Kungan, 268
  rehabilitation, 269
  relational ontology, 269
  spiritual Selves, 270
Anthropocentric modeling, 94
Anthropocentrism, 8, 29, 50, 92–95, 165
Anthropomorphism, 61, 278
Anthropomorphization, 161
Anti-clerical tendencies, materialists, 62
Aristotle's substantive, 188
Assemblages, 92
Asymmetry
  agility, 58
  amplification, 60
  anthropomorphism, 61
  conceptual framework, 61
  description, 58
  dualist assumption, 57
  feminist theorist, 57
  Haraway's "freedom", 58
  human-like qualities, 57
  human/nature dualism, 61

© Springer Nature Switzerland AG 2019   335
N. H. Kessler, *Ontology and Closeness in Human-Nature Relationships*,
AESS Interdisciplinary Environmental Studies and Sciences Series,
https://doi.org/10.1007/978-3-319-99274-7

Asymmetry (*cont.*)
  human-nature relationships, 58, 61
  human primacy, 57
  laboratory animals, 58
  Levinasian "face", 59
  perverse conception, 58
  portrait of animal agency, 59
  qualities and capacities, 60
  transformation, 60
  value-neutral, 57
  vestiges of dualisms, 60
A Thousand Plateaus, 36
Attachment, 233
Attachment theory, 96
Authoritarian-minded humans, 63

B
Barad, K., 35, 44, 50–56, 65, 69–73, 79–83,
       86, 92–94
Barad's agential realist ontology, 51
Barad's theories, 53
Becoming-animal, 36, 42
Beings, 17, 153
Biocommunication, 262
Bio/ecocentrism, 151
Biological evolution, 49
Biosphere reserve, 26
Biotechnological engineering, genetically
       modified organisms, 66
Bird banding data, 280, 281

C
Care ethics, 160
Cartesian dualism, 35, 44–46, 48, 49, 51, 58,
       62, 68, 82
Casting, 98
Chandler's statement, 84
Chickadee Woods, 163–165
Close relationships, 10–14
  assumption, 7
  indigenous cultures, 7, 8
  modern human
    child care center and mother racoon, 12
    children's perceptual capacities, 14
    gesture, 12
    intimacy, 12
    more-than-human, 10, 11
    "mutual sense of belonging", 13
    nature, 14
    post hoc alteration, 14
  more-than-human, 25
Closeness, 92

Cognitive bias, 250
Cognitive significance, 94
Common worlds pedagogy
  anthropocentrism, 166
  care ethics, 160
  ceta-morphism, 161
  childhood education, 159
  dualism, 162, 163, 166
  human-nature relationships, 159, 167
  intimacy, 163
  intra-action, 161
  monist materialism, 160, 162
  more-than-human beings, 160
  more-than-human forest selves, 166
  romanticism, 162
Communal theory, 96
Connection to Nature Scale (CNS), 128
Connectivity to nature
  anthropocentric view, 121
  beauty, 108
  definition, 121
  external connection, 122, 123
  helping behavior, 125
  human self, 124
  internalized connection, 123, 124
  merging, 125
  more-than-human beings, 126, 127
  psychological notion, 121
  reciprocal relationships, 110
  relational structure and dynamics, 127–129
  restorative and healing, 107
Consciousness, 53
Conservation psychology, 94
  definition, 107
  human-nature relationships, 107
  more-than-human beings, 109
Coole and Frost's characterization, 65
Critical animal studies (CAS)
  dualisms, 168
  equating animal with nonhuman, 173, 174
  human-nature relationships, 168
  sentience, 168–170
  veganism, 170–172
Critical plant studies, 168
Cultures
  history and social influences, 275
  human-nature relationships, 274
  "socially construct", 274

D
Darwin's natural selection, 64
De-anthropocentrizing ontologies, 56
Deep dualism, 164

Deleuze, G., 36–41, 43, 50, 59, 92, 95
Deleuzian becoming
  actual relational partners, 41
  affective contagion, 38
  A Thousand Plateaus, 36
  communication, 42
  contagion operates, 37
  depersonalized, 43
  description, 36
  emitted particles or molecules, 40
  explanatory mechanism, 37
  fleshy materialist, 38
  free-floating particles, 39, 40
  Huntington's intuition, 42, 43
  lack of *relationality*, 40
  material reality, 41
  materialism, 41
  microimitations, 38
  microrelationally, 43
  molar entity, 36
  molar *vs.* molecular, 36
  molecular communication, 38
  molecularity, 36
  molecules, corpuscles/particles, 39
  non-proximal becomings, 37
  quasi-material, 43
  schematics, 38
  set theory, 36
  spatially non-near set, 37
  virtual spacetime, 39
  zone of proximity, 36, 37, 39, 40
Depression, 234
Descriptively near sets, 36
Deterritorialization, 40
Digital technology, 66
Driesch's *entelechy*, 73
Dualisms, 241
Dualistically sourced assumptions, 76
Dyad, 52

E
Ecocentrism
  anthropocentrism, 151
  intrinsic and moral values, 152
  management strategy, 151
  systems-based approach, 152
Ecofeminism, 15
Ecofeminist dualisms, 47, 48
Ecofeminist *vs.* posthumanist selves, 78–80
Ecosystem management, 145
"Ecosystem services", 137
Egalitarianism, 152

Electrical basis of emotion, 237
  affecting another person, 252
  EEG, 251
  electrical resistance, 252
  electrochemical signals, 252
  honeybee research, 253
  internal and external emotions, 252
  romantic love, 252
  zoocentric orientation, 256
Electromagnetic bio-communication, 256
Eliminative materialism, 237
Emergence theory
  biological conditions, 296
  central nervous system, 297
  conscious life and experience, 294
  elements, 293
  fallacy of discoverable materiality, 295
  implausible, 295
  intuition and primary experience, 293
  intuitive-experiential means, 299
  material microproperties, 297
  material microstructures, 297, 300
  material ontology, 298
  material reality, 295
  mental activity, 298
  mental properties, 294
  microproperties, 295, 296
  ontological and epistemological fissures, 300
  physical and mental, 293
  physical components, 293
  process of emergence, 296
  scientific method, 299
  supervenience and physical realizationism, 292, 293
Empathy, 247
  animal (*see* Animal empathy)
  definition, 247
Environmental conservation, 26
Environmental effects, 149, 150
Environmental ontology, 3, 5
Environmental problems
  deforestation, 4
  faulty, 5
  individual interaction, 3
  large-scale, 3
  leaf-picking approach, 4
  relational Self, 4
  social-scientific analyses, 5
Epicurean materialism, 27
Epitome of humanism, 163
Etymology, 184
Exceptionalism, 44
Explanatory mechanism, 37

F

Faulty, 5
Feelings/emotions, 234, 235, 237–239
 attachment, 233
 happiness-inducing experience, 236
 inanimate, 272
 internal
  affective communication, 238
  grief, 238
  sharing of feelings, 238, 239
 material ontology
  depression, 234
  drug, 234
  electrical stimulation, 237
  eliminative materialism, 237
  schizophrenia, 235
 more-than-material origins, 235–237
 Ojibwa ontology, 273, 274
 posthumanist conceptualizations, 238
 real anger experience, 236
 stones, 273
Feeling-to-feeling interaction, 96
Feminist theorist, 57
Flattery, 257
Fleshy materialist, 38
Forest integrity, 165
Free-floating particles, 39, 40

G

Genetic information, 38
Geopolitics, 67
Grounding Knowledge, 274

H

Haraway's "freedom", 58
Harmony, 130
Heterogeneous elements, 40
Honeybee emotions, 249–251, 260
Horse in Colorado, 183
Horses as slaves, 105
Human activity, 147–150
Human agency, 83, 84
Human/animal Cartesian dualisms, 167
Human cultural identity, 113
Human exceptionalism, 18, 44–53, 56, 62, 85
Human-like brain, 55
Human-like qualities, 57
Human nature, 43
 anthropocentrism, 131
 economics disciplines, 134
 environmental education, 133
 innate human capacity, 135

more-than-human inhabitants, 133
 more-than-material emotional response, 135
 mutualism, 133
 pre-Enlightenment/Scientific
   Revolution, 136
 self-interest, 133, 135
 selfish and consumptive behavior, 132
 social and cultural influences, 134
Human-nature closeness, 96–99
Human/nature dualism, 9, 10, 35, 51, 61
 close relationships, 24, 25
 closeness as experienced, 25
 closeness as fabrication, 25
 instrumental, 24
 interhuman relationship models, 25
 material, 24
 passive, 24
 posthumanism, 24
 relational Self, 26
 self-reinforcing effect, 29
 subordinated nonhuman dependency, 23
Human-nature relationships, 96, 214–219,
   265, 266, 272
 act of relational *participation*, 197
 bird banding data, 281
 closeness, 331
 cognitive processing, 199
 community, 211–213
 cultural views, 18, 274
 curiosity, 210
 dualist objectification and
   instrumentalization, 210
 effects of relational knowledge's, 209
 ethical determination, 6
 fiction, 213
 foundation, 66
 human and more-than-human, 212
 human cognitive enterprises, 197
 human participant, 195
 immediacy, 197
 immunity to worldview differences, 208
 interdependence theory
   (*see* Interdependence theory)
 interhuman relational parallels, 91–93
 interpretation, 198
 justification, 195
 logic of domination, 209
 manipulations, 211
 material and more-than-material, 156
 mechanics, verification process, 197
 more-than-human, 209
 nonhierarchical multiplicity, 209
 objectivity and detachment, 211
 OED states, 208

Peircian feelings, 195
perceptions knowledge, 199
perturbations, 209
perversions, 211
poetic knowledge, 196, 202
post hoc processing, 195, 196
posthumanist, 198
relational inequality/asymmetry, 210
relational qualities and capacities, 209
reliability
    biotic machine, 215
    broken-off branches, 219
    experiment, 215
    fluidity and reciprocity, 217
    food, 215
    Hubbard Brook Experimental
        Forest, 214
    human-desired knowledge, 218
    human-nature encounter, 215
    human-nature relations, 219
    information and capacity, 216, 217
    level of physical intensity, 218
    more-than-human, 214
    ontological commitments, 216
    propositional knowledge, 217
    relational ontology, 218
    repetition-in-relationship, 218
religious fanatics, 197
territorial, 196
theories, 93–96
thoughts and feelings, 209
"Human progress", 149
Human sustainable activity, 151
Human values, 144
Hyperseparation, 44

I
Idealism, 8
"Imperialism of the self", 125
Inanimate feelings
    indigenous cultures, 277
    mountains, 279
    nature, 276
    poetry, 276
    relational ontology, 276
    wild nature, 277
Indeterminate vitality, 67
Indigenous human beliefs, 8
Individual-in-relation, 36, 54, 77, 81–83
Individualism, 29, 187
Individual' self, 73
Insect feelings
    cognitive bias, 250

honeybee, 249–251
materialist ontology, 251
mosquitoes, 250
Instrumentalism, 28, 29
Instrumentalization, 46
    more-than-human participants, 113
    natural resource and recreation, 114
    place dependence, 113
    place identity, 113
    valuation process, 114–115
Intensive and dynamic entity, 73
Interdependence theory, 94, 99, 100, 102,
        104, 105
    casting, 98
    duration, 98
    facilitation, 98
    human-nature relational
        characterizations, course require
            justification, 105
        communication of thoughts and
            feelings, 104
        description, 99
        MUSY, 100, 102, 105
        rainforest deforestation, 104
        self-possession, 104
        thoughts and feelings, by Walker and
            Blue, 99, 100
        U.S. Forest Service District Ranger, 100
    interaction of feelings, thoughts and
        actions, 96
    and interhuman closeness, 96–99
    personal conditions, 98
    physical environmental conditions, 98
    positive close relationship, 96
    relational conditions, 98
    "snap a line, 96
    social conditions, 98
    symmetry, 97, 98
Interhuman closeness, 96–99
Interhuman relational parallels
    and human-nature, 91–93
Interhuman relationship models, 25
Interpersonal closeness, 91
Intra-actions, 92–94

K
"Kook-fringe"
    Backster's approach, 261–262
    biocommunication, 262
    electromagnetic communication, 264
    emotional communication, 261
    plant harming experiment, 264
    polygraph, 264

"Kook-fringe" (*cont.*)
    relational Self, 263
    scientific approaches, 261, 263
    stress, 262

L
Latour's Action Network Theory, 84
Levinasian "face", 59
"Life-support systems", 137, 138
Logic of domination, 159
Love affair—of intimacy, 74

M
Marginalization, 94
Material, 63
    biological system, 167
    reciprocity, 164
    sustainability, 154
Materialism, 263
    agency, 49, 50, 52, 53
    Barad's agential realist ontology, 51
    biological evolution, 49
    the Brittlestar, 53–55
    Cartesian dualism, 49, 51
    conceiving matter, 52
    consciousness, 53
    directedness, 52
    dyad, 52
    ecocentrism, 151
    Epicurean, 27
    and equipment, 53
    human exceptionalism, 49, 52
    human-like brain, 55
    human/nature dualism, 51
    human teleology and spiritual
        experiences, 50
    justifications, 62–67
    and Midgley, 68–69
    naturalistic study of religion, 50
    posthumanist literature, 51
    primary experience, 28
    reality, 50
    reductive materialism, 27
    scattering, 56
    self-organization, 52
    self-transformation, 52
    selves/potential Selves, 51
    spiritual elements, 28
    theories, 49
    type, 48
Materially enacted, 55
Menominee's material practices, 155

Microimitations, 38
Milton's loved nature, 270–272
Mind-brain problem
    interlocking assumptions, 288
    literature, 288
    minds and material neurophysiology, 288
    nonhuman animals, 289
    physical realizationism, 290–292
    physical things, 288
    supervenience, 289–290
    Theory of Mind, 288
Molar entity, 36
Molecular communication, 38
Molecular interaction, 40
Molecularity, 36
Monist materialism, 30, 31, 162
Monist materialist, 160
More-than-human beings, 18
    ecocentrism, 153, 154
    feelings and thoughts, 331
    Menominee's relationship, 154
    sustainable development, 153
    thoughts and feelings, 333
More-than-human effects, 150
More-than-human emotions, 240, 258
More-than-human world
    definition, 26
    posthumanism, 26
More-than-material, 31, 32
More-than-material elements, 154–156
More-than-material ontology
    astonishment, 229
    beauty in nature, 227
    behavior of, 221
    capacity to astonish, 228
    elements of reality, 221
    emotions, 221
    evidence, 223–225
    Grand Canyon in Arizona, 228
    human attributes, 226
    human-nature relationships, 226
    kind of knowledge, 221
    material mechanism, emergence, 227
    materialism, 221
    materialism's support, 222
    more-than-human relational qualities and
        capacities, 222
    Peircian feeling and relational
        knowledge, 221
    relational interactions, 227
    times environmental education
        programs, 226
Multiple Use Sustained Yield (MUSY), 100,
    102, 105

Mutual and harmonious relationships, 146
Mutual influence, 94
Mutualism, 130, 133

N
Near sets, 36
Nerve-like electrical signaling, 255
New materialism, 35
Newtonian mechanics, 65
Non-proximal becomings, 37

O
Ojibwa ontology, 273, 274
Ontologically basic human-nature relations, 17
Ontologically basic relations
  care and closeness, 186
  characteristics, 189–192
  human-nature, 192–193
  human-nature relationships, 186
  more-than-human beings, 186, 187
  mutual dependence, 188
  posthumanism, 184
  Self, 184
  substance and relations, 188
  substantive and relational scales, 188
  substantivism, 187
"Ontological primitive", 188
Ontology of emotions
  eliminative materialism, 237 (*see also*
    Feelings/emotions)

P
Passive material more-than-human objects, 272
Peircian feelings
  capacity for, 199
  communication of external feeling, 200
  continuity of feeling, 201
  external experiences, 202
  human-nature relationships, 201, 202
  legitimacy, 202
  physical feelings, 201
  posthumanism, 201
  psychic feeling, 199
  tension/suspense, 200
Personal acquaintance knowledge
  elements of, 204
  food consumption, 204
  fracking process, 207
  Goodall's methods, 207
  interhuman relationships, 206
  more-than-human, 205

  observations/manipulations, 204
  poetic knowledge, 206
  post hoc conceptualization, 205
  sensory, emotional and intuitive
    capacities, 206
  types of knowledge, 205
Personification, 273
Perverse conception, 58
Physical lives, 66
Pitiless cosmic motions, 64
Place attachment theory, 16, 113–115, 118, 119
  absence of more-than-human beings, 120
  definition, 112
  instrumentalization
    more-than-human participants, 113
    natural resource and recreation, 114
    place dependence, 113
    place identity, 113
    valuation process, 115
  more-than-human reduction, 115–117
  relational structure and dynamics
    emotional bonds, 118
    place dependence, 119
    scale items, 119
    unilateral, 117
Place dependence, 113, 119
Place identity, 113
Plant-animal parallels, 256
Plant electrical activity
  vs. bar magnet, 254
  oscillatory signals, 254
  plant brain, 254
  respiration and photosynthesis, 254
Plant feelings
  chemical processes, 251
  vs. earthworms, 257
  electrical activity, 253–255, 258, 259
  flattery, 257
  "kook-fringe", 260
  material representation, 258
  more-than-material, 259
  plant evolution, 258
  sensations, 259
Plant-human communication, 261
Plant intelligence
  awareness/recognition, 315
  behavior, 315
  belief/desire architecture, 318
  "black box" explanation, 304
  cellular networks, 304
  chemical and electrical phenomena, 316
  decentralization, 304, 305
  elements, 305
  embodied cognition, 305

Plant intelligence (*cont.*)
   experimental evidence, 302
   growing belief, 301
   human-as-intelligent, 302
   human behavior, 302
   intelligence, 301
   intrinsic capacity, 303
   learning and memory, 316, 317
   mental activity, 316
   mental experience and consciousness, 317
   mental representation, 318
   microstructural, 303
   neural network, 303
   perception-action cycles, 301
   plant behavior, 302
   plant-as-intelligent, 302
   self/other recognition, 315
   T. repens, 316
Plant neurobiology, 255
Poetic knowledge, 202
Positivism, 242
Posthumanism, 17, 24, 244, 262
   agency without agents, 80–86
   asymmetry, 56–62
   definition, 35
   Deleuzian becoming, 36–44
   ecofeminist *vs.* posthumanist selves, 78–80
   human exceptionalism, 44–48
   human/nature dualism, 35
   human-nature relationships, 158
   individuality, 137
   intra-action, 157
   materialism's, 48–56
   materially embedded, 158
   more-than-human world, 26
   and new materialism, 35
   place attachment theory, 112
   relations of matter, 73–78
   relations of Selves, 73–78
   relations, *Relata* and the loss of the self,
      70–73
   scholarship, 35
   sustainability, 157
Posthumanists, 92
Potency, 65
Pragmatism, 15, 185
Preservationism, 155
Primary experience, 15, 28
Primary perception, 275
Pro-environmental behavior, 91
Propositional knowledge
   context of relational knowledge, 208
   epistemology, 207
   posthumanism, 208

   scientific method, 208
   types, 207
Pure proximity, 190

R
"Radical empiricism", 185
Radical differences, 163
Rainforest deforestation, 104
Reality, 50
Reciprocity, 108
Reductive materialism, 27
Rehabilitation, 91
Relational asymmetry, 57
Relational Self, 142
Relations, 17
Reterritorialization, 40
Romanticism, 162

S
Scattering, 56
Schematics, 38
Schizophrenia, 235
Scientific Revolution and Enlightenment
   periods, 46
Self
   description, 16
Self-in-relation, 93
Set theory, 36
Sharing of feelings, 239
Single material plane, 46
Social values, 143, 144
Sophrology, 235
Southeast Alaska national forest, 105
Spatially near sets, 36
Spatially non-near set, 37
Spiritual forces, 62
Spirituality, 155
"Strong sustainability", 147
Subject-based agency, 85
Substantivism, 187
Sustainability theories, 16, 131–136
   anthropocentric orientation, 140
   biocentric, 153
   definition, 141
   ecosystem management, 145
   environmental, 143
   environmental education, 133
   harmony and mutuality, 130, 131
   human activity, 147
   human desire, 141, 142
   human nature
      anthropocentrism, 131

economics disciplines, 134
innate human capacity, 135
more-than-human inhabitants, 133
more-than-material emotional
    response, 135
pre-Enlightenment/Scientific
    Revolution, 136
self-interest, 133, 135
selfish and consumptive behavior, 132
social and cultural influences, 134
human societies, 143
human values and desires, 145
individuality, 137
industrial capitalism, 129
loss of individuality, 139
material relations, 146
more-than-human beings, 130, 138
more-than-human relational partner, 140
more-than-material relations, 146
mutual/harmonious, 145, 146
mutualism, 133
natural and human-man capital, 137
pillars, 147
relational approach, 144
superficial/value-driven, 142
UNESCO-UNEP, 130
USEPA, 129, 130
watershed ecological economics, 141
Sustainable development, 16

T
Tachyphylaxis, 234
The Brittlestar, 53–55
The Divided Self, 235
The Myths We Live By (2004), 68
The Secret Life of Plants, 259
Thoughts, 306–312
    anthropocentrism, 320
    conscious awareness, 319, 320
    exchange of, 321–325
    explanatory theories
        acts of organization, 309
        anthropocentrism, 311

automatons, 310
embodied cognition theory, 306
human thinking, 308
material nervous systems, 311
materialist ontology, 307
non-cognitive emergence, 310
origin of intelligence, 310
physically intelligent response, 310
plant intelligence theorists, 312
plant thinking, 308
qualitative/quantitative, 310
field of plant neurobiology, 301
human intimacy, 287
human-nature relationships, 287
inanimate, 321
interhuman relationships, 322
lack of mind, 287
material neurophysiological structures, 301
mental experiences in plants, 312–315
more-than-human beings, 287
unexchanged/unexpressed, 287
Transformative agency, 84

U
US Environmental Protection Agency
    (USEPA), 129
U.S. Forest Service District Ranger, 100

V
Veganism, 170–172
Virtual spacetime, 39
Virtual technology, 66
Vital materialism in Driesch's wake, 63

W
Wild wolf management, 140
Wireless technology, 66

Z
Zone of proximity, 36, 37, 39, 40

Printed in the United States
By Bookmasters